Chirality in Natural and Applied Science

Chirality in Natural and Applied Science

Edited by

W.J. Lough
Institute of Pharmacy and Chemistry
University of Sunderland
UK

and

I.W. Wainer
National Institute on Aging
National Institutes of Health
Baltimore
USA

Blackwell
Science

CRC Press

© 2002 by Blackwell Science Ltd,
a Blackwell Publishing Company
Editorial Offices:
Osney Mead, Oxford OX2 0EL, UK
 Tel: +44 (0)1865 206206
Blackwell Publishing Asia Pty Ltd,
550 Swanston Street, Carlton South,
Victoria 3053, Australia
 Tel: +61 (0)3 9347 0300
Blackwell Wissenschafts Verlag,
Kurfürstendamm 57, 10707 Berlin,
Germany
 Tel: +49 (0)30 32 79 060

ISBN 0-632-05435-2

A catalogue record for this title is available
from the British Library

Published in the USA and Canada (only) by
CRC Press LLC
2000 Corporate Blvd., N.W.
Boca Raton, FL 33431, USA
Orders from the USA and Canada (only) to
CRC Press LLC

USA and Canada only:
0-8493-2434-3

The right of the Author to be identified as
the Author of this Work has been asserted
in accordance with the Copyright, Designs
and Patents Act 1988.

All rights reserved. No part of this
publication may be reproduced, stored in a
retrieval system, or transmitted, in any form
or by any means, electronic, mechanical,
photocopying, recording or otherwise,
except as permitted by the UK Copyright,
Designs and Patents Act 1988, without the
prior permission of the publisher.

First published 2002 by Blackwell
Science Ltd

Library of Congress
Cataloging-in-Publication Data
is available

Printed and bound in Great Britain by
MPG Books, Bodmin, Cornwall

For further information on
Blackwell Science, visit our website:
www.blackwell-science.com

Contributors

L.D. Barron	Department of Chemistry, University of Glasgow, University Avenue, Glasgow G12 8QQ, UK
L. Di Bari	Centro di Studio del C.N.R. per le Macromolecole Stereordinate ed Otticamente Attive, Dipartimento di Chimica e Chimica Industriale, Università di Pisa, 56126 Pisa, Italy
J. Greer	Pharmaceutical Discovery, Abbott Laboratories, Illinois, USA
W.A. König	Institut für Organische Chemie, Universität Hamburg, 20146 Hamburg, Germany
W.J. Lough	Institute of Pharmacy and Chemistry, University of Sunderland, Sunderland SR1 3SD, UK
A.J. MacDermott	Department of Chemistry, University of Cambridge, Lensfield Road, Cambridge CB2 1EW, UK
S.F. Mason	Department of Chemistry, King's College London, London WC2R 2LS, UK, and Department of History and Philosophy of Science, University of Cambridge, Cambridge CB2 3RH, UK
D.D. Miller	Department of Medicinal Chemistry, College of Pharmacy, University of Tennessee, Memphis, TN 38163, USA
K. Mislow	Department of Chemistry, Princeton University, Princeton, New Jersey 08544, USA
K. Mori	Department of Chemistry, Science University of Tokyo, Japan
P.N. Patil	Division of Pharmacology, College of Pharmacy, The Ohio State University, Parks Hall, 500 West 12th Avenue, Columbus, OH 43210, USA
C. Rosini	Dipartimento di Chimica Università della Basilicata a Potenza, 85100 Potenza, Italy
P. Salvadori	Centro di Studio del C.N.R. per le Macromolecole Stereordinate ed Otticamente Attive, Dipartimento di Chimica e Chimica Industriale, Università di Pisa, 56126 Pisa, Italy
D.J. Triggle	The Graduate School, State University of New York at Buffalo, Buffalo NY 14260-1200, USA

I.W. Wainer	Gerontology Research Center, National Institute on Aging, National Institutes of Health, Baltimore, MD, USA
C.J. Welch	Merck & Co., Inc., Rahway, NJ 007065, USA

Introduction
Retournons à Pasteur!

Kurt Mislow
Department of Chemistry, Princeton University, Princeton, New Jersey 08544, USA

In his autobiography, the geneticist François Jacob (1988) describes his first visit to the tomb of Louis Pasteur in the Pasteur Institute:

> On the walls to either side of the tomb were marble panels on which were carved, like Napoleon's victories in the Invalides, Pasteur's victories. Instead of the battles of Austerlitz, Jena, and Friedland, one read of molecular dissymmetry; fermentation; so-called spontaneous generation; studies on wine; silkworm disease; studies on beer; virulent diseases; vaccines; prophylaxis against rabies... How could one not admire this scientific odyssey, this way of vaulting from one domain to another, of going from chemistry and crystallography to the study of living things, from the diseases of beer to those of man?

The year 1998 was the sesquicentennial of Pasteur's first scientific triumph. Herschel's finding that there is a causal relationship between the handedness of hemihedral quartz crystals and their sense of optical rotation (Herschel 1822), with concurrent studies by Biot and Fresnel, had prepared the ground for Pasteur's momentous discovery, when he was only 26 years old, that crystalline hemihedry and optical rotation are similarly correlated in the tartrates (Pasteur 1848, 1850, 1853a). Pasteur's findings, which he presented on May 22, 1848 before the Paris Academy of Sciences, connected chirality on the macroscopic scale with chirality on the molecular scale and thus led to his insight that the optical activity of the tartrates is a manifestation of "dissymétrie moléculaire" (Pasteur 1861).

Pasteur's scientific breakthrough marked the beginning of modern stereochemistry. Since that time, stereochemistry has experienced enormous intellectual growth, but molecular dissymmetry—or what, adopting Lord Kelvin's coinage (Kelvin 1894), is now commonly called molecular chirality—remains at the heart of this and allied branches of science. Over one hundred and fifty years after Pasteur's brilliant discovery, molecular chirality is still a topical issue of broad scientific interest. Thus, a spate of articles, reviews, books, and international conferences and symposia have in recent years dealt with the role of chirality in chemistry, including applications to pharmaceutical research (see, for example: Collins *et al.* 1992; Sheldon 1993; Crossley 1995), and three new journals have been specifically devoted to this topic: Chirality, by Wiley–Liss in 1989, Tetrahedron: Asymmetry, by Pergamon Press in 1990, and Enantiomer, by Gordon and Breach in 1996.

It is a measure of Pasteur's genius that he pinpointed the single most essential feature of molecular chirality, well before the advent of structural theory and the asymmetric carbon atom of Jacobus Henricus van't Hoff (Pasteur 1861):

> Are the atoms of the right [tartaric] acid grouped on the spirals of a dextrogyrate helix, or placed at the summits of an irregular tetrahedron, or disposed according to some particular asymmetric grouping or other? We cannot answer these questions. But it cannot be a subject of doubt that there exists an arrangement of the atoms in an asymmetric order, having a non-superposable [mirror] image. It is not less certain that the atoms of the left acid realize precisely the asymmetric grouping which is the inverse of this.

As Pasteur recognized, whatever the precise arrangement of atoms in the molecule, the chirality of that arrangement is the necessary and sufficient condition for enantiomorphism and for its manifestation as optical activity. No further details and no recourse to structural theory were needed to arrive at this conclusion, which was based purely on a symmetry argument and which F. M. Jaeger called Pasteur's Law (Jaeger 1923; see also Jaeger 1930). Thus, the location and nature of the bonds between the atoms are immaterial—all that matters is whether the spatial arrangement of the atoms in the molecule, taken as a whole, is chiral under the conditions of observation.

The explosive development of structural theory since Pasteur's day has yielded the tremendous diversity of chiral molecular structures that fill the chemical journals and stereochemistry textbooks at the end of this century—yet all, without exception, conform to Pasteur's law. The growing pains of conceptual developments in stereochemistry, and much of the attendant confusion, could to a large extent have been avoided if more attention had been paid to Jaeger's exhortation "Retournons à Pasteur!" (Jaeger 1923; see also Jaeger 1930). Indeed, it might be argued that had the present-day collection of these structures been available at the end of Pasteur's life (1895), it would have been possible, with the help of the ball-and-stick models which had by then been developed, to visualize the chirality of most of these organic and inorganic molecules and to predict their optical activity—even of such unconventional entities as chiral molecular knots, links, and graphs (Mislow 1996).

There are, of course, molecules whose structural chirality could not have been envisaged by Pasteur's contemporaries. Among these are molecules whose chirality depends on differences between isotopic masses, such as the innumerable molecules that owe their chirality to the presence of naturally abundant stable isotopes. The minute percentages of CH_3CHDOH in ethanol and of $(^{12}CH_3)(^{13}CH_3)CHOH$ in 2-propanol are examples in this class. None of this could have been imagined before the discovery of isotopes, early in the twentieth century. Equally beyond the ken of a nineteenth-century chemist would have been the notion that molecular chirality could be achieved by irradiation of an appropriately structured achiral molecule with circularly polarized light resulting in an asymmetric distribution of electrons in the photoexcited state (Schippers & Dekkers 1983; Miesen *et al.* 1994).

Yet it was Pasteur, anticipating one of the most dramatic developments of modern physics, who first advanced the idea that our world is chiral ("L'univers est dissymétrique"; Pasteur 1874; see also Haldane 1960). This prescient conjecture was confirmed by the discovery, in 1956, that parity is not conserved for the weak interactions governing β-decay (Lee and Yang 1956) and hence that matter is inherently chiral (Latal 1991). Thus, atoms are chiral and in principle optically active, although the effect is tiny and has so far been observed only in heavy atoms (Pb, Bi, Tl, Cs). That this inherent chirality of matter might possibly have played a role in the origin and persistence of biomolecular homochirality in the natural world (i.e., the sole occurrence of L amino acids in proteins and of D sugars in nucleic acids and biologically important polysaccharides) was foreshadowed by Pasteur's speculation regarding the influence of what he characterized as "forces dissymétriques" on the evolution of homochirality in nature. In his words (Pasteur 1861):

> But the difference in properties of corresponding right and left substances when they are subjected to asymmetric forces, seems to me to be interesting in the highest degree on account of the ideas which it suggests to us in regard to the mysterious cause which presides over the asymmetric arrangement of the atoms in natural organic substances. Why this asymmetry? Why the one asymmetry rather than its inverse? ... Is it not necessary and sufficient to admit that at the moment of the elaboration of the primary principles in the vegetable organism, an asymmetric force is present? ... Do these asymmetric actions, possibly placed under cosmic influences, reside in light, in electricity, in magnetism, or in heat? Can they be related to the motion of the earth, or to the electric currents by which physicists explain the terrestrial magnetic poles? It is not even possible at the present time to express the slightest conjecture in this direction. But I regard as necessary the conclusion that asymmetric forces exist at the moment of the elaboration of natural organic products ...

While Pasteur was unable to specify the nature of the agency responsible for symmetry breaking at the biomolecular level, the fall of parity, nearly 100 years later, inspired speculations, supported by theoretical calculations, that the small but persistent chirality impressed on to prebiotic chemistry by PVEDs (parity-violating energy differences) between enantiomers might serve to explain the apparent mystery of biomolecular homochirality. This modern version of Pasteur's "force dissymétrique" does not lack appeal; there is, however, no experimental evidence for a causal relationship between PVEDs and the broken symmetry in biopolymers (Bonner 1996). Indeed, the postulation of such a force has been rendered unnecessary with the realization that the evolution of homochirality from abiotic sources can be simply accounted for on the basis of statistical arguments (Bolli *et al.* 1997; Siegel 1997; see also Green and Garetz 1984).

Pasteur's awareness of a "difference in properties of corresponding right and left substances when they are subjected to asymmetric forces" was based on the pathbreaking observation that is best described in his own words (Pasteur 1861):

> I referred to some observations ... connected with the comparison of the physical and chemical properties of the corresponding right and left [tartaric acid] isomers. I have already insisted on the perfect identity of all their properties,

excepting always the inversion in their crystalline forms and the opposite sense of their optical deviations. The physical aspect, lustre of the crystals, solubility, specific weight, simple or double refraction, all these things are not merely alike, similar, nearly allied, but identical in the strictest sense of the word...Taking this into account, the identity of properties above described ... exists constantly ... whenever these substances are placed in contact with any compound of the class with superposable image, such as potash, soda, ammonia, lime, baryta, aniline, alcohol, ethers,—in a word, with any compounds whatever which are non-asymmetric, non-hemihedral in form, and without action on polarized light ... If, on the contrary, they are submitted to the action of products of the second class with non-superposable image,— asparagine, quinine, strychnine, brucine, albumen, sugar, etc., bodies asymmetric like themselves,—all is changed in an instant. The solubility is no longer the same. If combination takes place, the crystalline form, the specific weight, the quantity of water of crystallization, the more or less easy destruction by heating, all differ as much as in the case of the most distantly related isomers.

Pasteur, in short, had discovered not only enantiomers but also the principle of discriminating between them (Pasteur 1861):

Here, then, the molecular asymmetry of a substance obtrudes itself on chemistry as a powerful modifier of chemical affinities. Towards the two tartaric acids, quinine does not behave like potash, simply because it is asymmetric and potash is not. Molecular asymmetry exhibits itself henceforth as a property capable by itself, in virtue of its being asymmetry, of modifying chemical affinities.

As an application of this principle, Pasteur showed that the enantiomers of tartaric acid could be separated by means of the diastereomeric salts formed with (+)-cinchotoxine and (+)-quinotoxine (Pasteur 1853b). This was the second example of enantiomeric resolution. Pasteur subsequently observed that (+)-tartaric acid was fermented in a broth of microorganisms ("matières albuminoïdes") whereas (−)-tartaric acid was not; the finding that such organisms react enantiospecifically led him to the inexorable conclusion that living matter is asymmetric. His interest thereafter turned increasingly to microbiology.

We have seen that Pasteur established stereochemistry as a new branch of science. Addressing his audience in 1860, he concluded (Pasteur 1861):

You have understood, as we proceeded, why I entitled my exposition, "On the Molecular Asymmetry of Natural Organic Products". It is, in fact, the theory of molecular asymmetry that we have just established, one of the most exalted chapters of the science. It was completely unforeseen, and opens to physiology new horizons, distant but sure.

The contents of this volume are ample testimony that Pasteur's great work has opened new horizons, not only in physiology but in all branches of the natural sciences. It is a privilege to pay this modest tribute to one of the greatest among scientific pioneers.

REFERENCES

Bolli M., Micura R., & Eschenmoser A. (1997) *Chem. Biol.*, **4**, 309–20.
Bonner W. A. (1996) *Orig. Life Evol. Biosphere*, **26**, 27–46, and references therein.
Collins A. N., Sheldrake G. N., & Crosby J. (Eds) (1992) *Chirality in Industry*, Wiley & Sons, New York.
Crossley R. (1995) *Chirality and the Biological Activity of Drugs*, CRC Press, Boca Raton.
Green M. M. & Garetz B. A. (1984) *Tetrahedron Lett.*, **25**, 2831–4.
Haldane J. B. S. (1960) *Nature*, **185**, 87.
Herschel J. F. W. (1822) *Trans. Cambridge Philos. Soc.*, **1**, 43.
Jacob F. (1988) *The Statue Within*, pp. 247–248. Basic Books, New York. (Translated from: *La statue intérieure* (1987), p. 276. Éditions Odile Jacob, Paris.)
Jaeger F. M. (1923) *Bull. Soc. Chim. Fr.*, **33**, 853–89.
Jaeger F. M. (1930) *Spatial Arrangements of Atomic Systems and Optical Activity*, McGraw–Hill, New York.
Kelvin W. T. (1894) The Second Robert Boyle Lecture. In: *J. Oxford Univ. Junior Scientific Club*, No. 18, p. 25.
Latal H. (1991) Parity Violation in Atomic Physics. In: *Chirality: From Weak Bosons to the α-Helix* (Ed. by R. Janoschek), pp. 1–17. Springer, Berlin.
Lee T. D. & Yang C. N. (1956) *Phys. Rev.*, **104**, 254–8.
Miesen F. W. A. M., Wollersheim A. P. P, Meskers S. C. J., Dekkers H. P. J. M., & Meijer E. W. (1994) *J. Am. Chem. Soc.*, **116**, 5129-33.
Mislow K. (1996) *Croat. Chem. Acta*, **69**, 485–511, and references therein.
Pasteur L. (1848) *Ann. Chim. Phys.*, **24**, 442–59.
Pasteur L. (1850) *Ann. Chim. Phys.*, **28**, 56–117.
Pasteur L. (1853a) *Ann. Chim. Phys.*, **38**, 437–88.
Pasteur L. (1853b) *Compt. Rend. Acad. Sci.*, **37**, 162–6.
Pasteur L. (1861) Recherches sur la Dissymétrie Moléculaire des Produits Organiques Naturels. In: *Leçons de Chimie Professées en 1860*, Lib. Hachette, Paris, pp. 1–48. This work, which consists of two lectures delivered at the invitation of the council of the Société chimique de Paris, was intended by Pasteur to provide an overview of his researches on molecular dissymmetry. All quotations in the present article are from the translation: L. Pasteur, Researches on the Molecular Asymmetry of Natural Organic Products, an Alembic Club Reprint, E. & S. Livingstone, Edinburgh, 1964. In general, this English version adheres quite faithfully to the French text, except for the consistent mis-translation of 'dissymétrie' and 'dissymétrique' as 'asymmetry' and 'asymmetric'.
Pasteur L. (1874) *Compt. Rend. Acad. Sci.*, **78**, 1515–8.
Schippers P. H. & Dekkers H. P. J. M. (1983) *J. Am. Chem. Soc.*, **105**, 145–6.
Sheldon R. D. (1993) *Chirotechnology*, Marcel Dekker, New York.
Siegel J. S. (1998) *Chirality*, **10**, 24-7.

Contents

Contributors	i
Introduction—retournons à Pasteur! K. MISLOW	iii
References	vii

1 Pasteur on molecular handedness—and the sequel — 1
S.F. MASON

Pasteur's optical resolutions	1
The chemical–physics background to Pasteur's 1848 discovery	3
Molecular structure theory and stereoselective chiral synthesis	11
Chance, necessity, and chiral fields in the evolution of biomolecular homochirality	15
Parity and its non-conservation	17
References	19

2 The origin of biomolecular chirality — 23
A.J. MACDERMOTT

Introduction	23
Homochirality—a hallmark of life	23
Homochirality—also a pre-condition for life?	24
The need for a chiral influence	24
False chirality—magnetic fields	28
True chirality—circularly polarized photons	29
True chirality—the w boson	31
True chirality—the z boson	33
Amplification mechanisms	35
PVEDS of biomolecules	37
So was it really the weak force?	41
Look in space!	42
Is the universe intrinsically handed?	46
References	49

3 Chirality at the sub-molecular level: true and false chirality — 53
L.D. BARRON

Introduction	53
Symmetry principles	56

Non-observables and symmetry operations	56
Symmetry of physical quantities	57
Symmetry in quantum mechanics	58
Parity	58
Time reversal	59
Charge conjugation	60
True chirality	61
Natural and magnetic optical activity	61
A new definition of chirality	62
Translating spinning electrons, photons and cones	63
Electric, magnetic and gravitational fields	64
Absolute asymmetric synthesis	65
Truly chiral influences	66
Falsely chiral influences	66
The breakdown of microscopic reversibility induced by a falsely chiral influence: enantiomeric detailed balancing	67
Unitarity and thermodynamic equilibrium	69
Chirality and symmetry violation	70
The fall of parity	70
Parity violation in atoms and molecules: true enantiomers	72
Violation of time reversal and *CP* violation	74
CP violation is analogous to chemical catalysis!	75
CPT violation	75
The mixed-parity states of a chiral molecule	76
The double well model	76
Two-state systems and parity violation	78
Parity violation and spontaneous parity breaking	80
Chirality and relativity	81
Chirality in two dimensions	82
Concluding remarks	83
Acknowledgements	84
References	84

4 The molecular basis of chiral recognition   87
J. GREER and I.W. WAINER

Introduction	87
Kinetic and thermodynamic aspects of enzymatic enantioselectivity	88
Kinetics of enzymatic reactions	88
Thermodynamics of enzyme reactions	89
Enzymatic enantioselectivity	89
Molecular chiral recognition mechanisms	91
The 'three-point' model	91
A re-examination of the 'three-point' model	92
Conformationally-driven molecular chiral recognition	94
The N-dechloroethylation of ifosfamide enantiomers	94
Interaction of peptides and larger molecules with chiral biopolymers	96
Binding of peptides and peptide-like moieties to a protein	96
How critical are the chiral centres for binding to the enzyme?	101

Modification of multiple centres	101
Modification of one centre at a time	103
Concluding remarks	105
References	105

5 Chirality in drug design and development — 109
D.J. TRIGGLE

Introduction	109
The stereoselectivity of drug–receptor interactions	110
Biological consequences of stereoselective drug action	113
The basis of stereoselectivity	116
Absorption	117
Distribution	117
Metabolism	118
Excretion	120
Stereoselectivity of drug classes	120
G protein-coupled receptors: β-adrenoceptor antagonists	120
Ion-channel receptors: voltage-gated Na^+ and Ca^{2+} channels	123
General anaesthetic receptors: general anaesthetics	128
Regulatory requirements for chiral drugs	128
Decision processes	130
Acknowledgements	131
References	131

6 Chirality in medicinal chemistry — 139
P.N. PATIL and D.D. MILLER

Historical aspects	139
Stereoselectivity of the biosynthesis of transmitters and related medicinals	141
Affinity and intrinsic activity	144
Steric aspects of drug action at the cholinergic neuroeffector junction	146
Muscarinic agonists	146
Nicotinic agonists	149
Acetylcholinesterase inhibitors	150
Cholinergic muscarinic blockers	151
Nicotinic cholinergic blockers	153
Steric aspects of drug action at the noradrenergic and dopaminergic neuroeffector junction	154
Inhibition of noradrenaline transport	155
Chirality at postjunctional receptors	157
Purity of enantiomers	157
Receptor active conformation	158
Conformation induction by enantiomers	160
Efficacy and isomeric activity ratios	161
The Easson–Stedman hypothesis	161
Deviation from Easson–Stedman hypothesis	162

Chirality of adrenoceptor blockers	164
Competitive reversible α-adrenoceptor blockers	164
Irreversible α-adrenoceptor blockers	167
Enantiomers of β-adrenoceptor blockers	168
Concluding remarks	170
Acknowledgement	171
References	171

7 Separation of chiral compounds—from crystallisation to chromatography — 179
W.J. LOUGH

Crystallisation	179
Russia, Israel, and NMR	181
Derivatisation	183
Commercialised chiral HPLC	184
Chiral HPLC beyond the eighties	189
Beyond HPLC	192
Meeting unfulfilled needs?	195
References	199

8 Electronic circular dichroism—fundamentals, methods and applications — 203
P. SALVADORI, L. DI BARI, and C. ROSINI

Introduction	203
Phenomenological aspects	204
Instrumentation—essential description of a dichrograph	208
Towards an understanding of optical activity and related phenomena	209
Preliminary considerations	209
Macroscopic characteristics of an optically active medium	209
The helix model	210
The nature of electronic transitions	211
Dipolar strength	214
Rotational strength	214
Symmetry properties of the rotational strength—spectroscopic foundation of optical activity	215
Classification of optically active molecules	216
Optical activity of dissymmetric chromophores	216
The independent systems approach	217
Coupled oscillators	219
Applications to structural chemistry	220
Empirical correlations	220
Semi-empirical rules	224
Helicity rules	224
Sector rules	224
Chirality rules	226

Non-empirical methods of analysis	227
Conclusions and future perspectives	233
References	234

9 Chirality in the Natural World: Chemical Communications — 241
K. MORI

Introduction	241
When was chirality found to be important in chemical communications?	243
Background—theory of olfaction in the 1960s and early 1970s	243
Pioneering work with insects	243
The bioactivity of pheromones depends on their chirality	244
How was the utmost importance of chirality in chemical communications recognized?	246
Synergistic response based on enantiomers—sulcatol	246
Inhibition by the wrong enantiomer—disparlure and japonilure	247
One enantiomer is active against males whereas the opposite enantiomer affects females—olean	247
Chirality plays a role even with an achiral pheromone	248
Current understanding of the importance of chirality in chemical communications—stereochemistry–bioactivity relationships among pheromones	249
Only one enantiomer is bioactive, and its antipode does not inhibit the action of the active stereoisomer	249
Only one enantiomer is bioactive, and its antipode inhibits the action of the pheromone	249
Only one enantiomer is bioactive, and its diastereomer inhibits the action of the pheromone	250
The natural pheromone is a single enantiomer, and its antipode or diastereomer is also active	250
The natural pheromone is an enantiomeric mixture, and both enantiomers are separately active	251
Different enantiomers or diastereomers are employed by different species	252
Both enantiomers are necessary for bioactivity	253
One enantiomer is more active than the other stereoisomer(s), but an enantiomeric or diastereomeric mixture is more active than that enantiomer alone	253
One enantiomer is active on males, the other on females	254
Only the *meso* isomer is active	254
The practical importance of chirality in pheromone-based pest control	255
Conclusion	255
Acknowledgement	255
References	255

10 Chirality in the natural world—odours and tastes — 261
W.A. KÖNIG

Introduction	261

The influence of chirality on biological activity	262
The relationship between olfaction and chemical structure	264
The relationship between odour and taste and chirality	264
Enantiomeric odour differences and their quantitative measurement	265
Analysis of the configuration and enantiomeric composition of volatile chiral compounds	266
Stereochemical analysis of odorous compounds	267
Monoterpene hydrocarbons	267
Terpene alcohols	270
Carbonyl compounds	273
Sesquiterpenes and related compounds	277
Sulphur compounds	279
'Electronic nose', sensor technology	280
Concluding remarks	281
References	281

11 Chirality in the natural world: life through the looking glass   285
C.J. WELCH

Introduction	285
Background	286
Special problems in the study of asymmetric organisms	286
The machinery of life: symmetric organisms from asymmetric building blocks	289
The asymmetry of microorganisms	291
Asymmetry in the plant world	291
Asymmetry of invertebrates	294
Asymmetry of fishes	296
Asymmetry of birds	297
Asymmetry of reptiles and amphibians	298
Asymmetry of mammals	298
Asymmetry in humans	299
Some recent advances in the developmental biology of animal asymmetry	300
Behavioural asymmetry	300
Summary and conclusion	300
Acknowledgements	301
References	301

Concluding Remarks—Chirality, Chemistry, the Future   303
THE EDITORS

References	304

Chapter 1
Pasteur on Molecular Handedness—and the Sequel

Stephen F. Mason

Department of Chemistry, King's College London, London WC2R 2LS, UK, and Department of History and Philosophy of Science, University of Cambridge, Cambridge CB2 3RH, UK

PASTEUR'S OPTICAL RESOLUTIONS

In the history of science Louis Pasteur (1822–95) is celebrated for a range of major contributions to chemistry and microbiology, of which the earliest of his discoveries, made in 1848, was the most fundamental. After completing his studies (1843–46) at the École Normale Supérieure in Paris, Pasteur stayed on as a teaching assistant for research (1846–48). He studied the crystal forms of the tartrates isolated from wine lees and within a few months achieved the first resolution of optically active isomers from an inactive mixture by hand sorting crystals grown from solutions of sodium ammonium paratartrate (racemic tartaric acid) into two sets, distinguished by a non-superposable mirror-image morphology (Pasteur 1848).

The crystals with hemihedral facets oriented to the right were morphologically identical with the crystals formed by the sodium ammonium salt of naturally occurring (+)-tartaric acid and, in aqueous solution, they gave the same specific optical rotation quantitatively, with the same sign qualitatively, clockwise or positive, to the right. In contrast crystals of the other set, with hemihedral facets oriented to the left, produced in solution a quantitatively similar specific optical rotation, but qualitatively with the opposite sign, anticlockwise or negative, to the left.

From the morphological difference between the two sets of crystals, Pasteur surmised by analogy with the crystal forms that the new substance isolated, (−)-tartaric acid, has a molecular shape which is the non-superposable mirror image of the molecular form of its isomer, (+)-tartaric acid. The term 'optical isomerism' had previously been used to describe substances of the same elemental composition and general properties, which differed only in that one was optically active and the other inactive, as for (+)-tartaric acid and paratartaric acid. They were frequently termed 'physical isomers'.

The concept of optical isomerism was now enlarged in 1848 to cover a pair of molecules with equal and opposite optical rotation in the fluid phase, the pair forming an inactive racemate from an equimolecular mixture of the two active forms. For the non-superposable mirror image relationship of molecular or macroscopic enantiomorphs ('opposite forms') Pasteur coined the term 'dissymétrie', often translated as 'asymmetry', although dissymmetry is consistent with all elements of pure rotational symmetry, as in helical or dihedral propeller-shaped molecules. The term 'chirality' (from the Greek *chir* = hand) for the relationship was introduced later, notably by Kelvin in his *Baltimore Lectures* of 1904.

Pasteur's chemical insights were confined to molecular shapes, for there were no agreed sets of atomic and molecular weights or atomic valences, and no fruitful theory of molecular structure before the 1860s (Mason 1976). During Pasteur's lifetime, all of the substances found to be optically active in solution, where the optical rotation necessarily had a molecular origin, were natural products, or derivatives of natural products, from which he conjectured that living organisms had access to a chiral natural force for the biosynthesis of their optically active products. Pasteur attempted to characterize the proposed universal chiral agency while a professor of chemistry at Strasbourg (1849–54) and at Lille (1854–7). The experiments he conducted on the project are recorded in his laboratory notebooks and mentioned in correspondence over this period. Pasteur felt that if he could discover the cosmic force of dissymmetry, he would become "the Galileo or Newton of biology" (Geison 1995a).

The negative outcome of his experiments at Strasbourg and Lille on the chiral natural force were not made public until 1883, when he reviewed his studies of hemihedral crystals and optical resolution, and his conceptions of cosmological handedness (Pasteur 1883a). The motions of the heavenly bodies in the solar and sidereal systems are not superposable on their mirror-image forms, Pasteur argued, and so plant photosynthesis through solar radiation cannot be even-handed. Following the work of Faraday, who had shown in 1846 that glass and other isotropic transparent media become optically active in a magnetic field, Pasteur in 1853 grew crystals normally with a symmetric holohedral habit in the field of a powerful electromagnet, with the expectation that dissymmetric hemihedral facets would evolve, but to no avail.

At Lille in 1854 he used a clockwork heliostat with a mirror to present plants with the semblance of the sun rising in the west and setting in the east, expecting an influence of the earth's diurnal rotation on the optical activity of plant natural products generated photosynthetically (Pasteur 1883b). The failure of these experiments led Pasteur to a less direct approach, employing living organisms with their surmised command of the chiral natural forces, and the optically active natural products thereby generated with retention of chiral potency, as laboratory reagents for the isolation of individual enantiomers from racemic mixtures. By these means he found two more methods of isolating one or both of the optically active enantiomers from inactive paratartaric acid.

In 1853 Pasteur discovered that paratartaric acid with an optically active plant-alkaloid base formed two distinct diastereomeric salts, which have different solubilities and so may be readily separated by fractional crystallization (Pasteur 1853). In pursuit of the universal chiral force, Pasteur irradiated alkaloid tartrates with polarized light, but to no effect. After heating cinchonine tartrate to 170°C for several hours, however, he was able to extract from the carbonaceous residue a new inactive isomer, *meso*-tartaric acid, which he described as 'untwisted' in molecular shape (Pasteur 1853).

In experiments from 1858 Pasteur discovered a third method for the optical resolution of racemates, showing that the use of ammonium paratartrate as the carbon source for the growth of the mould *Penicillium glaucum* led to the preferential consumption of (+)-tartrate. The growth medium became progressively laevorotatory to an optimum value, at which point Pasteur isolated ammonium (−)-tartrate from the medium (Pasteur 1858).

Pasteur returned to Paris as the director of scientific studies and administrator of the École Normale Supérieure (1857–67) and then director of a new laboratory of physiological chemistry at the École (1867–88). He also held a chair of chemistry at the Sorbonne (1867–74) during the construction of the new laboratory at the École. The completion of the laboratory was delayed by the lack of funding during the 1870 Franco–Prussian war, the ensuing collapse of the Second Empire of Louis Napoleon followed by the Paris Commune, and by administrative changes brought in subsequently. Pasteur prospered in scientific and social standing under the Second Empire (1852–70) and the Third Republic (1871–1919), ending his career as director (1888–95) of the Institut Pasteur in Paris, founded from the donations which poured in after his discovery of an anti-rabies vaccine.

By 1860, when the Paris Chemical Society invited him to give two review lectures on molecular dissymmetry, Pasteur had abandoned hope of physically characterizing the cosmic chiral forces. These handed agencies appeared to be uniquely the monopoly of living organisms in which "dissymmetric forces exist at the moment of the elaboration of natural organic products; forces absent or ineffectual in the reactions of our laboratories" (Pasteur 1860). Pasteur reiterated this thesis in subsequent review lectures, delivered in 1874 and 1883.

The formation of optically active products from inactive substrates was a demarcation criterion, distinguishing the biochemistry of living organisms from the chemistry of the laboratory. The sole way of harnessing the natural chiral force, it seemed, was the employment of living organisms and their optically active products as reagents in the laboratory, as illustrated by his methods for the optical resolution of racemic mixtures. Pasteur conjectured in 1883 that the spontaneous formation of crystals with left- and right-handed hemihedral facets by sodium ammonium paratartrate, the basis for his first optical resolution, accomplished in 1848, might well have been because of optically active organic dust or pollen which had settled on the surfaces of his crystallization dishes (Pasteur 1883c).

THE CHEMICAL–PHYSICS BACKGROUND TO PASTEUR'S 1848 DISCOVERY

During the first half of the 19th century, chemistry developed empirically with no coherent theory generally accepted as a reliable guide—there were only a few principles, rules, and laws, apparently of limited application, to aid further chemical discovery. One such principle, that a crystal was built up from space-filling units which microscopically had the same geometric form as the macroscopic crystal, had been evolved by the mineralogists and crystallographers of the 18th century. A given substance might well crystallize with different habits, or secondary forms, which cleaved down to a common polyhedral primitive form, retaining geometric shape on further cleavage.

René Just Haüy (1743–1822), professor of mineralogy at the Paris Museum of Natural History, conjectured in 1809 that if cleavage were continued to the ultimate limit of the *molécule intégrante,* it would be found that the crystal and its constituent molecules were morphologically "images of each other" (Haüy 1809). Haüy's principle was influential in French chemistry throughout the 19th century (Mauskopf 1976a). The principle led Ampère in 1814 to an alternative version of the hypothesis (1811) of Avogadro, who had inferred the diatomicity of the elementary gases from Gay-Lussac's law (1808) of gaseous combination by integer volumes, as in the formation of two volumes of hydrogen chloride from one volume of hydrogen and one of chlorine.

Ampère argued that the space-filling units forming a macroscopic crystal are necessarily three dimensional, so that all molecules, including those of the elementary gases, must be at least tetra-atomic, or consist of some larger, even number of atoms (Ampère 1814). At mid-century, the principle led Pasteur to the conclusion that the molecules of (+)- and (–)-tartaric acid have non-superposable mirror-image shapes, analogous to the macroscopic morphology of the corresponding crystals formed by the sodium ammonium salts (Pasteur 1848). Near the end of the century, the principle gave Pasteur's disciple, Le Bel, grounds for the conclusion that the four valences of the carbon atom are directed to the apices of a distorted tetrahedron, because crystals of carbon tetrabromide and carbon tetraiodide proved to be birefringent and so are anisotropic, whereas isotropic crystals with cubic symmetry would result if the molecular building units were regular tetrahedra (Le Bel 1890–94a).

Le Bel was unfortunate, for had he crystallized carbon tetrabromide at or above 48°C, the temperature at which the molecules rotate on their lattice sites and become effectively spherical, he would have obtained cubic crystals. Indeed Le Bel noted that "above a certain temperature" the crystals seem to become cubic, but he took the observation no further. Pasteur was correspondingly fortunate in crystallizing sodium ammonium paratartrate below 27°C, obtaining two enantiomorphous sets of crystals—above 27°C the salt crystallizes as a holohedral racemate with a close-packed lattice in which the two enantiomeric tartrate ions are paired through an inversion centre (Kuroda & Mason 1981).

Haüy's principle was extended from geometric form to molecular and crystal chemistry and physics by subsequent crystallographers, with divergent consequences. Haüy held that the primitive crystal form of each individual substance was characteristic and specific to that substance, but other crystallographers, including his own students, found that crystals of related series of substances, such as the sulfates of the transition metals, are isomorphic and form mixed crystals (Burke 1966a). The discovery became a guiding principle in 19th century chemistry as the law of isomorphism, put forward in 1819 by Eilhard Mitscherlich (1794–1863) at Berlin. The law specified that "the same number of atoms combined in the same manner produces the same crystalline form, which is independent of the chemical nature of the atoms and determined only by the number and the relative positions of the atoms" (Burke 1966a).

Mitscherlich used the law to determine the relative atomic weights of selenium and sulfur from the isomorphism of corresponding selenates and sulfates. Haüy never accepted the law of isomorphism, but he was mollified by the finding that the intrafacial and interfacial angles of isomorphous crystals of different substances are usually exhibit slightly different.

One of Haüy's former students, Gabriel Delafosse (1796–1878), who taught Pasteur crystallography from 1843 to 1846 at the École Normale Supérieure, studied the secondary features of crystals in relation to their particular physical properties and the character of their constituent substances. Delafosse correlated the surface striations of crystal faces with the inferred shape and internal arrangement of the constituent molecules, and he drew attention to instances where small secondary facets, additional to the primary form of a crystal, were accompanied by unusual electrical or optical crystal properties (Mauskopf 1976b).

Jewellers had long known that tourmaline crystals become electrically charged at the ends on heating, and attracted pieces of straw or paper. In a 1756 study of crystal pyroelectricity, Franz Ulrich Aepinus (1724–1802) at Berlin attributed the effect to heat driving the positive and negative electrical fluids along the grain of the crystal striations to poles terminating the crystal axis. Wilhelm Gottlieb Hankel (1814–99) at Halle found in 1843 that pyroelectric crystals lacked an inversion centre of symmetry and had hemihedral crystal facets—such crystals had a single axis, at the ends of which electric charges developed with a change in the ambient temperature.

The discovery by Étienne Malus (1775–1812) in 1809 that light is plane-polarized by reflection stimulated the development of crystal optics, and led to the optical classification of crystals as isotropic, uniaxial, or biaxial, according to the number of different refractive indices (one, two, or three) required to account for the propagation of plane polarized light along the geometric axes of the crystal. In 1812 Jean Baptiste Biot (1774–1862) at the Collège de France in Paris found that the plane of polarized light undergoes a rotation on transmission through a plate of quartz cut perpendicular to the trigonal (pseudo-hexagonal) geometric axis of this uniaxial crystal. Such an orthoaxial crystal section was expected to be isotropic for light incident in the direction perpendicular to the slice. For a given wavelength of the light, Biot found that the rotation of the polarization plane had a magnitude proportional to the thickness of the crystal plate and a sign, sometimes to the right and at other times to the left, dependent on the particular quartz crystal from which the slice was cut.

Biot was aware, following Haüy before him, that many natural quartz crystals have obliquely-inclined (plagihedral) facets on alternate corners (hemihedral) of the hexagonal prism geometric shape which Haüy assigned to the primitive form of quartz. These plagihedral and hemihedral facets reduce the holohedral hexagonal symmetry of the primitive form down to the trigonal symmetry of the secondary forms found by mineralogists in the natural world.

Haüy distinguished two principal secondary forms of the quartz crystal—one set had the hemihedral obliquely inclined facets oriented to the right whereas those facets were oriented to the left in the other set. In 1820 the polymath John Herschel (1792–1871), while working with his father, the King's Astronomer, William Herschel (1738–1822) showed that slices of the crystal cut perpendicular to the trigonal axis rotated the plane of polarized light clockwise for one morphological set, and anticlockwise for the other set. John Herschel concluded that the morphological handedness of the secondary forms of the quartz crystal and the clockwise or anticlockwise rotation of the plane of polarized light have a common origin in the particular asymmetric arrangement of the molecules in the quartz crystal, producing hemihedral facets through the intermolecular forces and optical rotation through the forces between the asymmetric lattice and the light ray (Mauskopf 1976c).

The apparent anomaly that seemingly isotropic quartz plates rotate the plane of polarized light, whereas analogous orthoaxial sections of other uniaxial crystals do not, was resolved by the principal architect of the transverse-wave theory of light, Augustin Fresnel (1788–1827). In 1822 Fresnel discovered circularly polarized light, in which the transverse wave vector of the radiation traces out either a right-handed or left-handed helical path around the direction of propagation. Circularly polarized light was envisaged as two superimposed rays of plane-polarized rays with mutually perpendicular polarization planes a quarter of a wavelength out of phase, advanced or retarded, according to the handedness of the circular radiation.

On the basis of this model Fresnel invented two methods of producing circularly polarized light, the quarter-wave plate and the Fresnel rhomb, and a third method, the Fresnel triprism, based on his theory of the rotation of plane polarized light exhibited by orthoaxial slices of the quartz crystal.

Fresnel proposed that quartz and other optically active substances have different refractive indices for left- and right-circularly polarized light. A plane polarized beam on entering an optically active medium is resolved into superimposed left- and right-circularly polarized rays, which are differentially retarded during transmission through the medium, producing a phase shift between the two circular components. The beam emerges from the medium plane polarized, reconstituted from the superimposed circular components, and the phase shift produced during transmission is expressed as a rotation of the polarization plane relative to the original plane at incidence.

Fresnel attributed the circular birefringence of an optically active medium to the helicoidal shape or arrangement of the molecules constituting the medium (Fresnel 1824). His theory applied equally to optically active substances in fluid phases, where the optical rotation necessarily has a molecular origin. From 1815 Biot made extensive studies of the optical activity of organic natural products in the liquid and vapour phase or in solutions, in quest of a molecular basis for optical rotation.

Delafosse lectured on these several recent developments in crystallography, chemistry and optics at the École Normale Supérieure when Pasteur was a student 1843–46, as shown by Pasteur's surviving lecture notes (Mauskopf 1976d). When Pasteur was working on his doctoral theses, one on chemistry and the other on physics, presented in August 1847, he was also influenced by Auguste Laurent (1807–53), who was at the École Normale from late 1846 to April 1847. Laurent worked on the substitution reactions of aromatic substances, arguing that such reactions left unchanged the basic structure of the "aromatic nucleus" (a term he coined), for many properties of the product were only slightly different from those of the substrate.

Crystals of chlorostrychnine were virtually isomorphous with those of strychnine itself, so the structures of the constituent molecules in the crystals of the two compounds must be closely similar; both substances were, moreover, identically poisonous and equally optically active (Laurent 1855a). Under the influence of Laurent, Pasteur changed the topics of research for his doctoral theses to studies of the isomorphism of arsenious and antimonious salts in chemistry, and to investigations of the optical rotatory power of liquids in physics (Geison & Secord 1988).

Pasteur's studies 1846–47 of the optical rotatory power of liquid natural products and of tartrate salts in solution brought him into contact with Biot, whose polarimeter he used at the Collège de France, and with Biot's recent account of an anomalous relationship between optical isomerism and isomorphism, discovered by Mitscherlich.

During the 1830s studies were performed on two compounds with the same chemical constitution, one of which was optically active and the other inactive. Mitscherlich in 1831 had compared the crystal forms of corresponding salts of (+)-tartaric acid and inactive paratartaric acid, finding that they differed in crystal morphology in all cases except one, the sodium ammonium salt; these crystals seemed to be wholly isomorphous. At the time he attached no special significance to this singular case, but with further studies over the years the exception seemed increasingly anomalous. Mitscherlich repeated and confirmed his earlier crystallographic investigation of the tartrate and paratartrate salts; in 1843 he sent a summary of his findings to Biot, who published Mitscherlich's results with his own commentary in the following year (Mauskopf 1976e).

Many chemists found these results astonishing. Thomas Graham (1805–69) at University College London wrote in his textbook *Elements of Chemistry* (1842, 1850) that optical isomerism "defeated every attempt at explanation". "But what is most surprising", Graham continued, the sodium ammonium salts of (+)-tartaric acid and inactive paratartaric acid form crystals which "not only coincide in the proportion of their water and other constituents, and in the composition of their acids, but also in external form, having been observed by Mitscherlich to be isomorphous. A nearer approach to identity could scarcely be conceived than is exhibited by these salts" (Graham 1842, 1850).

Pasteur found himself well placed in 1847–48 for an attack on what was generally agreed to be a fundamental problem, the nature of optical isomerism, now prominent through the isomorphism anomaly identified by Mitscherlich. In his later years Pasteur often quoted the aphorism that chance favours the prepared mind in scientific discovery. His scientific outlook had been well prepared in his student years, 1843–46, by the lectures of Delafosse on the relationship between crystal morphology and the shape of the constituent molecules, and the connection of hemihedral crystal facets with particular crystal properties; by those on pyroelectricity by Hankel in 1843; and by those on the oppositely handed hemihedral facets of the two main secondary forms of quartz, noted by Haüy, and correlated with equal and opposite optical rotatory power by Herschel in 1822.

Pasteur was further influenced while working for his doctoral theses 1846–47 by Laurent's ideas on the architecture of molecules, with structural resemblances reflected in approximate crystal isomorphism and similar physical properties, including optical activity, and Biot's finding that optical rotatory power in the fluid phase seemed to be confined to biological natural products.

Pasteur began research in 1847–48 by repeating previous crystal studies of the tartrates, and he found that all the salts of (+)-tartaric acid he examined crystallized with hemihedral facets, a feature which had escaped Hervé de La Provostaye earlier, 1841–42. Pasteur concluded in his "law of hemihedral correlation" that all optically active substances afford crystals characterized by hemihedral facets, referring to the 1843 claim of Hankel that crystals with hemihedral facets are pyroelectric. Salts of the inactive paratartaric acid Pasteur examined gave symmetrical holohedral crystals, lacking the property of pyroelectricity, except for the anomalous case of the sodium ammonium salt discovered by Mitscherlich.

The hemihedral facets of the sodium ammonium paratartrate crystals implied that they were optically active tartrates in some sense. Pasteur noted that the facets of only about a half of these crystals had the same orientation as those of the (+)-tartrate salt crystals examined previously and, after separation, this set of crystals gave the same (+) optical rotation in solution. The facets of the other moiety of the crystals had a complementary or mirror-image orientation, and this group of crystals on isolation resulted in the converse (–) optical rotation in solution. Because optical rotation in the fluid phase implied that the rotatory effect is molecular, (+)-tartaric acid and the newly discovered (–)-tartaric acid, with a chemically identical constitution, must differ in molecular assembly, Pasteur surmised, distinguished by their non-superposable mirror-image shapes, like the morphological secondary forms of the corresponding macroscopic crystals (Fig. 1.1).

Pasteur's discovery gave a new and more concrete meaning to optical isomerism, now seen to describe a pair of dissymmetric molecules, chemically identical, and differing only in oppositely signed optical rotation in the fluid phase, and in oppositely handed molecular shape (enantiomers). The proposed chemical identity did not last for long, because in his second method of optical resolution Pasteur found in 1853 that (+)- and (–)-tartaric acid formed salts of different solubility with a plant alkaloid base.

Pasteur withdrew his "law of hemihedral correlation" in 1856 on finding that amyl alcohol produced by fermentation consists of an optically active and an inactive isomer. The two isomers proved to be readily separable by crystallizing the barium salt of the monoester formed by amyl alcohol with sulfuric acid (barium sulfamylate). The crystals formed by the two amyl alcohol isomers proved to be isomorphous and devoid of hemihedral facets (Pasteur 1856).

In the report of his first method of optical resolution, Pasteur acknowledged in 1848 the influence of previous workers, Haüy, Delafosse, de La Provostaye, Biot, Mitscherlich and Hankel, but he did not mention that Haüy had described left- and right-handed hemihedral forms of quartz, that Biot had found some orthoaxial sections of quartz to be dextrorotatory and others to be laevorotatory, nor that Herschel had established a one-to-one correlation between the sign of the optical rotation of the section and the hand of the hemihedral morphology of the quartz crystal from which the section was cut. Pasteur remedied the omissions in his two lectures on molecular dissymmetry in 1860 but, after acknowledging the ideas and the influence of Laurent in his doctoral theses of 1847, Pasteur never mentioned Laurent again with approval.

Laurent was primarily concerned with the development of mainstream chemistry, particularly with the interpretation of the structures and reactions of organic substances. He extended the traditional analogies between crystal form and the structure of the constituent molecules, and looked for "approximate isomorphism" between the crystals of a parent substance and those of a substituted derivative, assuming that substitution reactions produce no profound structural change. Laurent had the vision that, while it is not possible to deduce the molecular structure of a substance from its reactions, it is feasible to predict the likely reactions of a hypothesized structure and thereby test the validity of that structure. Laurent's posthumous *Méthode de chimie* (1853) was widely influential. The English translator of the book, William Odling (1829–1921), held "the generalities of Laurent to be in our day as important as those of Lavoisier were in his" (Laurent 1855b).

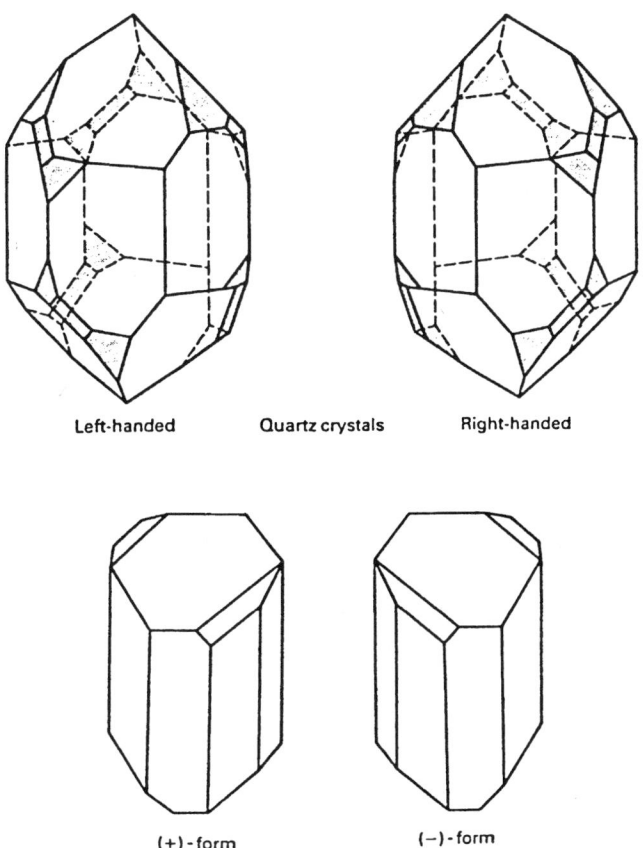

Fig. 1.1 The enantiomorphous crystal sets of quartz and of sodium ammonium tartrate. The minor crystal facets of a quartz crystal follow either a left- or a right-handed screw sequence viewed on end, along the direction of the threefold crystal axis. A section of the quartz crystal cut perpendicular to the trigonal axis is laevorotatory for plane polarized light propagated along the crystal axis if the crystal is morphologically left-handed, or dextrorotatory for the right-handed morphological form. Racemic sodium ammonium tartrate crystallized below 27°C forms two sets of crystals, similarly distinguished by their non-superposable mirror-image morphology, and by their oppositely signed optical rotation. The optical activity of the tartrate crystals occurs in solution, where the rotatory effect arises from the structure of the individual molecules, and not that of a crystal lattice, as in quartz.

Laurent was not only an outstanding chemist but also a fierce critic, particularly of the establishment figure of Jean Baptiste Dumas (1800–84), who held chairs of chemistry at the Paris Écoles and then the Sorbonne in the 1830s and 1840s. Dumas entered political life as Minister of Agriculture and Commerce and then Minister of Education after the revolutionary changes of 1848 under Louis Napoleon, who made Dumas a senator during the Second Empire 1852–70. Laurent was a radical liberal who supported the reforms of 1848, but opposed the imperial aspirations of Louis Napoleon. Laurent aspired to a chemistry chair in Paris, but never secured such an appointment; he died prematurely in 1853, age 46, a victim of tuberculosis, like Fresnel before him.

With the scientific acclaim that greeted his first method of optical resolution of 1848, Pasteur felt inclined to adopt Dumas as his role model and to decouple from Laurent. In a letter to Dumas in 1852 Pasteur deprecated the early influence Laurent had exercised on him as a student, and looked to Dumas for patronage (Geison & Secord 1988). Pasteur approved of the *coup d'etat* in December 1851 whereby Louis Napoleon dissolved the Constituent Assembly and declared himself Emperor Napoleon III. Dumas introduced Pasteur to the Emperor, who promoted Pasteur to commander of the Legion of Honour in 1868. An imperial decree of July 1870 awarding Pasteur a pension and making him a senator for life remained unfulfilled after the defeat of France in the Franco–Prussian war and the ensuing abdication of Louis Napoleon later in the year. Pasteur sustained imperial values under the Third Republic, but when he stood for election to the Senate as a conservative patriot in 1875 he was overwhelmingly defeated by the republican candidates (Geison 1995b).

After approximately 1850 Pasteur took little interest in the internal development of mainstream chemistry, except to defend his demarcation criteria of molecular dissymmetry and optical activity distinguishing the chemistry of the laboratory from the chemistry of life. Synthetic organic chemists continually encroached on his criteria, and they were actively repulsed. In 1850 Pasteur found that aspartic acid from asparagus is optically active, and in the same year Dessaignes synthesized aspartic acid by heating ammonium fumarate, an optically inactive substance. Pasteur was so surprised that he visited Dessaignes in Vendôme to obtain a sample of the synthetic aspartic acid, and he then showed that the synthetic compound was optically inactive, as he had expected.

After he had isolated *meso*-tartaric acid in 1853, Pasteur held that potentially dissymmetric substances existed in four forms, as he had now established for the tartrates—the optically active (+) and (–) enantiomers; the equimolecular racemic mixture of paratartaric acid, optically inactive by external compensation; and the untwisted *meso* form, optically inactive by internal constitution.

In his 1860 lectures on molecular dissymmetry Pasteur claimed that chemists at best could synthesize untwisted *meso* forms in the laboratory, and that the synthetic aspartic acid made by Dessaignes must be such a *meso* form. The synthetic aspartic acid could not be the racemic form, Pasteur held, "for in that case not only would *one* active have been made from an inactive one, but there would have been *two* such prepared, one dextro- and the other laevo-rotatory" (Pasteur 1860).

Pasteur clearly did not examine at all closely the material given him by Dessaignes, for synthetic aspartic acid is a racemate, readily resolved into its enantiomers through the different solubilities of its alkaloid salts, or by the crystallization of its amide, asparagine, when the enantiomers spontaneously crystallize as individual (+)- and (–)-rotatory crystals. He was soon obliged to concede that chemists could synthesize racemates, for Perkin and Duppa reported in 1860 an *ab initio* synthesis of tartaric acid from succinic acid, which had been prepared itself from ethylene (Perkin & Duppa 1860). Pasteur obtained a sample of the synthetic tartaric acid and found it was not the *meso* form, but the racemic paratartrate, which he resolved into (+)- and (–)-tartaric acid.

In later review lectures of 1875 and 1883 Pasteur retreated to the position that chemists cannot synthesize optically active compounds directly from inactive substances in the laboratory by standard procedures. As he himself had demonstrated, chemists must employ for such syntheses either living organisms or dissymmetric natural products biosynthesized through the universal chiral force uniquely at the command of those organisms. The spontaneous optical resolution of racemates by crystallization arose simply from optically active contaminants of biological origin in the crystallization medium. The distinction between the laboratory and life was not fundamental, however. "Not only have I refrained from posing as absolute the existence of a barrier between the products of the laboratory and those of life", Pasteur affirmed, "but I was the first to prove that it was merely an artificial barrier, and I indicated the general procedure necessary to remove it, by recourse to those forces of dissymmetry never before employed in the laboratory" (Pasteur 1883c).

Pasteur linked the biochemical autonomy of the various species of microbe, each with its own shape and form under the microscope, and each producing its own specific products from a given substrate, to the microbial command of the universal chiral force of nature. In 1862 Pasteur was awarded the Alhumbert Prize by the Paris Academy of Sciences for his studies demonstrating that bacterial spores are omnipresent in the atmosphere, and that the cases of spontaneous generation of living organisms claimed by Félix Pouchet and others could be attributed to contamination by airborne microbes (Geison 1995c).

Pasteur felt in 1883 that new life forms might well be induced if the cosmological forces of dissymmetry could be harnessed. "Life is dominated by the dissymmetric forces presented to us in their enveloping and cosmic action", Pasteur asserted, "What can one say of the development of organic species if it becomes feasible to replace cellulose, albumin, and their analogues in the living cell by their optical enantiomers? The outcome would be in part the discovery of spontaneous generation, if such is within our power" (Pasteur 1883c). There was something of the iatrochemical tradition of Paracelsus (1493–1541) and van Helmont (1577–1644) in Pasteur's natural philosophy—the chemical autonomy of fermentation and disease came from omnipresent and persistent germs or seeds, invested with singular powers which, although uncharacterized as yet, could hardly be denied.

MOLECULAR STRUCTURE THEORY AND STEREOSELECTIVE CHIRAL SYNTHESIS

After the international conference of chemists at Karlsruhe in 1860, sets of atomic weights and valences of the chemical elements became generally agreed. This development facilitated the construction of molecular models, based on the number and possible orientations of the valence bonds between atoms.

August Kekulé (1829–96), influenced by Laurent's *Chemical Method*, proposed from 1865 a flatland structure theory for aromatic molecules, based on the hexagonal ring model for benzene. Kekulé's theory was well supported by aromatic substitution reactions, which afforded the number of isomeric benzene derivatives, and their geometric type, predicted by the ring model. The structure theory of aliphatic substances remained less successful, with optical isomerism a major problem, until the theory became three-dimensional.

In 1874 Joseph Achille Le Bel (1847–1930) in Paris and Jacobus Henricus van't Hoff (1852–1911) at Utrecht independently proposed that the four valences of the carbon atom are directed to the vertices of a tetrahedron, so that four different groups bonded to a carbon atoms give rise to two enantiomorphous structures, differing only in that one is the non-superposable mirror image of the other (van't Hoff 1874; Le Bel 1874). This model accounted for all the soluble optical isomers then known, and van't Hoff went on to predict types of optical isomerism as yet undiscovered.

Pasteur did not mention van't Hoff in his review lectures of 1875 and 1883, but he often alluded to Le Bel, recognizing him as a disciple. Indeed Le Bel took up Pasteurian themes in his research. In 1874 Le Bel looked for dissymmetric forces that would overcome the usual production of racemates in organic synthesis: he suggested the use of optically active catalysts in thermal reactions, and he proposed the chiral agency of circularly polarized light to favour the formation of one of one optical isomer over the other in photochemical reactions. Le Bel, in looking for optical isomers, was inclined to regard as three-dimensional molecules generally considered to be planar. He adopted Ladenburg's trigonal prism structure for benzene, and considered a potential antiplanar structure for ethylene, pointing out that the 1:2 disubstituted derivatives would be resolvable into optical isomers in both cases.

Over a brief period, 1890–94, Le Bel adopted the aberrant view that the four valences of carbon have an orientation of lower symmetry than regular tetrahedral, because crystals of carbon tetraiodide and carbon tetrabromide are birefringent. On the principle of Haüy, that a crystal and its constituent molecules are morphological images, a regular tetrahedral molecule would be expected to crystallize in an isotropic form with cubic symmetry. Accordingly Le Bel supposed that ethylene must be a non-planar molecule and that its 1:2 derivatives would be resolvable into optical isomers. In 1892 Le Bel attempted to resolve citraconic acid by Pasteur's third method, employing microbial cultures to consume preferentially one of the enantiomers as a carbon source. He found that the growth medium did indeed become optically active, but in 1894 he showed that the optically active substance was (−)-methylmalic acid, formed by the stereospecific addition of the elements of water across the carbon–carbon double bond of citraconic acid, catalysed by a microbial enzyme (Le Bel 1890–94b) (Fig. 1.2).

Van't Hoff followed more closely the tradition of Laurent and Kekulé, developing detailed expectations of optical and geometrical isomerism for saturated and unsaturated organic molecules from the tetrahedral orientation of the carbon bond valences. Van't Hoff's predictions of the number and type of stereoisomers resulting from a chain of n bonded chiral carbon atoms, A–$[CXY]_n$–B, were initially tested and subsequently used as a guide by Emil Fischer (1852–1919) in his studies of sugars (1884–1908) and peptides (1908–19) (Freudenberg 1966a). Fischer found, as van't Hoff had foreseen, that there are 2^n stereoisomers, all optically active, if A and B are inequivalent groups. The 2^n stereoisomers are made up of $2^{(n-1)}$ chemically distinct diastereomers, each consisting of a pair of mirror-image enantiomers, taken to be chemically equivalent in reactions with achiral reagents before Fischer's time.

CH₃   COOH
  \   /
   C
  / \
 H   COOH

citraconic acid

       HO   COOH
        \   /
     H— C
         |
     CH₃—C
        / \
       H   COOH

(−)-methylmalic acid

Fig. 1.2. The structure of citraconic acid, which Le Bel (1892) attempted to resolve optically on the assumption that the atoms of the ethylene molecule are not coplanar. The use of a microbial culture for the attempted resolution gave an optically active product, identified by Le Bel (1894) as (−)-methylmalic acid, formed by the stereospecific addition of the elements of a water molecule across the carbon–carbon double bond of the substrate, catalysed by a microbial enzyme.

On ascending the series from a sugar with n chiral carbon atoms, $HOCH_2$–$[CHOH]_n$–CHO, Fischer found that the yields of the two diastereomeric sugar products with $(n+1)$ chiral centres were markedly unequal. Only achiral reagents were employed in the ascent. The addition of hydrogen cyanide to the aldehyde group was followed by the hydrolysis of the cyanohydrin adduct to the corresponding hydroxy acid, and then the reduction of the acid lactone with sodium amalgam to generate the homologous aldose. But in the ascent from the pentose, arabinose ($n=3$), to the two hexose homologues, mannose and glucose ($n=4$), Fischer showed that mannose is the major product, and glucose is produced in small yield, so small that it had escaped detection in the earlier work of Kiliani. Such stereoselectivity in the synthetic reactions of a chiral molecule with achiral reagents became more marked with chiral reactants, and even stereospecific in reactions mediated by enzyme catalysts or by microbes.

Fischer linked the various members of the sugar series by chemical methods of ascent and descent, and cross-correlations between the ketose and aldose series, to establish his D and L configurational convention, formally related back to an assumed absolute configuration for the parental triose, D-(+)-glyceraldehyde ($n=1$), by Rosanoff (1906). Subsequently Fischer's choice of configurational convention was found to be stereochemically correct by the anomalous scattering method in X-ray crystallography, introduced in 1951 by Bijvoet and his coworkers (Bijvoet et al. 1951) (Fig. 1.3).

Aided by the convention for relative configurations, Fischer compared the discriminatory reactions of the sugars with microbial and enzyme systems. He found that of fourteen different sugars tested with pure strains of yeast only four monosaccharides are fermentable, three from the aldohexose series (D-glucose, D-galactose and D-mannose) and one ketohexose (D-fructose). These hexose sugars have a common stereo configuration at carbon-5 (numbered from the aldehyde group), which specifies the D configuration through chemical correlations back to the unique chiral centre of D-(+)-glyceraldehyde. But the D configuration, although necessary, was not a sufficient condition for reaction, because D-talose proved to be a non-fermentable aldohexose. Individually all of the chiral carbon atoms of D-talose have the same stereo configuration as one or more of the fermentable hexose sugars, but the particular sequence of chiral centre configurations is distinctive, and incompatible with the enzyme catalysts of the fermentation process (Fig. 1.4).

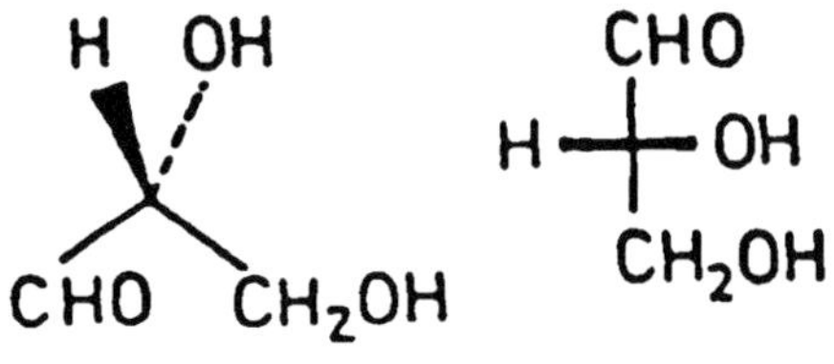

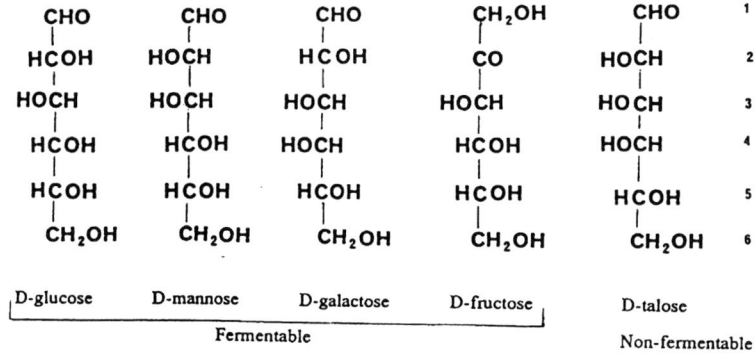

Fig. 1.3. The absolute stereochemical configuration of D-(+)-glyceraldehyde (left), and the corresponding projection structure, according to Fischer's convention that groups linked to a chiral carbon atom lie above a plane passing through the carbon atom if the bonds are depicted as horizontal lines, or behind the plane if the bonds are represented by vertical lines.

Fig. 1.4. The hexose monosaccharides found by Emil Fischer (1894) to be fermentable with pure yeast strains (D-glucose, D-mannose, D-galactose, and D-fructose), or non-fermentable (D-talose).

Fischer went on to study the action of particular enzyme systems on the two types of glucoside found in natural products, or synthesized in the laboratory. The treatment of D-glucose with methyl alcohol and a trace of acid as a catalyst generates two isomers, α- and β-methyl-D-glucoside, through the introduction of an additional chiral centre by semi-acetal formation at the aldehydic group of the sugar. Fischer discovered that emulsin, first isolated from almonds by Liebig and Wöhler in 1837, hydrolyses only a β-D-glucoside, whereas maltase, obtained with invertase from yeast by Berthelot in 1860, is active solely with an α-D-glucoside as a substrate. Both enzyme preparations were found to be inactive to the corresponding enantiomeric L-glucoside substrates and to other semi-acetal sugar derivatives.

From these and analogous studies, Fischer concluded that "the difference frequently assumed in the past to exist between the chemical activity of living cells and of chemical reagents, in regard to molecular asymmetry, is non-existent". There is a progressive scale of chiral discrimination from the stereoselectivity of achiral reagents in the synthetic reactions of the sugars to the stereospecificity of fermentations or the enzyme-catalysed hydrolyses of the glycosides. In the latter case, the enzyme and its specific substrate have complementary structures: "the enzyme and the glucoside must fit each other like a lock and key, in order to effect a chemical reaction on each other". The scale of chiral discrimination from stereoselectivity to stereospecificity offers "a simple solution to the enigma of natural asymmetric synthesis". Starting with a single enantiomer, or even with an enantiomeric excess in an otherwise racemic mixture, synthetic reactions lead inevitably to a dominant diastereomeric product favoured by the steric congruence of the reaction intermediates: "once a molecule is asymmetric, its extension proceeds also in an asymmetric sense". Neither a *vis vitalis* internal to the organism, nor Pasteur's universal chiral force of nature, are required to account for the predominance characterized by Fischer of the D series of sugars and the L series of amino acids in the homochiral biochemistry of living organisms. The new perspective left unsolved, as Fischer appreciated, the problem of the origin of the primordial enantiomer from which stereoselective synthesis began (Freudenberg 1966b).

CHANCE, NECESSITY, AND CHIRAL FIELDS IN THE EVOLUTION OF BIOMOLECULAR HOMOCHIRALITY

During the late 1890s, when Fischer's new perspective began to emerge, a controversy arose in the columns of *Nature* on the origin of biomolecular optical activity; this was sparked off by the Presidential address of Francis Japp (1848–1925) on 'Stereochemistry and Vitalism' to the chemistry section of the British Association meeting at Bristol in 1898 (Palladino 1990). Japp, an Aberdeen organic chemist of Dutch Catholic descent, argued that a directive vital force had come into play at the moment of organic creation, "a force of precisely the same character as that which enables the intelligent operator, by the exercise of his Will, to select one crystallized enantiomorph and reject its asymmetric opposite" (Japp 1898).

Japp found no supporters for his vitalism in the ensuing *Nature* debate over the months following. Most of his critics favoured a "chance" origin of the prebiotic enantiomer from which biochemical homochirality had evolved, but a minority attributed the ancestral enantiomer to asymmetric physical causes. Japp replied to his critics that all known physical causes are necessarily symmetrical, just as 'chance' favours equally the production of either of the two mirror image enantiomers.

Percy Frankland (1848–1946), a Birmingham chemist who had delivered the Pasteur Memorial Lecture to the Chemical Society of London in 1897 and had published, with his wife, a biochemist, a scientific biography of Pasteur, contributed to the *Nature* debate the proposal that polarized solar radiation had been a chiral agency generating the primordial enantiomer (Frankland 1898; Frankland & Frankland 1898). Here Frankland was referring to recent studies in France of the theory of chiral forces, and the discovery of the differential absorption of left- and right-circularly polarized light by coloured optically active substances.

In 1894 Pierre Curie (1859–1906) at the École de Physique et Chimie in Paris analysed the general symmetry relations between a cause and its effect in physical phenomena (Curie 1894). Curie showed that, while an effect might have a higher symmetry than its cause, any asymmetry in an effect must be present also in the cause.

Each of the natural forces that Pasteur had considered to be potentially dissymmetric was found by Curie to be individually symmetric to mirror plane reflection. But a collinear combination of two complementary types of force, Curie demonstrated, one a rotatory (axial) force and the other a linear (polar) vector force, provides two possible chiral force fields, with the antiparallel combination affording the non-superposable mirror-image enantiomorph of the corresponding parallel combination. Rotatory and linear motion combined give a helical motion, which is right-handed if the axial and polar vectors are parallel, or left-handed if they are antiparallel. Similarly a magnetic (axial) field and an electric (polar) field lose their individual mirror-plane symmetry in a collinear union. When the two components oscillate at a common frequency in an electromagnetic field, the two enantiomorphous combinations are represented by left- and right-circularly polarized light.

In 1874 Le Bel had conjectured that circularly polarized light constitutes a chiral agency favouring the formation in photochemical reactions of one member of two mirror-paired optical isomers. Two decades later Curie provided a theoretical basis for the surmise, which was soon strengthened and given a practical basis by the discovery in 1895 of circular dichroism, the absorption counterpart of optical rotation (circular birefringence) over transparent wavelength regions.

In the year Pasteur died, Aimé Cotton (1869–1951), at the Paris École Normale, where Pasteur had studied half a century earlier and worked from 1857 to 1888, made the discovery that, at an absorption wavelength, an optical isomer in homogeneous solution differentially absorbs left and right circularly polarized light. For the enantiomeric molecule at the same wavelength the differential absorption is equal in magnitude but opposite in sign, analogous to the optical rotation at a wavelength in the region of optical transparency (Cotton 1895).

The 'Cotton effect', as circular dichroism came to be termed, implied that the irradiation of a photo-labile racemate in solution with left- or right-circularly polarized light at an absorption wavelength would enrich the mixture in the enantiomer with the smaller absorption coefficient for the circular component employed. This expectation led to a minor but enduring tradition in which circularly polarized light from the sun or other celestial source was invoked as the physical cause of an original enantiomeric excess, which evolved through autocatalytic stereoselective reactions to biomolecular homochirality (Bonner 1972).

Cotton himself searched for the expected photoresolution of racemates by means of circularly polarized radiation, but without success. The first partial optical resolution of a racemate in solution by means of monochromatic circularly polarized light was accomplished in 1929 by the Heidelberg physical chemist Werner Kuhn (1899–1963), who went on to demonstrate that broad-band 'white' circularly polarized radiation, while photochemically active, cannot discriminate between the enantiomers of a racemate (Kuhn 1930). Enantioselective photolysis or photosynthesis requires the irradiation of the racemic substrate with circularly polarized light tuned to the wavelength of a particular circular dichroism absorption maximum, so that one enantiomer absorbs more radiation and undergoes a greater photochemical change than the other.

Broad-band circularly polarized radiation lacks chiral discrimination, for the circular dichroism absorption bands of every chiral molecule alternate in sign along the wavelength ordinate, and the frequency-weighted band areas sum to zero over the electromagnetic spectrum as a whole. His derivation of the rule for the vanishing sum of the optical rotatory power of an enantiomer over all radiation frequencies led Kuhn to the mainstream view of the period, that the prebiotic origin of the initial biomolecular enantiomer had been wholly a matter of chance. In his later years, Kuhn saw chance also at work in the spontaneous racemization of the structural and functional biomolecules of an organism over time; he considered this racemization process to be responsible for the biological process of ageing (Kuhn 1958).

PARITY AND ITS NON-CONSERVATION

During the 1920s it became clear that the electromagnetic and other chiral fields characterized by Curie are even-handed over a time and space average, and that their proposed role as originators of biomolecular chiral homogeneity requires special and non-typical conditions. Furthermore, even granted special conditions, there was no basis other than chance in the particular choice of the series of the D sugars and the L amino acids as the foundation of the biomolecular chiral homogeneity of terrestrial organisms, as opposed to the enantiomeric series of L sugars and D amino acids.

The apparent even-handedness of the forces of nature was elevated in 1927 to the principle of the conservation of parity by Eugene Wigner, then in Berlin. Wigner postulated that all physical causes and the laws linking them to the effects produced are invariant to spatial inversion through a co-ordinate origin (the parity operation) or, what is equivalent, they are unchanged by a mirror-plane reflection (Wigner 1927). Among theoretical physicists, "conservation of parity quickly became a sacred cow", sometimes with bizarre exemplifications, as in the 1927 report to *Nature* from Bohr's Institute in Copenhagen that, among Danish cattle, there are approximately equal numbers of right-circular and left-circular cud-chewers, providing a pastoral manifestation of universal parity conservation (Frauenfelder & Henley 1975).

The natural forces assumed to conserve parity included the strong and the weak nuclear interaction, accounting for α- and β-radioactivity, respectively. Studies of electron polarization and scattering in 1928 and 1930 using β-emitting radionucleides as beam sources gave puzzling results which, in retrospect at least, showed that parity is violated in the weak interaction (Franklin 1979). The accumulation of other anomalies over the years led ultimately to the conclusion in 1956 that parity is not conserved in the weak interaction, and tests designed to detect parity violation soon confirmed this (Mason 1988).

The β-decay experiments establishing that parity is not conserved in the weak interaction demonstrated that fundamental particles have an intrinsic handedness or helicity. The electrons emitted in β-decay are inherently left-handed, with the spin axis preferentially orientated antiparallel to the linear momentum direction, whereas the corresponding antiparticles, β-positrons, are right-handed, with a parallel alignment of the spin axis and momentum direction. Although parity itself is not conserved in the weak interaction, the combination (CP) of parity (P) and charge conjugation (C, the conversion of a particle into the corresponding oppositely charged antiparticle) is conserved, to a close approximation. The inclusion of time-reversal (T) gives the stricter and more complete principle of CPT conservation.

According to the principle of CP conservation, the negatively charged electron and the positively charged positron are CP mirror image forms; the hydrogen atom, composed of an electron and a proton, has a CP enantiomer made up of a positron and an antiproton. As chiral entities to the CP mirror, all atoms are expected to be optically active, giving an optical rotation proportional to the sixth power of the atomic number. This expectation was pursued through the 1970s, and progressive experimental refinements culminated in the early 1980s with the recording of an optical rotatory power for heavy metal atoms in the gas phase (bismuth, lead, thallium and caesium) in agreement with the sign and magnitude calculated (Mason 1988).

Again, according to the principle of CP conservation, a chiral molecule composed of normal solar-system atoms, such as L-alanine, has a true CP mirror image enantiomer made up of antiatoms in a counter-world of antimatter, whereas the terrestrial enantiomer with the standard particle composition (D-alanine) has properties dependent upon the weak nuclear force which are inequivalent, because of the violation of the simple parity equivalence. The inclusion of the electroweak interaction (the unified neutral weak force with electromagnetism) in *ab initio* quantum mechanical calculations of the binding energy of the electronic ground state show that the L amino acids, the L polypeptides, and common D sugars are slightly more stable than their respective terrestrial enantiomers (Mason 1988).

The neutral electroweak interaction has operated at all times and places, and it provides an advantage factor of the correct sign to the chiral molecules on which the biomolecular homochirality of terrestrial organisms is based, the L amino acids and the D sugars, characterized by Fischer. The classical chiral fields specified by Curie require special initial conditions for either sign of chiral discrimination, the northern or the southern hemisphere, dawn or dusk, a particular hand of circularly polarized light at a chosen wavelength, and so on. The electroweak advantage for the L amino acids and the D sugars is small: it is equivalent to an enantiomeric excess of approximately one million molecules of the energetically favoured isomer in one gram mole of the racemate in thermodynamic equilibrium at earth surface temperature. Mechanisms have been proposed for the amplification of a small enantiomeric excess of this order to homochirality on the early earth (Mason 1991).

There is evidence of a significant enantiomeric excess of L amino acids in the extraterrestrial carbonaceous meteorites which originated from the organic substances, ices and grains in the giant interstellar molecular cloud from which the solar system was formed some 4.6 billion years ago. After the fall in 1969 of the Murchison meteorite in Victoria, Australia, analyses of the amino-acid content of its carbonaceous material demonstrated an enantiomeric excess of the L isomer of the natural protein amino acids.

The L enantiomer excess was generally ascribed to terrestrial contamination of the carbonaceous material after the fall of the meteorite, because the α-hydrogen atom of the protein α-amino acids is relatively labile in the presence of water. The carbonaceous meteorites are hydrated to some extent, and the lability of the α hydrogen atom results in the racemization of the protein amino acids in an aqueous environment over time. The half-life for the racemization of protein amino acids in fossil bones, teeth and shells ranges from a few thousand to a million or so years at earth surface temperatures from the tropics to the Arctic, whereas the α-methyl analogues of these amino acids are resistant to racemization (Bada 1991).

Later chiral analyses of the organic substances in the Murchison carbonaceous meteorite focussed on four of the α-methyl-α-amino acids which are relatively abundant in the carbonaceous material but are unknown in the biomolecules of terrestrial organisms or (for isovaline) of limited occurrence. These four non-protein amino acids exhibit an L-enantiomeric excess over the range from 2 to 9 per cent. It is concluded that a chiral influence on chemical evolution was active at least 4.6 billion years ago in the materials of the proto-solar system, before the origin of terrestrial life around 3.8 billion years before the present (Cronin & Pizzarello 1997). The magnitude of the enantiomeric excess reported would probably serve as an adequate basis for the sequence of stereoselective reactions leading to biomolecular homochirality envisaged by Fischer at the turn of the 19th century (Freudenberg 1966b).

After the detection of parity violation in the weak interaction, Haldane (1960) commented that the discovery had vindicated Pasteur's concept of cosmic dissymmetry. The comment requires the qualification that the weak nuclear force has neither the magnitude nor the biosynthetic role that Pasteur envisaged. The neutral electroweak interaction, however, has the chiral signature corresponding to the L amino acid and D sugar series characterized by Fischer in the biomolecular homochirality of terrestrial organisms, and the L signature is now extended to the prebiotic extraterrestrial amino acids.

The universal handedness perceived by Pasteur was limited to the solar system and the constellations of the fixed stars, to hemihedral crystals, and to the products of biosynthesis. Cosmological asymmetry and handedness now extends from the overwhelming predominance of particles over antiparticles, arising from the violation of approximate CP conservation within the exact principle of CPT conservation, to the assemblies of galactic clusters and superclusters, where the left-handed spirals are favoured. Surveys of the winding directions of the spiral galaxies demonstrate a significant excess of galaxies rotating anticlockwise (S-type, left-handed) over those rotating clockwise (Z-type, right-handed) as they recede from the earth. The enantiomorphous excess of S-rotational forms amounts to some 9.8 per cent for our Local Supercluster of galaxies (Mason 1988).

REFERENCES

Ampère A. M. (1814) Lettre sur la détermination des proportions dans lequelles les corps se combinent d'apres le nombre et la disposition respective des molecules dont leurs particules intégrantes sont composées. *Ann. Chim.*, **30**, 43–86. English translation by S. H. Mauskopf (1969) *Isis* **60**, 61–74.

Bada J. L. (1991) Amino acid cosmogeochemistry. *Philos. Trans. R. Soc. London*, **B333**, 349–58.

Bijvoet J. M., Peerdeman A. F. & van Bommel A. J. (1951) Determination of the absolute configuration of optically active compounds by means of X-rays. *Nature*, **168**, 271–2.

Bonner W. A. (1972) Origins of molecular chirality. In: *Exobiology* (Ed. by C. Ponnamperuma) pp. 170–234. North–Holland, Amsterdam.

Burke J. G. (1966a) *Origins of the Science of Crystals*, pp. 107–46. University of California Press, Berkeley and Los Angeles.

Burke J. G. (1966b) *Origins of the Science of Crystals*, p. 131. University of California Press, Berkeley and Los Angeles.

Cotton A. (1895) Absorption inégale des rayons circulaires droit et gauche dans certaines corps actifs. *Compt. Rend. Acad. Sci.*, **120**, 989–91.

Cronin J. R. & Pizzarello K. (1997) Enantiomeric excesses in meteoritic amino acids. *Science*, **275**, 951–5.

Curie P. (1894) Sur la symétrie dans les phénomènes physiques. *J. Phys (Paris)*, **III**, 393–416; in: the *Oeuvres de Pierre Curie* (1908) pp. 118–41. Gauthiers–Villars, Paris.

Frankland P. F. (1898) Asymmetry and vitalism. *Nature*, **59**, 30–31.

Frankland P. F. & Frankland G. C. (1898) *Pasteur*, Cassel, London.

Franklin A. (1979) The discovery and the nondiscovery of parity nonconservation. *Stud. Hist. Philos. Sci.*, **10**, 201–57.

Frauenfelder H. & Henley E. M. (1975) *Nuclear and Particle Physics*, p. 359. Benjamin, Reading, Mass.

Fresnel A. (1824) Considerations Théoriques sur la polarization de la lumière. *Bull. Sci. Soc. Philomathique*, 147–58. (English translation in Lowry T. M. (1935) *Optical Rotatory Power*, pp. 13–19. Longman, Green & Co., London.)

Freudenberg K. (1966) Emil Fischer and his contribution to carbohydrate chemistry. *Adv. Carbohydr. Chem.*, **21**, (a) 1–38; (b) 34–38.

Geison G. L. (1995) *The Private Science of Louis Pasteur*, Princeton University Press, Princeton, NJ; (a) pp. 37, 139; (b) pp. 44–45; (c) pp. 110–42.

Geison G. L. & Secord J. A. (1988) Pasteur and the Process of Discovery: The Case of Optical Isomerism. *Isis*, **79**, 6–36.

Graham T. (1842) *Elements of Chemistry*, pp. 157–8. Bailliere, London.

Graham T. (1850) *Elements of Chemistry*, p. 183. Bailliere, London.

Haldane J. B. S. (1960) Pasteur and cosmic asymmetry. *Nature*, **185**, 87.

Haüy R. J. (1809) *Tableau Comparatif des Résultats de la Crystallographie et de l'Analyse Chimique relativement à la Classification des Minéraux*, Paris, p. xvii.

Japp F. R. (1898) Stereochemistry and vitalism. *Nature*, **58**, 452–60, 616–18.

Kuhn W. (1930) The physical significance of optical rotatory power. *Trans. Faraday Soc.*, **26**, 293–308.

Kuhn W. (1958) Possible relation between optical activity and ageing. *Adv. Enzymol.*, **20**, 1–29.

Kuroda R. & Mason S. F. (1981) Crystal and Molecular Structures of Dextrorotatory and Racemic Sodium Ammonium Tartrate. *J. Chem. Soc. Dalton Trans.*, 1268–73.

Laurent A. (1855) *Chemical Method* (English translation by W. Odling), Cavendish Society, London; (a) p. 203; (b) p. viii.

Le Bel J. A. (1874). English translation in: Benfey O. T. (Ed.) (1963) *Classics in the Theory of Chemical Combination*, paper 9, pp. 161–71. Dover, New York.

Le Bel J. A. (1890–94). In: Delépine M. M. (1949) *Vie et Oeuvres de Joseph-Achille Le Bel*, Dupont, Paris; (a) pp. 85–97; (b) pp. 93–97 & 114–16.

Mason S. F. (1976) The Foundations of Classical Stereochemistry. *Top. Stereochem.*, **9**, 1–34.

Mason S. F. (1988) Biomolecular homochirality. *Chem. Soc. Rev.*, **30**, 347–59.

Mason S. F. (1991) *Chemical Evolution: Origins of the Elements, Molecules and Living Systems*, pp. 280–3. Clarendon Press, Oxford.

Mauskopf S. H. (1976) Crystals and Compounds: Molecular structure and composition in nineteenth century French science. *Trans. Am. Philos. Soc.*, **66** (3), (a) 5–82; (b) 51–55; (c) 62–64; (d) 70; 66–68.

Palladino P. (1990) Stereochemistry and the nature of life: mechanist, vitalist, and evolutionary perspectives. *Isis*, **81**, 44–67.

Pasteur L. (1848) Mémoire sur la relation qui peut exister entre la forme cristalline et la composition chimique, et sur la cause de la polarization rotatoire. *Compt. Rend. Acad. Sci.*, **26**, 53–38; in: Pasteur Vallery-Radot (Ed.) (1922) *Oeuvres de Pasteur*, Vol. I, *Dissymétrie Moléculaire*, pp. 61–64. Masson, Paris.

Pasteur, L. (1853) in *Oeuvres*, **I**, pp.258–62.

Pasteur, L. (1856) in *Oeuvres* **I**, pp.275–9, 284–8.

Pasteur L. (1858) in *Oeuvres*, **II**, pp. 25–28, 129–30.

Pasteur L. (1860) in *Oeuvres*, **I**, pp.314–44. (Page 342 is quoted. English translation, *Researches on Molecular Asymmetry, by Louis Pasteur (1860)*, Alembic Club Reprint No. 14, Edinburgh, 1948, p. 43.)

Pasteur L. (1883) in *Oeuvres*, **I**; (a) pp. 369–86; (b) pp. 375–6; (c) p. 377.

Perkin W. H. & Duppa B. F. (1860) Synthesis of tartaric acid. *J. Chem. Soc.*, **13**, 102.

Rosanoff M. A. (1906) On Fischer's classification of stereo-isomers. *J. Am. Chem. Soc.*, **28**, 114–21.

Van't Hoff J. H. (1874). English translation in: Benfey O. T. (Ed.) (1963) *Classics in the Theory of Chemical Combination*, paper 8, pp. 151–60. Dover, New York.

Wigner E. (1927) Einige Folgerungen aus der Schrödingerschen Theorie für die Termstrukturen. *Z. Phys.*, **43**, 624–52.

Chapter 2
The Origin of Biomolecular Chirality

Alexandra J. MacDermott
Department of Chemistry, University of Cambridge, Lensfield Road, Cambridge CB2 1EW, UK

INTRODUCTION

The question of the *origin* of biomolecular chirality—*why* life is based on L amino acids and D sugars and not their mirror images—has been an unresolved puzzle since Pasteur's epoch-making hand resolution of tartaric acid in 1848. Pasteur himself believed that biological homochirality came about because the universe was intrinsically handed in some way, but he failed to identify the source of this handedness during his lifetime (S.F. Mason, this volume). In this chapter some possible chiral influences are reviewed and the question is asked whether the parity-violating weak force is the cosmic dissymmetry that Pasteur was looking for.

HOMOCHIRALITY—A HALLMARK OF LIFE

A characteristic hallmark of life is its *homochirality*—biology uses only one enantiomer and not the other, with animals made of proteins based exclusively on L amino acids, coded for by DNA based exclusively on D-deoxyribose. Although a few 'unnatural' D amino acids and L sugars do occur with specific roles, as in bacterial cell walls (but not bacterial protein or nucleic acids) and antibiotics (Ulbricht 1981; Meister 1965), these examples in fact illustrate the dominance of the L amino acids and the general rule of homochirality—D amino acids are used in cell walls precisely because the enzymes of predators only digest L amino acids; furthermore, the genetic code does not code for these D amino acids, which are synthesized by enzymatic inversion of the L form (Spach & Brack 1983).

Homochirality is essential for efficient metabolism, like the universal adoption of right-hand screws in engineering. Fischer (1894) recognized this in his stereochemical 'lock and key' hypothesis—a chiral molecule can 'feel' the difference between the enantiomers of another chiral molecule, just as a left foot can feel the difference between left and right shoes and will only fit comfortably into the left. But molecules do not always interact preferentially with other molecules of the same hand as with feet and shoes—the preference depends on the situation, *e.g.* a right hand prefers to *shake* another right hand (standing facing each other) but prefers to *hold* a left hand (standing side by side).

It is chirally discriminating interactions of this kind that produce the connection between the L amino acids and the D sugars (Wolfrom *et al.* 1949, Melcher 1974), making L amino acid/D sugar life (or equally D amino acid/L sugar life) preferable to L amino acid/L sugar or D amino acid/D sugar life. In this way, whichever hand is selected in ancestral biomolecules will dictate the handedness of the rest of biology through diastereomeric connections.

The importance of homochirality in biological systems is underlined by the fact that the 'wrong' hand often has destructive effects. If an enzyme incorporates even one D amino acid, it will be the wrong shape, and will therefore be inactivated, and molecules of the 'wrong' hand have been implicated in the processes of ageing and carcinogenesis (Ulbricht 1962, 1981); the body therefore contains special enzymes called D-amino-acid oxidases to eliminate amino acids of the wrong hand. A classic case was the 1960s sedative drug thalidomide, which was sold as a racemic mixture but only one hand caused the limb deformities in babies born to mothers taking the drug during pregnancy (Blaschke *et al.* 1979).

HOMOCHIRALITY—ALSO A PRE-CONDITION FOR LIFE?

It is clear from the above that homochirality is certainly at least a *consequence* of life, and it is now believed it might also be a *pre-condition* for life. This is because polymerization to give the necessary long-chain homochiral polymers (*e.g.* all-L polypeptides or all-D nucleic acids) will not go efficiently in racemic solution even when directed by a homochiral template. Joyce *et al.* (1984) showed that poly(C)-directed formation of poly(G) is strongly chiroselective, in that only the D monomers get incorporated into the growing poly(G) chain, but that the reaction is severely inhibited by the presence of the L monomer, and this *enantiomeric cross-inhibition* effect has been confirmed in a recent study by Orgel (1996).

An almost homochiral monomer solution thus seems to be needed for efficient polymerization, so a homochiral pre-biotic chemistry might be a pre-condition for life. If so, the origin of biomolecular chirality is not just a matter of curiosity, it is one of the central problems of the origin of life—a chiral influence might be needed to get life started, and the nature of that chiral influence will determine which hand was selected.

THE NEED FOR A CHIRAL INFLUENCE

Where we have chirality we must have homochirality for an efficient biochemistry—but is life necessarily chiral? The answer does seem to be "yes", because only the smallest and simplest molecules are achiral. In any homologous series chirality is normally only absent for the very lowest members, and becomes increasingly common with increasing size, *e.g.* for the alkanes, C_nH_{2n+2}, chirality becomes possible at $n = 5$; for buckminsterfullerenes the smallest chiral molecule is C_{28}, and there are often several different chiral isomers of the higher homologues. Molecules large enough and complex enough to support life are thus almost certain to be chiral, and therefore homochiral. A possible 'achiral' replicator based on glycerol instead of deoxyribose has been proposed (Schwartz & Orgel 1985), but even here it is only the backbone that is achiral—it becomes chiral as soon as it coils up into a base-paired double helix analogous to DNA, so at some stage the choice must be made between left and right hand helix.

But is a chiral influence really needed? Although the view that homochirality (produced by some physical influence) must *precede* life is now widely held, there are those who consider that homochirality emerged *with* life, without the need for a chiral influence. This view is exemplified in the recent work of Eschenmoser (Bolli *et al.* 1997). It has already been seen that chiroselective polymerization requires a homochiral monomer solution because of enantiomeric cross-inhibition. Eschenmoser reasoned, however, that *diastereomeric* cross-inhibition might not be as fatal as enantiomeric cross-inhibition. He therefore investigated a system in which the 'monomers' were, in fact, tetramers, in this example not of normal RNA, but of 'p-RNA', which has pyranose instead of furanose rings. Monomers or oligomers of RNA analogues cannot polymerize on their own without enzymes unless the terminal phosphate group is artificially 'activated', *e.g.* by formation of the 2′,3′-cyclophosphate.

Eschenmoser found that activated tetramers of p-RNA oligomerized over several weeks to form octamers, 12-mers and 16-mers in good yield. He used mixtures of the normal all-D tetramers with diastereomeric versions containing L nucleotides, *e.g.* DDDL, DDLD, DLDL, *etc*. He found that the polymerization was indeed chiroselective, in that only DDDD tetramers were incorporated into the polymers—and, most importantly, the polymerization was only slightly inhibited by the presence of the diastereomeric tetramers.

However, Eschenmoser does not seem to have included LLLL tetramers in his reaction mixture, so he has not overcome the enantiomeric (as opposed to merely diastereomeric) inhibition problem. Nevertheless, his experiment is obviously a step in this direction, and he extrapolates it to a thought-experiment in which the enantiomeric inhibition problem *has* been overcome, and a racemic mixture of all possible diastereomeric tetramers (including LLLL and DDDD as well as DLDL, LLLD, *etc*) auto-oligomerizes chiroselectively to all-L and all-D polymers. But this would still give equal amounts of all-L and all-D polymers. With starting tetramers of random sequence, however, the number of possible n-mers is 4^n, which greatly exceeds Avogadro's number even when n is quite modest.

As an example, the number of possible sequences exceeds the number of actual molecules by a factor of ten million for 50-mers ($4^{50} = 10^{30}$) using 1-mole samples (10^{23} molecules), and for 30-mers ($4^{30} = 10^{18}$) using picomole samples (10^{11} molecules). This means that in any finite sample the all-L n-mers that are formed constitute only a tiny subset of the 4^n possible sequences; and if the polymerization is random, the all-D n-mers will constitute a different tiny subset from the all-L n-mers. In other words, although there are equal numbers of all-L and all-D polymers, it is not a racemic mixture as they have different sequences. Some sequences will then do better than others in a subsequent evolutionary process—they might be more resistant to decomposition, or they might act more efficiently as templates for the assembly of further copies of themselves—and these selected sequences will be homochiral (if we assume that no sequence appears in both the all-D and all-L subsets) and will happen to be either D or L, thus fixing the handedness of biology.

One might feel that Eschenmoser has not really overcome the enantioselective cross-inhibition bottleneck to the origin of life, but his subtle concept of *de-*

racemization by random polymerization of finite samples is undoubtedly an idea of great elegance that might well have swung opinion against the need for a pre-biotic chiral influence were it not for the recent discovery of striking evidence for pre-biotic homochirality in the form of excesses of L amino acids in the Murchison meteorite (Cronin & Pizzarello 1997, Engel & Macko 1997).

Most meteorites come from asteroidal parent bodies. Like comets, asteroids are not considered promising sites for life itself; rather they contain pre-biotic organic raw materials from the pre-solar cloud—although asteroidal material has undergone a certain amount of subsequent processing and is less primordial than cometary material. There have been hints of enantiomeric excesses ever since the Murchison meteorite fell in 1969: Engel and Nagy (1982, 1983) repeatedly published findings of excesses of L amino acids, but these were almost universally dismissed as terrestrial contamination despite the fact that in one paper (Engel *et al.* 1990) isotope ratios were reported apparently confirming the extra-terrestrial origin of their amino acids. The relative abundances of stable isotopes are slightly different on each body in the solar system because of different fractionation processes reflecting each body's history and often its mass (very light isotopes might escape). Ratios such as D/H, $^{13}C/^{12}C$, $^{15}N/^{14}N$ and those of the various oxygen and noble gas isotopes together provide a fingerprint of a sample's origin.

Cronin & Pizzarello (1997) adopted a different approach to avoiding the contamination problem—they focused on the non-terrestrial α-methyl α-amino acids. These also had the advantage of being non-racemizable (having a methyl group instead of the α-hydrogen prevents racemization by the usual method of proton removal), and this was important to Cronin because he believed he might be looking for a very ancient enantiomeric excess, possibly created by a chiral influence on the pre-solar cloud. Sure enough, he found excesses of the L form, ranging from 2.8 to 10.4%, in four α-methyl amino acids—isovaline, α-methylnorvaline, α-methyl-isoleucine and α-methyl-*allo*-isoleucine—but not in norvaline, which can racemize because it lacks the α-methyl group.

Cronin's landmark discovery has shifted the paradigm so that people are now ready to accept an ancient enantiomeric excess as evidence for the operation of a pre-biotic chiral influence on chemical evolution rather than a sure sign of terrestrial contamination. In this new climate, Engel's latest paper (Engel & Macko 1997) has been widely acclaimed—he found even larger enantiomeric excesses, with D/L ratios of 0.5 to 0.6 in various amino acids that do occur on Earth, including alanine, with the isotope ratios showing the same extra-terrestrial signature for *both* enantiomers, thus excluding terrestrial contamination.

The reason that the Murchison enantiomeric excesses are being hailed as evidence for the operation of a pre-biotic chiral influence rather than the signature of life itself is not just the age of the meteorite (4.5 billion years, *i.e.* as old as the solar system itself, and long pre-dating life on Earth, which is thought to have started 3.8 billion years ago). Neither is it the current hostile conditions on meteorites and their parent asteroids, because conditions there might have been more favourable to life during a warmer aqueous past—isotopic evidence (Cronin & Chang 1993) suggests that the parent asteroids were formed from volatile-rich interstellar organic matter; aqueous processing then occurred on the parent body before the volatile compounds were lost to space, producing the present hostile environment.

The most telling evidence against a biotic origin for the Murchison excesses is the *total lack of fractionation*. Biotic chemistry is characterized by the *selection* of useful molecules, which eventually become dominant to the exclusion of non-useful molecules. Analysis of numerous classes of Murchison organic (Cronin & Chang 1993), however, reveals none of this fractionation characteristic of life—all isomers are present, including branched chains, in the amounts to be expected from laboratory syntheses, with the amounts of higher homologues dropping off with increasing number of carbon atoms.

There have, however, been reports (Hoover 1997) of sausage-like alleged 'microfossils' in Murchison, some of which are similar in shape to those found by McKay *et al.* (1996) in Mars meteorite ALH84001. Many would consider even the morphology of these 'microfossils' extremely unconvincing by comparison with McKay's—the blobs and 'sausages' could be almost anything, and there is not, at present, much other evidence that they could be associated with life, although some of the blobs seem to be carbon-rich. This is in sharp contrast to McKay's work on ALH84001, in which the morphological evidence is supported by several strands of chemical evidence and possible biomarkers—none individually conclusive but extremely suggestive when found all together in one place.

In the probably unlikely event that the Murchison 'microfossils' proved to be real, however, it would tie in with the recent suggestion (also not generally accepted) that the carbonaceous chondrite class of meteorites, to which Murchison belongs, all came from Mars (Brandenburg 1997) and not from asteroids. It would be most surprising to find life itself on asteroids or comets, but it is widely accepted that life might have evolved on both Earth and Mars (on which conditions were very similar when life was evolving on Earth 3.8 billion years ago) only to go extinct on Mars when it became cold and dry 3.5 billion years ago. An open mind must thus be kept, but the current evidence is overwhelmingly against life itself in Murchison, and the enantiomeric excesses are in all probability pre-biotic, the result of some physical chiral influence.

Many physical chiral influences have been proposed since Pasteur, but many—including Pasteur's own favourite—are unfortunately not viable because they have the wrong symmetry. There was great confusion about what constituted a valid chiral influence until Barron (1982, 1986) clarified the subject with his seminal work on 'true' and 'false' chirality (L. D. Barron, this volume). The hallmark of 'true' chirality is natural optical rotation, which is a parity-odd, time-even pseudo-scalar. Barron realised that absolute asymmetric synthesis could only be induced by something with this same symmetry, *i.e.* a parity-odd, time-even influence.

Most allegedly chiral influences are in fact 'falsely chiral'—parity-odd, time-odd—and so, apparently, of the wrong symmetry. Truly chiral influences are very rare, so Barron's discovery at first appeared to greatly restrict the possibilities for a chiral influence on pre-biotic chemistry. But Barron later realised that although only a truly chiral influence can lift the degeneracy of enantiomers—and therefore only a truly chiral influence can induce asymmetric synthesis under thermodynamic control—the situation might be different under kinetic control (Barron 1987). Under non-equilibrium conditions the changing entropy confers a preferred time direction which destroys any time-reversal symmetry (Birss 1966), so that true *and* false chirality (both parity-odd) might now lead to asymmetric synthesis. False chirality cannot lift the degeneracy of enantiomeric eigenstates, but it might produce unequal barrier heights and therefore unequal reaction rates for enantiomers.

Candidate chiral influences have been reviewed by MacDermott & Tranter (1989), Bonner (1991) and Keszthelyi (1995). Most give only a very minute enantiomeric excess, as we shall see, so these excesses would have to be amplified by autocatalytic mechanisms (see below) to give the homochirality that might be necessary to get life started.

FALSE CHIRALITY—MAGNETIC FIELDS

Because life is intrinsically a non-equilibrium phenomenon, falsely chiral influences could have considerable potential in the quest for the origin of biomolecular chirality. Pasteur's preferred influence, a magnetic field, is however parity-even, time-odd, and thus on its own not even falsely chiral, let alone truly chiral. But when the magnetic field (parity-even, time-odd) is combined with a collinear electric field (parity-odd, time-even), a falsely chiral parity-odd, time-odd influence is obtained overall. Curie (1894) first suggested this as a chiral influence, and various experiments have claimed asymmetric synthesis both in this way (Gerike 1975) and analogously from stirring (parity-even, time-odd) in the Earth's gravitational field (parity-odd, time-even) (Dougherty 1980, 1981).

It is, however, very doubtful that any of these experiments have actually demonstrated the rather special conditions under which false chirality might lead to an enantiomeric excess. The reaction must be under kinetic control, and there must be a *breakdown of microscopic reversibility*, *i.e.* unequal barrier heights for forward and backward reactions (see L. D. Barron, this volume). This means that in forming L or D molecules from an achiral precursor A, the A $\rightarrow$ L and L $\rightarrow$ A barrier heights are unequal (as are A $\rightarrow$ D and D $\rightarrow$ A); also the A $\rightarrow$ L and A $\rightarrow$ D barrier heights are unequal (hence the possibility of asymmetric synthesis); but the A $\rightarrow$ L and D $\rightarrow$ A heights are equal, as are the A $\rightarrow$ D and L $\rightarrow$ A heights (Fig. 5, L. D. Barron, this volume), giving *enantiomeric* microscopic reversibility.

A falsely chiral influence can only have this effect in practice if there is an appropriate mechanism. It is very hard to see what mechanism could enable the effect to be felt as a result of stirring in a gravitational field, despite this influence having the correct symmetry (it is not clear how gravity could influence a chemical reaction even if one ignores its incredible weakness compared with even the weak force). In collinear electric and magnetic fields, however, the effect could be felt by reactions involving a circulation of charge in a plane perpendicular to the magnetic field.

Barron (1987) gives the conrotatory interconversion of chiral enantiomeric cyclobutenes as an example. In a magnetic field there is a preference for one direction of charge circulation over the other; but for this to give an enantiomeric excess, all the molecules must be oriented the same way, so the function of the electric field is simply to orient the molecules. It might seem that collinear magnetic and electric fields are an unlikely influence to find on the early Earth, but the electric field would not be required if the molecules were already oriented (perhaps on some unusual kind of surface) and the Earth's magnetic field alone might then suffice.

Because no clear experimental demonstration has yet been made of a breakdown of microscopic reversibility, and no calculations have yet been performed, the size of the enantioselective effect is unknown, although the effects of magnetic fields on chemical reactions are in general small and large fields are required to produce detectable effects (*e.g.* 1 T or more as compared with the Earth's magnetic field of 5×10^{-5} T).

A rare truly chiral influence is provided by the Barron–Thiemann effect—a magnetic field (parity-even, time-odd) parallel to a (not necessarily polarized) light beam (parity-even, time-odd), which is overall parity-odd, time-even (Barron & Vrbancich 1984; Teutsch & Thiemann 1986). Preliminary experiments seemed to demonstrate this effect in the synthesis of hexahelicene (Thiemann, private communication; Teutsch 1988), more L molecules being obtained with the field parallel to the light beam, and more D with the field anti-parallel. The enantiomeric excess, at 0.07%, was, however, just at the limits of detectability, and there has been some difficulty reproducing the result.

Attempts were later made to observe enantioselective destruction as opposed to synthesis of hexahelicene; no effect was observed with a 308 nm laser and 0.3 T (3 kG) magnet (MacDermott & Tranter, unpublished result), but a tentative positive result was obtained with a 1.1 T (11 kG) magnet (Thiemann, private communication). Although a magnetic field parallel to a light beam has the right symmetry, it is not at all clear what the mechanism of these enantioselective effects might be, even if they are real. The efficacy would anyway be reduced in nature—sunlight does not reach the Earth's surface as a laser-like beam—and the directional characteristics of the sun's rays are markedly reduced (although not completely eliminated) by atmospheric scattering. The Barron–Thiemann effect is, however, at least an example of a magnetic field effect which gives some hint of a positive result in the laboratory.

The Barron–Thiemann effect has now been observed in the form of magneto-chiral dichroism of five parts in 10^3 at 0.9 T in the fluorescence of chiral europium complexes (Rikken & Raupach 1997).

TRUE CHIRALITY—CIRCULARLY POLARIZED PHOTONS

A classic truly chiral influence is, of course, the circularly polarized photon, the enantioselective properties of which are well established in the laboratory (Mason 1982, 1983). Sunlight is, indeed, 0.1% circularly polarized because of multiple aerosol scattering of sunlight in the Earth's atmosphere, but unfortunately it is equally and oppositely so at dawn and dusk, with the right circularly polarized component in excess at sunrise (Angel et al. 1972; Wolstencroft 1985).

The overall even-handedness of this effect could be broken if the proverbial 'warm little pond' were on a steep mountain slope facing either west or east, and so receiving respectively only left circularly polarized evening light or right circularly polarized morning light. Laboratory studies of asymmetric photolysis of racemic amino acids with circularly polarized 200–230 nm radiation (Florey et al. 1977; Norden 1977) suggest that morning light would cause enantioselective photolysis of a pool of racemic amino acids which left the natural L enantiomer in excess.

The Kuhn–Condon sum rule for rotational strengths,

$$\sum_a R_{oa} = 0$$

might, however, pose problems for photochemical enantioselection with a broad-band source such as sunlight—the enantioselective effect of one circular dichroism (CD) band will be cancelled by that of another band of opposite sign, so that any effect averages to zero over the spectrum as a whole. Photochemical enantioselection therefore works best with almost monochromatic light tuned to the wavelength of a major CD band maximum, a situation unlikely to occur in nature. One could, however, imagine certain special cases of molecules in which only one absorption band actually led to reaction, with light from CD bands of opposite sign having no effect on the molecule.

Sunlight might be even-handed on a space and time average, but extra-solar astronomical sources of circularly polarized radiation, although even-handed overall, are *not* even-handed in their effect on individual solar systems. Synchrotron radiation from rotating objects such as neutron stars and magnetic white dwarfs is of opposite circular polarization from opposite poles, and the idea is that it could selectively eliminate one hand or the other in the pre-solar dust cloud according to which side of the neutron star or white dwarf the cloud happened to pass as it traversed the galaxy. When the cloud later condensed to form the solar system, chiral molecules with an enantiomeric excess could then be delivered to Earth by comets. This effect was first suggested for neutron stars by Bonner (1991) and Greenberg *et al.* (1994).

Greenberg conducted laboratory simulations of the action of (unpolarized) UV radiation on the dust grains in clouds (Greenberg *et al.* 1994; Greenberg 1997), and obtained a range of interesting pre-biotic molecules including (racemic mixtures of) chiral molecules, *e.g.* glycerol, glyceric acid, glyceramide, and traces of serine and alanine. He then used data on the luminosity of neutron stars to estimate the total dose of circularly polarized radiation that molecules in the pre-solar cloud would receive as it passed a neutron star at 40 pc, and found that this could be simulated by 50 h irradiation of a racemic mixture with a laboratory source of circularly polarized light, which produced enantiomeric excesses of 10% or more in his experiments. Data on the number density of neutron stars in the galaxy suggested that approximately 10% of pre-solar clouds should pass sufficiently near a neutron star to develop such a significant enantiomeric excess.

Greenberg proposes that chiral molecules produced in the pre-solar cloud, and subsequently acquiring an enantiomeric excess from neutron stars, could then be delivered to Earth by comets after collapse of the cloud to form the solar system. It is well documented that many organics were indeed delivered to Earth by comets.

Racemization should not occur because it is also well documented that cometary molecules did not undergo excessive heating during formation of the solar system, and the low density of comets means that the Earth's atmosphere would cushion the fall of the cometary particles and avoid heating, thus ensuring delivery of the primordial enantiomeric excess intact to Earth. Cometary dust particles are, furthermore, very porous and fluffy—if they fell into the Earth's oceans water would enter the pores which would then serve as tiny 'warm little ponds' with high concentrations of organics (thus overcoming the 'concentration problem' in theories of the origin of life in larger ponds or oceans) and plenty of convenient surface for catalysis.

Greenberg's estimate of the total dosage of circularly polarized radiation from neutron stars involves several assumptions, however. In particular it is not clear that the strong circular polarization of pulsars (*i.e.* neutron stars) in the radiofrequency region extends to the photochemically important optical and UV regions—although the linear polarization of pulsars in the optical region is high (Smith *et al.* 1988), their circular polarization in this region is small or zero (Cocke *et al.* 1971). Strong optical circular polarization is, however, observed for magnetic white dwarfs in binary systems (Chanmugam, 1992) and so could be more promising. It is not completely clear at present, however, whether these various astronomical sources are sufficiently abundant, sufficiently luminous, and with sufficiently strong circular polarization in the correct spectral region.

It is worth noting that the concern is not only that the sources might be insufficiently abundant, but that they might be *too* abundant: in this circumstance the pre-solar cloud would pass randomly on one side or the other of *several* sources during its lifetime of a few tens of millions of years before dispersing or collapsing, thus producing on average no net chiral effect.

Astronomical sources of this kind can have a significant effect on the pre-solar cloud only (as opposed to the primitive Earth), because the gas is transparent and the whole cloud is irradiated; after the cloud collapsed the newly formed planets would, of course, be opaque to radiation so that only those few molecules on the surface were irradiated. The effect of astronomical sources would, furthermore, be swamped by the (overall even-handed) effect of circularly polarized sunlight. Astronomical sources affecting the pre-solar cloud were previously dismissed (Wolstencroft 1985) in favour of sunlight affecting the Earth's surface because no mechanism was then known for safe transfer of pre-solar chiral molecules intact to Earth. Although Greenberg's cometary delivery model now makes astronomical sources an extremely promising possibility, as with sunlight they could fall foul of the Kuhn–Condon sum rule.

TRUE CHIRALITY—THE W BOSON

An important truly chiral influence is provided by the weak force, which is one of the four forces of nature—the electromagnetic force (carried by the photon), the weak force (carried by the W and Z bosons), the strong force (carried by mesons and gluons), and gravity (carried by gravitons). The weak force is the only one of the four that can tell the difference between left and right—it is said to *violate parity*. Elementary particles themselves have a handedness, which is felt only by the weak force. Fermions such as electrons exist in two states of opposite helicity (Aitchison & Hey 1982) which are interconverted by parity, and correspond to spin and momentum vectors parallel (right-handed) or anti-parallel (left-handed).

The two helicity states participate equally in the parity-conserving electromagnetic, strong and gravitational interactions, but participate *unequally* in the weak interaction, to an extent proportional to v/c, the velocity of the fermions relative to that of light. This parity violation by the weak force was first predicted by Lee and Yang (1956) and rapidly confirmed by observations of the handedness of β-decay electrons (Wu *et al.* 1957).

Radioactive β-decay involves the decay of a neutron into a proton, an anti-neutrino, and an electron:

$$n \rightarrow p + e^- + \bar{v}_e$$

As a neutron is two down quarks and an up quark (udd), whereas a proton is two ups and a down (uud), this is really the decay of a down quark (charge $-1/3$) into an up (charge $+2/3$) by emission of a W^- boson, which then decays into an electron and an anti-neutrino:

$$d \rightarrow u + W^- \rightarrow u + e^- + \bar{v}_e$$

(the force-carrying bosons can be regarded as fermion–anti-fermion pairs—the photon as an electron-positron pair, e^-e^+, and the W^- as $d\bar{u}$, $e^-\bar{v}_e$, *etc.*). The β-electrons have predominantly left-handed spin polarization, whereas the corresponding process in the anti-world,

$$\bar{n} \to \bar{p} + e^+ + \nu_e \quad \text{or} \quad \bar{d} \to \bar{u} + e^+ + \nu_e$$

produces β-positrons with predominantly right-handed spin polarization.

In the matter world β^+-emission can also arise from p → n via a W^+:

$$p \to n + e^+ + \nu_e$$

but this is much less common than normal β^--emission from n → p via a W^- because the neutron is heavier than the proton—the decay p → n can occur only if enough energy is supplied by the rest of the nucleus. Although a few β^+-emitters are known, especially in the context of β^+-decay of daughter nuclei formed when α-particles strike a parent nucleus, most β-decay is of the β^- variety, involving emission of (predominantly left-handed) β-electrons.

The excess of left-handed β-electrons raises the possibility of enantioselective β-radiolysis as a symmetry-breaker in pre-biotic chemistry, as first suggested by Ulbricht & Vester (1962). A theoretical study by Hegstrom (1982, 1987) estimates the potential enantiomeric excess to be approximately 10^{-10}–10^{-11} for the radiolysis of amino acids ($Z = 6$) but ca 100 times larger for molecules containing heavy atoms with $Z \approx 100$.

The effect occurs, as with enantioselective photolysis, by the preferential destruction of one hand. There are many natural β-emitters which could affect pre-biotic chemistry, including ^{238}U, ^{235}U and, especially, ^{40}K, which is ubiquitous in terrestrial organisms (Noyes et al. 1977). But the short-range (ca 1 cm) of the β-electrons of ^{40}K requires us to assume an early and quite intimate association between the pre-biotic amino acids and potassium; this problem is circumvented if ^{14}C provides a source of handed β-rays from within the pre-biotic molecules themselves (Noyes et al. 1977), but the only experiment so far on the self-radiolysis of ^{14}C-labelled amino acids yielded negative results (Bernstein et al. 1972).

As far as we are aware no reproducible bioenantioselective effects have yet been obtained from β-radiolysis in the laboratory (Bonner 1984, 1991; van House et al. 1984; Keszthelyi 1995). But some grounds for optimism come from an experiment by Campbell & Farago (1985), who reported differential absorption by camphor vapour of beams of left and right helically polarized 5-eV electrons, with an unexpectedly large asymmetry factor of 10^{-4} with an electron beam of 0.5% excess left helicity.

Campbell and Farago's experiment used electrons in a narrow energy range, however. In reality β-electrons are emitted with a spectrum of energies, ranging from zero to hundreds of keV. The lower energy end is likely to be more enantioselective even though the magnitude of the helicity is less at low energies (it is proportional to the value of v/c at the moment of production of the β-electron, and not, as frequently misunderstood, to the instantaneous value at later times when it decelerates). This is because the most energetic β-electrons simply blast both L and D molecules apart indiscriminately. Just as the enemy of enantioselective photolysis with a broad band source is the Kuhn–Condon sum rule, there might also be an analogous sum rule for β-radiolysis, because Campbell & Farago (1985) found resonances for polarized electrons analogous to circular dichroism bands for circularly polarized photons. Again there might, however, be specific molecules for which one absorption band leads to biosynthesis whereas the others do not.

A recent news report (Service 1999) described a very exciting experiment by R. N. Compton in which bombarding a solution of sodium chlorate with (left-handed) electrons from β^--emitters produced an excess of right-handed crystals, whereas use of (right-handed) positrons from β^+-emitters produced an excess of left-handed crystals. The opposite results with β^- and β^+ rays are very convincing evidence that this is the long-awaited first definitive demonstration of chiroselection by β-rays.

TRUE CHIRALITY—THE Z BOSON

The weak charged currents mediated by the W boson are only involved in flavour-changing reactions, *e.g.* d → u as in β-decay. But the weak neutral currents mediated by the Z boson do not involve any flavour change, and are ever-present and felt by all particles. Because of electroweak unification (Weinberg 1967; Salam 1968), every ordinary electromagnetic interaction in atoms and molecules is accompanied by a much smaller weak neutral current interaction.

Although there are normally equal numbers of left- and right-handed electrons (except where they are produced by a parity-violating weak process such as β-decay), the parity-violating weak neutral current does not treat them equally—it interacts preferentially with the left-handed electrons, and so sees electrons not as 'racemic' but as more left-handed than right. Although parity (P) is therefore violated, CP (parity plus charge conjugation) is conserved, because although a left-handed electron participates preferentially in the weak interaction (compared with a right-handed electron), a right-handed positron participates preferentially (compared with a left-handed positron) to the same extent.

As far as the weak force is concerned, therefore, electrons have an intrinsic left-handedness and positrons have an intrinsic right-handedness—and the Pasteur-like dissymmetry that pervades the universe is the fact that it is made of matter and not anti-matter. It must be emphasized that electrons are not 'really' left-handed (there being normally equal numbers of left and right-handed electrons)—it is just that the weak force can 'see' the left-handed ones better.

A consequence of this handedness of elementary particles is that all atoms and achiral molecules have a very slight electroweak optical rotation (Sandars 1980; Fortson & Wilets 1980), which has been measured experimentally (Emmons *et al.* 1983, 1984). As a result of this electroweak optical activity, the rotational strengths of two enantiomeric molecules, although oppositely signed, are no longer equal in magnitude, because the contribution of the common constituent atoms has the same sign for both enantiomers.

This means that the circular dichroism is not quite the same for the two enantiomers, and so photochemical enantioselection is possible even with unpolarized light, especially because this effect of the weak interaction is present constantly (not just at dawn and dusk) and so may be cumulative over extended time periods. The resulting enantiomeric excess is expected to be *ca* 10^{-14} for light-atom biomolecules, and 10^{-7} for their complexes with heavy atoms (Mason & Tranter 1985).

Another consequence of the handedness of elementary particles is that left- and right-handed chiral molecules are not true enantiomers but diastereoisomers—the true enantiomer of an L amino acid is the D amino acid made of anti-matter, as first pointed out by Barron (1981). Being diastereoisomers, left and right-handed molecules differ slightly in all properties—in NMR chemical shifts, for example (Barra et al. 1987; Robert, 1997) and, most importantly, in energy. This parity-violating energy difference between enantiomers (Hegstrom et al. 1979) arises from weak neutral current interactions, mediated by the Z boson, between electrons and neutrons. These interactions impart a parity-violating energy shift (PVES), E_{pv}, to the energy of a chiral molecule, and an equal and opposite shift, $-E_{pv}$, to that of its enantiomer, giving a parity-violating energy difference (PVED) of $\Delta E_{pv} = 2E_{pv}$.

To calculate the PVED (MacDermott & Tranter 1990) we start with the Hamiltonian for the parity-violating weak neutral current interaction between electrons and neutrons:

$$\hat{H}_{pv} = -(\Gamma/2)\sum_a \sum_i N_a \{\mathbf{s}_i \cdot \mathbf{p}_i, \delta^3(\mathbf{r}_i - \mathbf{r}_a)\}_+$$

where the sums over i and a are over all electrons and nuclei, respectively, in the molecule, and N_a is the neutron number of nucleus a. This beautifully elegant expression summarizes the physical origin of the PVES. The term $\mathbf{s} \cdot \mathbf{p}$ represents the projection of the spin on to the direction of momentum, thus touching directly on the left-handedness of the electron. $\hat{H}_{pv}$ is of opposite sign for enantiomers—$\mathbf{p}$ changes sign under parity, being a polar vector, whereas $\mathbf{s}$ remains the same, being an angular momentum and therefore an axial vector. The delta function expresses the contact nature of the weak force. The smallness of the PVES arises from the smallness of the constant Γ, which contains the very small weak coupling constant.

The PVES E_{pv} is evaluated by taking the expectation value of $\hat{H}_{pv}$ over the ground-state wave functions. But because $\mathbf{p}$ is imaginary, $\hat{H}_{pv}$ is also imaginary, and therefore its expectation value will be zero (Hermitian operators have real eigenvalues, so if the operator is imaginary its eigenvalue must be zero). But if the ground-state wave function is corrected for the effect of spin–orbit coupling, the PVES no longer vanishes. One can think of the spin–orbit coupling as serving to produce an actual excess of either left or right-handed electrons by favouring one orientation of the spin over the other (Hegstrom & Kondepudi 1990). The equation obtained is:

$$\Delta E_{pv} = 2\sum_j^o \sum_k^u P_{jk} (\varepsilon_j - \varepsilon_k)^{-1}$$

where:

$$P_{jk} = \mathrm{Re}\langle\psi_j|\hat{V}_{pv}|\psi_k\rangle \cdot \langle\psi_k|\hat{V}_{so}|\psi_j\rangle$$

is the 'parity-violating strength'. The sums are over all occupied (o) MOs j and all unoccupied (u) MOs k, ε_j and ε_k are the energies of the MOs ψ_j and ψ_k, respectively, $\hat{V}_{pv}$ is a one-electron version of the parity-violating Hamiltonian and $\hat{V}_{so}$ is a spin–orbit coupling operator.

The parity-violating strength P_{jk} is closely analogous to the rotational strength:

$$R_{OA} = \mathrm{Im}\langle\psi_O|\hat{\mu}|\psi_A\rangle \cdot \langle\psi_A|\hat{\mathbf{m}}|\psi_O\rangle$$

in optical activity. $\hat{V}_{pv}$ and $\hat{\mu}$ are both parity-odd (being polar vectors), whereas $\hat{V}_{so}$ and $\hat{\mathbf{m}}$ are both parity-even (being axial vectors), with the result that P_{jk} and R_{OA} are oppositely signed for enantiomers.

The PVED has been evaluated by *ab initio* methods (Mason & Tranter 1985; MacDermott & Tranter 1989, 1990) using GAUSSIAN or CADPAC to obtain the MO's ψ_j and corresponding energies ε_j. The magnitude of ΔE_{pv} is typically *ca* 10^{-20} hartree, equivalent to 10^{-38} J per molecule, 10^{-15} J per mole, or $10^{-17} kT$ at room temperature, as exemplified by the first calculation of the PVED, by Hegstrom *et al.* (1980), who considered a chirally twisted ethylene molecule. Although tiny, these energy differences represent a promising possibility for electroweak enantioselection because they are present in all chiral molecules everywhere at all times, thus providing a uniform background chiral bias.

Before further discussion of PVEDs that have now been calculated for a large range of important biomolecules, it is appropriate to consider the most important aspect of the origin of chirality—the mechanism necessary to amplify the small initial enantiomeric excesses from the various chiral influences to homochirality.

AMPLIFICATION MECHANISMS

The possible amplification mechanisms fall into two classes: (a) the Yamagata cumulative mechanism, applicable to crystallization or polymerization of optically *labile* or achiral monomers; and (b) the Kondepudi catastrophic mechanism, applicable to optically *non-labile* molecules.

An example of the Yamagata (1966) cumulative mechanism is the crystallization of quartz, which consists of helical crystals made of achiral silica units. During crystal growth an achiral unit A can add on to a growing crystal of either hand:

$$L_{n-1} + A \leftrightarrow L_n$$

$$D_{n-1} + A \leftrightarrow D_n$$

But, owing to the PVED, the corresponding free energy changes are unequal,

$$\Delta G_L \neq \Delta G_D$$

which means that A will add preferentially to one hand of crystal rather than the other, resulting in a fractional excess at each of N stages of crystallization which is small but multiplicative, leading cumulatively to an excess of one hand. This is equivalent to the PVED of the crystal being N times the PVED of the individual units within the crystal,

$$\Delta E_{pv} \text{ (}N\text{-unit crystal)} = N \Delta E_{pv} \text{ (one unit)}$$

resulting in a greatly amplified enantiomeric excess in the macroscopic crystal. Under conditions of kinetic as opposed to thermodynamic control, a similar amplification effect can occur because of parity-violating enantiomeric differences in activation energy for addition of A to L_{n-1} or D_{n-1}:

$$\Delta G_L^\ddagger \neq \Delta G_D^\ddagger$$

The first catastrophic amplification mechanism was that of Frank (1953), based on a kinetic scheme involving autocatalysis and enantiomeric antagonism, *i.e.* the presence of one enantiomer encourages production of itself, but inhibits production of its enantiomer. The Kondepudi mechanism (Kondepudi & Nelson 1983, 1984, 1985) is a more sophisticated development of Frank's original scheme and belongs to the mathematical class of 'hypersensitive bifurcations'.

This sort of catastrophic amplification by spontaneous chiral symmetry-breaking—in which the system 'spontaneously' goes from a near-racemic state to a homochiral one—can occur only in a non-equilibrium situation. It cannot occur in a system closed to the inflow of energy and matter—such a system will proceed toward thermodynamic equilibrium, which corresponds to only a tiny enantiomeric excess from any very small chiral influence. In an open-flow reactor system, however, spontaneous symmetry-breaking can occur. Kondepudi envisages a lake, fed by an input of achiral substances A and B, with an output of enantiomers X_L and X_D and other products. The enantiomers form reversibly from the substrates, both directly (k_1) and autocatalytically (k_2), while cross-inhibition between the enantiomers results in their irreversible conversion (k_3) to products P:

$$A + B \underset{k_{-1}}{\overset{k_1}{\Leftrightarrow}} X_{L(D)}$$

$$X_{L(D)} + A + B \underset{k_{-2}}{\overset{k_2}{\Leftrightarrow}} 2X_{L(D)}$$

$$X_L + X_D \overset{k_3}{\to} P$$

The scheme can accommodate unequal reaction rates for the two enantiomers, and can be extended (Kondepudi 1987) to include racemization, thermal fluctuations, and other factors such as asymmetric destruction rates of the two enantiomers by β-radiolysis or other environmental influences. With an input of A and B maintaining the substrate concentration at a constant or slowly increasing level, and a corresponding output of products, the system attains a dynamic quasi-steady state far removed from thermodynamic equilibrium.

As the input concentration is increased, the balance between autocatalysis and enantiomeric antagonism becomes metastable, and the system bifurcates into one homochiral channel or the other. Without a chiral influence, the choice of homochiral reaction channel is arbitrary, but at the transition point the system becomes hypersensitive to small chiral perturbations or fluctuations, which might cause the bifurcation to become determinate.

So why doesn't the Kondepudi mechanism simply amplify random fluctuations in concentration of the L and D enantiomers, because these will be much larger than enantiomeric excesses from small chiral influences? The answer lies in 'noise-averaging' over the very long time-scales that characterize these amplification reactions (Hegstrom & Kondepudi 1990). Although fluctuations will favour first one enantiomer and then the other, they cancel because they are random, and over a long period the systematic effect of the chiral influence, although small, will dominate simply because it is always in the same direction.

Large, well-mixed volumes also serve to eliminate the net effect of random fluctuations. Kondepudi showed that the minimum enantiomeric excess that can be amplified without being overcome by thermal fluctuations is 10^{-17}, and that this can be amplified to homochirality in 10^4 years in a small lake of volume 4×10^9 dm^3 (1 km square and 4 m deep) assuming modest concentrations of ca 10^{-3} M and realistic reaction rates of 10^{-10} mol dm^{-3} s^{-1}.

Although typical PVEDs of 10^{-17} kT are only just large enough to be amplified, the amplification time and lake volume required are very sensitive functions of the PVED. An increase in the PVED of one order of magnitude reduces the amplification time-scale by four orders of magnitude for a given lake volume, or alternatively reduces the required lake volume by two orders of magnitude for a given amplification time (Kondepudi 1987). Thus, for the 1 km × 1 km × 4 m lake, the amplification time would be reduced from 10^4 years to just one year if the PVED were increased to 10^{-16} kT.

The enantiomeric excess of 10^{-11} or so from β-radiolysis could be amplified in a few tens of litres over a period of months, which could be checked on a laboratory scale—although the enemy of terrestrial laboratory amplifications is of course biological contamination, which might be difficult to eliminate and could give a greater chiral bias than a small chiral influence. Another enemy of amplification is racemization, but Kondepudi's mechanism has been extended to include this (Kondepudi 1987), and can withstand a racemization half-life as low as 10^2 years, compared with typical values of 10^5–10^6 years for most amino acids.

Computer simulations of Kondepudi amplification have produced promising results (Buhse et al. 1993). The autocatalysis and enantiomeric antagonism features of the Kondepudi mechanism are observed for many polymerization reactions essential to life, and some enantiomeric enrichment has been demonstrated in the laboratory (Darge & Thiemann 1974; Brack & Spach 1979, 1980; Thiemann & Teutsch 1990; Bolm et al. 1996), although from initial enantiomeric excesses much larger than that from the PVED.

PVEDS OF BIOMOLECULES

Because only the PVED of ancestral biomolecules is relevant (the handedness of modern biomolecules is fixed not by their own PVED but by diastereomeric connection with their ancestors), we have to consider what came first in the origin of life. Scenarios fall into three main categories: (1) nucleic acids first, (2) proteins first, (3) clay minerals first.

The 'proteins first' scenario was once popular because amino acids are relatively simple molecules which form readily (e.g. in interstellar clouds and in the famous Miller–Urey experiment involving an electric discharge through a mixture of ammonia and methane). Nucleic acids are, by contrast, complex molecules consisting of three parts—sugar, phosphate and base—with ribose very difficult to form pre-biotically and phosphate incorporation a well-known problem. It was also thought in the past that nucleic acids could not replicate on their own without enzymes. But the advent of the 'RNA world', in which RNA molecules replicate by themselves in the laboratory, has swung general opinion behind the 'nucleic acids first' scenario.

The apparent difficulty in forming nucleic acids pre-biotically is circumvented if Cairns-Smith's clay mineral replicators came first. These are made from readily available materials, and more complex molecules such as nucleic acids could then form more easily on their surface, initially assisting the survival of the clay genes, but later usurping their function in Cairns-Smith's (1985) 'genetic take-over' hypothesis.

Mason and coworkers have calculated the PVEDs of molecules from all three classes of possible ancestral replicator by *ab initio* methods (MacDermott & Tranter 1989; MacDermott 1995). The natural L amino acids L-alanine, L-valine, L-serine and L-aspartate were all found to be more stable than their 'unnatural' D enantiomers, in their solution conformations and also the α-helix and β-sheet conformations, by 10^{-17} kT (Mason & Tranter 1984,1985). Turning to the natural D sugars, it was found that D-glyceraldehyde, the parent of the higher sugars, is indeed PVED-stabilized, by about 10^{-17} kT (Tranter 1986).

β-D-deoxyribose, in the C2-*endo* conformation found in DNA, is similarly PVED-stabilized, but β-D-ribose in the C3-*endo* form found in RNA is *less* stable than its enantiomer (Tranter et al. 1992). This latter result seemed discouraging until it was realized that the precursor of β-D-nucleotides is not β-D-ribose but α-D-ribose (Prieur 1994) (*via* an S_N2 Walden inversion at C1), and it was later shown that α-D-ribose *is* PVED-stabilized. Although β-D-ribose on its own has a positive PVES, the addition of phosphate and base causes the PVES to go negative, and both B DNA and A RNA are PVED-stabilized in their right-hand helical form, by 10^{-17} kT per nucleotide unit.

Of course ribose, being difficult to synthesize, might not be pre-biotic, a possible more primitive replicator being the glycerol-based polymer (essentially RNA with C2 removed) proposed by Bada & Miller (1987) and Schwartz & Orgel (1985). This is based on achiral monomer units, and owes its chirality purely to its helical conformation. The right-hand helix is again PVED-stabilized, in both B and A conformations, by 10^{-17} kT per nucleotide unit (MacDermott & Tranter 1989a; MacDermott 1995).

It is therefore possible that, instead of the handedness of the D sugars dictating the right-handedness of the DNA helix, it is, in fact, the other way round—a large excess of right-handed double helical glycerol-based replicators could arise by the Yamagata mechanism during polymerization of the achiral monomers, then steric considerations in any later insertion of C2 to give ribose would dictate production of the D form.

Thus the most important biomolecules—amino acids and sugars—have PVEDs of *ca* 10^{-17} kT, which is only just amplifiable by the Kondepudi mechanism. If the PVED were a little larger, Kondepudi amplification would be very much easier. The PVED is proportional to the sixth power of the atomic number (largely because of the importance of spin–orbit coupling), so the presence of second-row heavy atoms should produce much larger PVEDs. This does not happen with the phosphorus atom in nucleic acids—because of the electron-withdrawing effect of the four oxygens, the phosphorus atom is very electropositive in phosphates and this results in too little electron density on the phosphorus to feel the potentially larger parity-violating effect (MacDermott & Tranter 1989a).

Clearly more electronegative heavy elements such as sulfur would be better candidates for a large PVED, so MacDermott and Tranter turned next to thio-substituted DNA analogues (MacDermott et al. 1992). Whereas the normal DNA backbone with $-O-PO_2^--O-$ linkages has a PVED of 10^{-17} kT per nucleotide unit, thio-substitution of the side oxygens to give $-O-PS_2^--O-$ resulted in a PVED of 10^{-16} kT per unit, and thio-substitution in the helix itself gives more dramatic PVED-stabilization of 10^{-15} kT per unit for $-S-PS_2^--O-$ and $-S-PSO^--O-$ and 10^{-14} kT per unit for $-S-S-CH_2-$ links. These last are particularly exciting results—as discussed above, a PVED of 10^{-17} kT is only just amplifiable by the Kondepudi mechanism, taking 10^4 years in a 4×10^9 dm^3 pond, whereas the new PVEDs of 10^{-14} kT are amplifiable in 1 year in a small 4×10^5 dm^3 pond.

Thio-substituted links might have large PVEDs, but can they actually form a viable double helix, and are they relevant to the origin of life? They certainly do form a base-paired double helix, as shown by the recent use of many of the sulfur-linked DNAs as *antisense inhibitors* (Cohen 1989). If an oncogene, for example, results in an undesirable gene product, one can use a piece of DNA of complementary base sequence ('antisense') to bind to the corresponding m-RNA transcript to form a DNA–RNA hybrid double helix; this prevents that piece of m-RNA from being translated into undesirable proteins. But if one tries to introduce a normal antisense oligonucleotide into a cell, nucleases break it down, so it is necessary to alter the DNA backbone to make it nuclease-resistant. A thio-substituted backbone is found to be ideal because it forms a double helix with little change of geometry.

Thio-substituted DNA analogues could also be relevant to the origin of life because one of the problems is that the Miller–Urey experiment assumed a reducing atmosphere whereas the primitive Earth probably had a neutral atmosphere; but deep-sea volcanic vents *do* provide a reducing environment, and the sulfurous gases present could make a thio-substituted replicator highly plausible. Moreover, the well known problem of phosphate incorporation makes a non-phosphate sulfur-linked ancestral replicator even more likely.

Turning now from polymers to crystals, L-quartz consists of right-hand threefold helices of silica tetrahedra. As a chiral mineral made of achiral units it can undergo Yamagata amplification, and indeed a 1.4% excess of L-quartz has been reported in a large collection of crystals from all over the world (Palache *et al.* 1962). Calculations by MacDermott *et al.* (1992) showed that L-quartz is PVED-stabilized by 10^{-17} kT per SiO$_2$ unit. This was again disappointingly small for the heavy Si atom—but silicates have the same problem as phosphates in that the electron-withdrawing effect of the oxygens leaves the Si atoms very electropositive.

The PVED does not, however, need to be larger than 10^{-17} kT to account for the alleged 1% excess of L-quartz—according to the Yamagata mechanism

$$\Delta E_{pv}(\text{crystal}) = N\Delta E_{pv}(1 \text{ unit})$$

so $\Delta E_{pv}(\text{crystal}) = 10^{-2}$ kT (corresponding to a 1% enantiomeric excess) can be obtained from $\Delta E_{pv}(\text{unit}) = 10^{-17}$ kT if $N = 10^{15}$, which corresponds to a realistic small crystal of side 0.1 mm (Tranter 1985).

These results for quartz thus predict almost exactly the reported 1% excess of L-quartz. But are the reported enantiomeric excesses real? Although the original reports (Vistelius 1950; Palache *et al.* 1962) seem to show an excess of L in several separate samples from different locations all over the world, a later report (Frondel 1978) disputes the earlier statistics and shows a (statistically insignificant) excess of D quartz. More fundamentally, it is possible that any quartz enantiomeric excesses, even if real, might reflect not the PVED but *chiral nucleation*—if one small crystal happens to be L, it acts as a seed and causes other crystals nearby to be L also. The seed might, furthermore, break up if the solution is stirred, producing lots more L seeds which then become spread around the solution, with the result that the whole solution crystallizes in the L form.

This has been observed in solution in the famous 'Kondepudi effect' (Kondepudi *et al.* 1990) (not to be confused with the Kondepudi chiral amplification mechanism discussed earlier). Kondepudi obtained equal numbers of L and D sodium chlorate crystals from an unstirred solution, but if the solution was stirred he always got a large excess of either L or D (but the direction of stirring did not affect the hand of the crystals). The explanation is that stirring causes the seed which forms first to break up and spread throughout the solution.

The crucial question as far as natural chiral crystals are concerned is, therefore, whether there is any process analogous to stirring which would lead to chiral nucleation under geological conditions; if there is not, an enantiomeric excess could reflect the PVED, but if there is, any enantiomeric excess will merely reflect the hand of the seeding crystal. Because of the PVED, however, there will be slightly more L seeds than D, so one should find chirally nucleated samples with an excess of L quartz slightly more often than samples with an excess of D. So we are now counting not the excess of L crystals in any one pot, but the excess of pots with an excess of L. The size of this excess of L pots depends on the size of the nucleating crystal—from the thermodynamic version of the Yamagata mechanism

$$\Delta E(\text{seed crystal}) = N\Delta E(1 \text{ unit})$$

which will give an enantiomeric excess of the more stable form in the seed crystals (and a corresponding excess can also be obtained under kinetic control because of the corresponding enantiomeric difference in crystallization rates).

It is known that for quartz $\Delta E(1 \text{ unit})$ is $10^{-17} kT$, so the crucial question is how large is N for the nucleating crystal. Kondepudi *et al.* (1995) suggest that the side of the nucleating crystal is *ca* 0.5 mm. This would correspond to $N \approx 10^{17}$. If this is true, it would seem that there could indeed be a detectable enantiomeric excess in the seed crystals and, therefore, a detectable excess in the number of pots with a chirally nucleated enantiomeric excess of the PVED-determined form. Kondepudi's estimate of the size of the nucleating crystal might, of course, be too large, but even if N were only 10^{15} it would still be feasible to detect the resulting excess of samples with an excess of the PVED-determined form.

It would, therefore, be worth examining much larger samples of natural chiral crystals—especially those containing heavy atoms, bearing in mind the Z^6 dependence of the PVED—to see if there is any enantiomeric excess. It is worth looking because of the great efficiency with which any chiral bias in minerals can be transferred to biology through chiral surface catalysis. For example, L-quartz absorbs L amino acids preferentially from a racemic mixture, with 1% selectivity (Kavasmaneck & Bonner 1977); if this were combined with a 1% excess of L-quartz it would produce an overall electroweak enantioselectivity of 10^{-4}, which is dramatically larger than the PVED of individual molecules and would be correspondingly more easily amplified. Even a considerably smaller excess of L-quartz would still produce a substantial enhancement of the electroweak effect as compared with individual molecules.

More realistic candidates than quartz itself, both for pre-biotic surface catalysis or as Cairns-Smith's clay mineral ancestral replicators, are aluminosilicate quartz analogues, some of which contain heavy metal cations, for which PVEDs might be much larger. Other chiral crystals have been studied to see if heavy atoms produce a larger chiral bias; *e.g.* Mason (1986) observed a 6.8% enantiomeric excess in sodium uranyl acetate crystals. Laboratory crystallizations have suggested that potassium silicotungstate crystallizes predominantly in a dextrorotatory habit (Wyrouboff 1896; Soret 1899; Copaux, 1910), but it is almost impossible to prevent chiral contamination in such studies (Amariglio *et al.* 1968).

The weak force seems to predict the correct hand whatever came first—nucleic acids or proteins—in the origin of life. For nucleic acids first, D-glyceraldehyde and α-D-ribose are more stable, so that the D sugars and D ribonucleotides would be selected, and the right-hand helical backbone of DNA is also PVED-stabilized. For proteins first, the L amino acids are usually more stable.

Recent calculations (Bakasov *et al.* 1998; Lazzeretti & Zanussi 1997, 1998) using larger basis sets, and other improvements such as configuration interaction, have yielded PVEDs for amino acids that agree in sign with with those of Mason and Tranter, but are an order of magnitude larger. These encouraging results make it likely that the PVEDs described above for other molecules might also be revised upwards in the future, making for much easier pre-biotic amplification.

SO WAS IT REALLY THE WEAK FORCE?

The Murchison enantiomeric excesses seem on current evidence to be powerful evidence of a pre-biotic chiral influence. Eschenmoser's scheme for deracemization without a chiral influence is ingenious, but the time taken to produce the long polymers needed for deracemization and then to select a homochiral sequence by an evolutionary process might be long, possibly longer than Kondepudi amplification time-scales for some chiral influences. So it is possible that life based on Kondepudi-amplified homochirality could get going more quickly.

The question of whether Kondepudi amplification could select biomolecular handedness by chance by amplification of a random fluctuation without a chiral influence hinges on the time-scale of the amplification compared with that of the fluctuations. It is, however, characteristic of hypersensitive bifurcations that they are extremely slow, and it is unlikely that an amplification process could exist that was fast compared to the fluctuations. If it did, large enantiomeric excesses would surely be seen quite often in the laboratory, whereas in fact racemic mixtures are overwhelmingly the norm and asymmetric synthesis is difficult to the extent that the pharmaceutical industry has had to spend huge sums trying to find the sophisticated systems needed to achieve it.

Circularly polarized radiation from astronomical sources would seem to be the largest chiral influence if Greenberg's (1997) figures for pulsars are correct; but if the much weaker magnetic white dwarfs proves to be a more realistic source the effect could be more similar in size to the other possible chiral influences. The Kuhn–Condon sum rule might restrict the efficacy of astronomical sources, and also of sunlight and β-rays, although there may be specific molecules for which this is less of a problem.

Magnetic-field effects involving false chirality might at first seem unlikely because of the small size of the Earth's magnetic field, but with Kondepudi amplification they are no less plausible than β-rays or the PVED; although magnetic field reversals do occur, their time-scale is longer than likely amplification times.

The PVED represents the smallest of the candidate chiral influences, but it would have delighted Pasteur because it appears to vindicate his concept of a *universal* dissymmetry. It has the advantage of applying to all molecules everywhere at all times, while all the other influences rely on specific local conditions, such as the presence of β-emitters, polarized radiation or magnetic fields, and specific mechanisms such as photochemical reactions or reactions susceptible to magnetic field effects. Although the chiral bias from the PVED is very small (10^{-17}–10^{-14}), its effect could be greatly enhanced (10^{-4}) *via* chiral mineral catalysis.

In truth there is as yet insufficient knowledge of the origin of life on Earth, and in particular of what molecules came first, to draw any conclusions about which chiral influence was responsible for selecting biomolecular chirality. It might, however, be possible to discover if it was really the weak force by looking in space.

LOOK IN SPACE!

Chance or local chiral influences such as circularly polarized sunlight and influences involving magnetic fields will give different hands on different planets. The solar-system-wide chiral influence of circularly polarized radiation from astronomical sources will give the same hand throughout any one solar system but different hands in different solar systems. The universal chiral influence of the weak force will give the same hand for a given molecule throughout the universe when acting through the PVED; and when acting through β-emitters it will usually give the same hand but may occasionally give the opposite hand when β-positron emitters are involved instead of the much more common β-electron emitters. By looking in space, therefore, it might be possible to discover some clue about which chiral influence was responsible for life's handedness—consistently finding the same hand on different planets would tend to favour the weak force.

But, much more importantly, even finding the 'wrong' hand—as opposed to a racemic mixture—on another planet would be of enormous significance as it could be the chiral signature of life, and a search for extra-terrestrial life could be approached as a search for **extra-terrestrial h**omochirality, SETH (MacDermott *et al.* 1996). Although life necessarily implies homochirality, homochirality *on its own* does not, however, necessarily imply life itself, as the Murchison case shows. But *any* deviation from racemic-ness is clearly a very significant and unusual finding, indicative at the very least of pre-biotic chemistry with the operation of a chiral influence. Biotic and pre-biotic homochirality can be distinguished by taking into account other evidence as in the Murchison case (MacDermott 1997); where other biotic indications such as micro-fossils, biomarkers, non-equilibrium isotope ratios, *etc.*, are inconclusive individually (as for the Mars meteorite, for example), finding homochirality *as well* could be the clincher for life itself.

Considering first pre-biotic homochirality, comets are probably the best place to look for relict enantiomeric excesses caused by neutron stars, because they are relatively unprocessed relics of the pre-solar cloud. Asteroids (and meteorites therefrom) are similar to comets, but have undergone some subsequent aqueous processing, which could have provided an opportunity for any enantiomeric excess from neutron stars or the weak force to be amplified by the Kondepudi mechanism (which probably would not be very effective in the gaseous conditions of the pre-solar cloud).

Clues about the origin of the Murchison enantiomeric excesses can be obtained from Cronin's amino acids with two chiral centres, α-methyl-isoleucine and α-methyl-*allo*-isoleucine, the two enantiomeric pairs of $CH_3CH_2C^{**}H(CH_3)$-$C^*(CH_3)(NH_2)COOH$, for both of which the L form is in excess (Cronin & Pizzarello 1997). For α-methyl-*allo*-isoleucine the D and L forms are (2R,3S) and (2S,3R) respectively, whereas D and L α-methyl-isoleucine are (2R,3R) and (2S,3S) respectively—in other words 'D' and 'L' refer to R and S at C2 (C*).

It is generally believed (Cronin & Chang 1993) that amino acids in meteorites were formed from interstellar ketones *via* the Strecker cyanohydrin synthesis RR'CO $\rightarrow$ RR'C(NH$_2$)COOH during aqueous processing on the parent body. The ketones are known to be interstellar from isotope ratios, and from the predominance of branched chain isomers (leading to many chiral centres), suggesting gas-phase radical reactions. The evidence for later Strecker synthesis of the α-amino acids is that the corresponding α-hydroxy acids (also formed in the reaction) occur in comparable amounts. It is therefore most likely that the C3 (C**) asymmetric centre is interstellar whereas the C2 (C*) centre was introduced on the parent body.

If Cronin's enantiomeric excesses originated in the C3 centre from pre-solar action of a neutron star, producing (say) more 3R than 3S, then Strecker synthesis would result in L > D for α-methyl-*allo*-isoleucine and D > L for α-methyl-isoleucine. Cronin obtained L > D for *both*, however, showing that the main enantiomeric excess is in fact in the C2 centre (although there is possibly also a smaller enantiomeric excess in the C3 centre). It would therefore seem that neutron stars (which are effective only in the pre-solar cloud) cannot be the cause of the enantiomeric excess at C2. Cronin points out that the neutron star origin could be saved if the α-methyl amino acids had a separate origin from their α-H analogues (for which there is possibly some circumstantial evidence (Cronin & Pizzarello 1997), and were formed entirely in the interstellar medium.

Alternatively, if a Strecker origin on the parent body is accepted, the enantiomeric excess at C2 could be an amplified electroweak excess. In general the L amino acids are favoured by the weak force, and preliminary calculations (MacDermott & Bayliss, unpublished) suggest that this preference for L extends probably to α-methyl-isoleucine and very probably to α-methyl-*allo*-isoleucine, the uncertainty being because of the unavailability of reliable X-ray crystallographic structural data on these non-terrestrial amino acids.

Finding enantiomeric excesses on comets, which are more primordial, would strengthen the Murchison evidence for pre-biotic homochirality. The COSAC GC–MS experiment on the lander of the Rosetta mission to comet Wirtanen (launch 2003, arrival 2011) will carry chiral columns. In chiral gas chromatography (GC) the column is coated with a chiral substance to which enantiomers adhere to differing extents and so take different times to pass through the column, producing two peaks. Without a chiroptical detector, however, it is difficult to be sure that two peaks really correspond to L and D enantiomers. To circumvent this problem, the COSAC columns will come in pairs, one with a homochiral coating and the other with a racemic coating of the same substance. Enantiomers can be recognized by the appearance of two peaks after passage through the homochiral column where there was only one peak through the racemic column; if the two peaks also have identical mass spectra, they are almost certainly enantiomers.

One of the most ubiquitous extra-terrestrial molecules is HCN. It readily forms polymers, ranging in colour from orange to black, which are believed by Matthews *et al.* (1997) to be responsible for the orange haze seen on Titan, Jupiter and several other solar system bodies, as well as the black crust on comet Haley. Many of these polymers are chiral, with a polyamidine structure based on repeating $(HCN)_3$ units containing chiral centres. These chiral HCN polymers are particularly interesting, because they are readily hydrolysed to give polypeptides, and can also act as condensing agents in the production of nucleic acids.

This could enable parallel synthesis of polypeptides and polynucleotides, thus explaining the association of proteins and nucleic acids in life, and avoiding the problem of which came first—they came together! It is possible that the polymerization process could provide a Kondepudi-type amplification, so it could be worth looking for pre-biotic homochirality in HCN polymers on comets and other orange, brown and black solar system bodies. The active photochemistry on Titan could produce branched-chain chiral hydrocarbons (Brack & Spach 1987) such as 2,3-dimethylpentane, 3-methyl-1-pentene, and many more; the polymerization of these might similarly lead to chiral amplification and pre-biotic homochirality.

Comets and Titan might be good laboratories in which to demonstrate the emergence of pre-biotic homochirality, but they are not generally regarded as favourable sites for life itself. On Mars, however, and possibly Europa, one might find the homochiral signature of extant or extinct life. Europa might have a sub-surface ocean which could harbour life. Life might have evolved on Mars 3.8 billion years ago, when conditions were similar to those on Earth, and then gone extinct when Mars went cold and dry 3.5 billion years ago.

The enemy of relict homochirality is of course racemization, but Bada & McDonald (1995) have shown that this is greatly retarded under dry or frozen conditions compared with wet conditions—typical amino acid racemization half lives of 10^6 years in wet conditions rise to 10^{13} or more years in dry or frozen conditions at Martian temperatures. The homochiral signature of past Martian life might, therefore, still be preserved if there was no prolonged contact with liquid water after extinction. It is also possible that life might have hung on in isolated pockets, as in the dry valleys of Antarctica.

The most direct way of detecting enantiomeric excesses arising from extinct or extant life is by measurement of optical rotation, and a novel cigar-sized space polarimeter called the SETH Cigar has been proposed for this purpose (MacDermott *et al.* 1996), and would ideally complement a Rosetta-type chiral GC–MS.

Pasteur would no doubt hope to find L amino acids on Mars as on Earth. But if D amino acids were found on Mars it would be known that this represented an *independent* start to life—a very profound conclusion, suggesting that life can start easily, if it started twice in one solar system (although it would of course invalidate the weak force and neutron star theories of the origin of homochirality, at least for this solar system). If, however, L amino acids were found on Mars it would not be certain that this represented an independent start to life (there might have been exchange of material between the two planets), and although it would be at least consistent with global rather than local theories of the origin of homochirality, it would not be known whether it was the universal chiral influence of the weak force or the mere solar-system-wide effect of the neutron star. But if the same hand was consistently found on many different extra-solar planets this would strongly support the weak force as a universal symmetry-breaker.

The search for **extra-terrestrial homochirality** (SETH) could become a search for **extra-solar homochirality** (SEXSOH) (MacDermott 1997), using future space telescopes of the Darwin series. The Darwin space interferometer (Leger *et al.* 1996) (under consideration by ESA for launch between 2009 and 2017) aims to catch light from extra-solar planets around other stars with a view to detecting infra-red spectroscopic signatures of life—CO_2 would indicate a planet similar to the telluric planets (Venus, Earth, Mars) in our solar system; H_2O would further indicate that the planet was habitable; and O_3 (indicative of O_2) would indicate that it *was* inhabited by photosynthetic organisms.

Light from an Earth-like planet covered, Amazon rain-forest-style, with leaves would, moreover, be characterized by a slight circular polarization as a result of reflection circular dichroism (differential absorption of left and right circularly polarized light) from the highly optically active chiral chlorophyll molecules which absorb in the red visible region. Wolstencroft (1996) showed that the circular polarization of light reflected from leaves in the laboratory is *ca* 1%. This is too small to be detected by the Darwin Mark I telescope, but could be detected by including a small polarimeter on a follow-up Darwin Mark II telescope with larger mirrors for possible launch in the 2020s.

Although O_3 is generally regarded as a good signature of life, it is conceivable that it could be mimicked by hitherto unsuspected non-biotic photochemistry, so if the planet were also optically active the case for life would be greatly strengthened. It is worth noting that the problem of biotic *vs* pre-biotic homochirality does not arise with SEXSOH—*bulk* optical activity can *only* arise from global extant life. Isolated 'warm little ponds' with pre-biotic homochirality or caches of fossils from extinct life will not produce enough optical activity to be detectable from space.

SEXSOH should enable the solar-system-wide effect of neutron stars to be distinguished from the universal effect of the weak force. But would neighbouring solar systems have been affected by the same neutron star? This is not a problem because although it is true that stars are generally born in clusters, which would indeed all be affected by the same passing neutron star, the young stars later disperse in different directions so that our present-day neighbours are not our sibling solar systems, many being in fact of very different ages. The neutron star or white dwarf that might have affected our solar system will by now be very far away.

Darwin Mark I is intended to observe planets around *ca* 300 nearby stars; more will be accessible to Darwin Mark II. This should be more than enough to ascertain whether biomolecules in general have the same hand in other solar systems as on Earth, as would be predicted if the weak force determined life's handedness. But finding the 'same hand' as on Earth is only meaningful if the biomolecules found are the same as occur on Earth; if totally different biochemistry is found elsewhere it will be difficult to draw any conclusions about the origin of homochirality.

Many people subscribe to the principle of mediocrity—because our sun is such an ordinary star, our Earth must support a very ordinary and typical biochemistry. Others give more weight to the role of *contingency* in evolution (Gould 1989, 1993)—even on a very typical planet there might be many different possible scenarios, and what actually happens may be contingent on random factors (*e.g.* what if the impact that killed the dinosaurs had not occurred ...). Others argue that there are actually not that many possible scenarios, in the early stages at least, and that the nucleic acids, for example, have totally unique structural features (Prieur 1994) that will lead to their selection. Also, the first replicator will be the one that can form most quickly—and the attainment of homochirality may be an important factor here.

There is indeed considerable evidence that life must have emerged very quickly because there is only a very short time window for it to do so. The early Earth was subject to frequent major impacts that would have frustrated the development of life; these impacts only ceased *ca* 3.80 to 3.85 billion years ago (determined from crater ages on the Moon), and yet life seems to have started definitely by 3.5 billion years ago (from microfossil evidence in rocks from Western Australia; Schopf 1993) and very probably by 3.8 billion years ago (from life-like isotope ratios in the Isua rocks (Mojzsis *et al.* 1996)—life produces non-equilibrium $^{12}C/^{13}C$ ratios, with enrichment of the light isotope *via* the kinetic isotope effect).

Schwartzman *et al.* (1993) have, furthermore, suggested that high surface temperatures on the early Earth held back the development of life—it seems that life emerged *as soon as* the surface temperature had dropped sufficiently to prevent the relevant biomolecules from decomposing, and that each successive new stage, such as cyanobacteria, then eukaryotes, then metazoa, also appeared *as soon as* the temperature had dropped sufficiently.

All this leads increasingly to the conclusion that life is not something very rare and improbable, but something that starts very easily, using readily available materials—in fact elemental abundances suggest that life on Earth has simply used the elements that were available in sea water (Williams & Frausto da Silva 1996). Available evidence increasingly suggests that starting materials are similar in other solar systems. Many young stars, such as β-Pictoris, have a protoplanetary disk representing a solar system in the making. Butner *et al.* (1997) have recently examined the infra-red spectra of several β-Pictoris-like stars, and found that they bear a startling resemblance to the infra-red spectra of comets, suggesting that the dust around these stars has undergone processing similar to that which occurred in our own solar system.

This might mean that even if many solar systems do have a totally different biochemistry, there will still be enough with a similar biochemistry to our own to assess whether or not most of them have the same hand as on Earth and thus whether it was indeed the weak force that determined their chirality. There is also the possibility that different chiral influences might have operated on different planets, *e.g.* the same hand as on Earth might usually be found, indicating the universal effect of the PVED, but with the exceptions being planets with a particularly large magnetic field or an unusually high concentration of radioactive isotopes.

IS THE UNIVERSE INTRINSICALLY HANDED?

But is the weak force really the universal chiral influence that Pasteur was seeking? He would surely be pleased if SEXSOH revealed molecules of consistently the same hand in other solar systems, confirming the electroweak origin of biomolecular chirality. But would this really mean that the universe is intrinsically handed, as he believed?

To answer this, it is necessary to investigate *why* the weak force is 'left-handed'. The weak force favours left-handed fermions, L amino acids and right-handed DNA because the universe is made of matter; in an anti-matter universe the weak force favours right-handed fermions, D-amino acids and left-handed DNA. But *why* is the universe made of matter and not anti-matter? The answer lies in CP-violating processes in the early universe (Close 1983). Normally processes and their 'anti-processes' have equal probability, *e.g.* the decay of a neutron by emission of a predominantly left-handed electron,

$$n \to p + e_l^- + \overline{v}_e$$

and the CP-related decay of an anti-neutron by emission of a predominantly right-handed positron,

$$\overline{n} \to \overline{p} + e_r^+ + \overline{v}_e$$

occur with equal frequency—CP is said to be conserved. But for the K meson the decay

$$K^0 \to e_r^+ + \pi^- + \overline{v}_e$$

is 7 parts in 1000 more frequent than its anti-process

$$\overline{K}^0 \to e_l^- + \pi^+ + v_e$$

This *CP-violation* is still not fully understood but is probably because of the weak force, as the other forces do not even violate P let alone CP. A similar inequality in the decay rates of the X and $\overline{X}$ bosons could generate an excess of matter in the early universe. The X boson is a particle predicted by Grand Unified theories, which unite the electroweak and strong forces and allow quark–lepton interconversion; such exotic particles would have been quite common in the early universe. Lots of matter (and equal amounts of anti-matter) was produced in the heat of the Big Bang by pair creation from photons, *e.g.*

$$\gamma \to X + \overline{X}$$

The X-bosons then decay into two quarks or a lepton and an anti-quark, thus producing mainly matter:

$$\overline{q} + l \xleftarrow{1-a} X \xrightarrow{a} q + q$$

But there are equal numbers of anti-X bosons, $\overline{X}$, which decay predominantly into anti-matter:

$$q + \overline{l} \xleftarrow{1-b} \overline{X} \xrightarrow{b} \overline{q} + \overline{q}$$

So equal amounts of matter and anti-matter are still produced overall. If, however, it is supposed that the branching ratios for the two alternative decays are not the same for X and $\overline{X}$, *i.e.* $a \neq b$ so that the $X/\overline{X}$ system is CP-violated like $K^0/\overline{K}^0$. If $a > b$ there will eventually be slightly more quarks than anti-quarks. The bulk of the quarks annihilate with the anti-quarks, and the leptons with the anti-leptons, giving a huge number (N_γ) of photons (which now form the 3K background radiation). If, however, $a > b$ there will be a few quarks left to form a small number (N_B) of baryons (qqq, *e.g.* protons, uud) and electrons, which later combine to form atoms. The matter existing today is thus only a tiny remnant of that originally produced, as evidenced by the huge photon to baryon ratio, $N_\gamma/N_B = 10^9$ (obtainable from the intensity of the 3K background).

So does CP-violation mean that matter and anti-matter are 'really' different? The answer is emphatically no, provided that CPT is conserved, because CP-violation is then 'false' and not 'true' (see L. D. Barron this volume; Barron 1994). The CPT theorem states that even if C (charge-conjugation, $+ \leftrightarrow -$), P (parity, $x \to -x, y \to -y, z$

→ $-z$) and T (time-reversal, $t \to -t$) are individually violated, the three operations taken together are conserved.

Thus if CPT is conserved, CP-violation implies time-violation, and thus CP-violation is like false chirality, being CP-odd, time-odd. Just as a P-odd, T-odd falsely chiral influence cannot lift the degeneracy of P-enantiomers (L and D molecules), so a CP-odd, T-odd influence cannot lift the degeneracy of CP-enantiomers (matter and P-enantiomeric anti-matter). Like false chirality, CP-violation can affect only the *rates* of processes (as in the K meson and X boson examples), not the eigenstates.

So matter and anti-matter remain firmly degenerate, and CP-violation can only give an excess of matter when conditions are very far from equilibrium, as in the rapidly expanding early universe. But if CPT were also violated, CP-violation would be 'true' (CP-odd, T-even) and therefore able to lift the degeneracy of CP-enantiomers—in which circumstance matter and anti-matter really would be different.

But how did the universe enter its present CP-violated condition that favours matter and left-handed fermions? Here the realms of speculation are entered, but the answer might lie in an early 'fall from grace' during the era of inflation and symmetry-breaking during the first 10^{-35} s (Guth 1989). The universe began in a state of perfect symmetry with all four forces one, and presumably also with no distinction between left and right; it then underwent a phase transition to its current state of broken symmetry with the four forces very different and the weak force 'left-handed'. An appropriate analogy is a ball sitting at the top of a perfectly symmetrical hill; the symmetry is then spontaneously broken when the ball rolls down randomly to one side or the other.

Perhaps the complete story of the origin of biomolecular chirality might involve successive selection by all three types of influence—chance, false chirality and true chirality. The L amino acids might indeed have been selected by the truly chiral influence of the weak force—and if so the L amino acids were selected only because of the matter-dominance of the universe, which came about through the false chirality of CP-violation in X boson processes. The start of it all, however, was, presumably, the universe's early phase transition, in which it fell *by chance* into its present 'left-handed', matter-favouring, CP-violated condition. If so, the universe had no intrinsic handedness to begin with, even though it is now well and truly chiral in the spirit of Pasteur.

Die-hard proponents of an intrinsic handedness might want to believe in CPT-violation to salvage a determinate rather than chance selection of the 'left-handed' CP-violated condition. But such an intrinsic handedness as an initial condition is not really very attractive even to a dedicated disciple of Pasteur. There can be no initial conditions if one wants a 'free-lunch' vacuum fluctuation-type universe emerging out of *absolutely* nothing, not just almost nothing (Atkins 1992). It is more attractive to assume an initial perfect symmetry, so that if there was CPT violation it would arrive like the CP-violation—in a chance 'fall from grace'. The newly chiral weak force has ample opportunity for determinate selection thereafter.

REFERENCES

Aitchison I. J. R. & Hey A. J. G. (1982) *Gauge Theories in Particle Physics*. Adam Hilger, Bristol.
Amariglio A., Amariglio H. & Duval X. (1968) *Ann. Chim.*, **3**, 5.
Angel J. R. P., Illing R. & Martin P. G. (1972) *Nature*, **238**, 389.
Atkins P. W. (1992) *Creation Re-visited*. W. H. Freeman.
Bada J. L. & McDonald G. D. (1995) *Icarus*, **114**, 139.
Bada J. L. & Miller S. L. (1987) *BioSystems*, **20**, 21.
Bakasov A., Ha T.-K. & Quack M. (1998) *J. Chem. Phys.*, **109**, 7263.
Barra A. L., Robert J. B. & Wiesenfeld L. (1987) *BioSystems*, **20**, 57.
Barron L. D. (1981) *Mol. Phys.*, **43**, 1395.
Barron L. D. (1982) *Molecular Light Scattering and Optical Activity*. Cambridge University Press.
Barron L. D. (1986) *Chem. Phys. Lett.*, **123**, 423.
Barron L. D. (1987) *Chem. Phys. Lett.*, **135**, 1.
Barron L. D. (1994) *Chem. Phys. Lett.*, **221**, 311.
Barron L. D. & Vrbancich J. (1984) *Mol. Phys.*, **51**, 715.
Bernstein W. J., Lemmon R. M. & Calvin, M. (1972) *Molecular Evolution* (Ed. by L. Rohlfing & A. I. Oparin). Plenum, New York.
Birss R. R. (1966) *Symmetry and Magnetism*. North Holland, Amsterdam.
Blaschke G., Kraft H. P., Fichentscher K. & Kohler F. (1979) *Arzneim. Forsch.*, **29**, 1640.
Bolli M., Micura R. & Eschenmoser, A. (1997) *Chem. Biol.*, **4**, 309.
Bolm C., Bienewald F. & Seger A. (1996), *Angew. Chem. Intl. Ed. Engl.*, **35**, 1657.
Bonner W. A. (1984) *Orig. Life*, **14**, 383.
Bonner W. A. (1991) *Orig. Life Evol. Biosphere*, **21**, 59.
Brack A. & Spach G. (1979) *J. Mol. Evol.*, **13**, 35 and 47.
Brack A. & Spach G. (1980) *J. Mol. Evol.*, **15**, 231.
Brack A. & Spach G. (1987) *BioSystems*, **20**, 95.
Brandenburg J. E. (1997) *Proc. SPIE*, **3111**, 69.
Cairns-Smith A. G. (1985) *Seven Clues to the Origin of Life*. Cambridge University Press.
Campbell D. M. & Farago P. S. (1985) *Nature*, **318**, 52.
Chanmugam G. (1992) *Ann. Rev. Astron. Astrophys.*, **30**, 143.
Close F. E. (1983) *The Cosmic Onion*. Heinemann.
Cocke W. J., Muncaster G. W. & Gehrels T. (1971) *Astrophys. J.*, **169**, L119.
Cohen J. S. (Ed.) (1989) *Topics Mol. Struct. Biol.*, **12**.
Copaux H. (1910) *Bull. Soc. Mineral. Fr.*, **33**, 167.
Cronin J. R. & Chang S. (1993) *The Chemistry of Life's Origins* (Ed. by J. M. Greenberg *et al.*), p. 209. Kluwer Academic.
Cronin J. R. & Pizzarello S. (1997) *Science*, **275**, 951.
Curie P. (1894) *J. Phys. (Paris)*, **3**, 393.
Darge W. & Thiemann W. (1974) *Orig. Life*, **5**, 263.
Dougherty R. C. (1980) *J. Am. Chem. Soc.*, **102**, 380.
Dougherty R. C. (1981) *Orig. Life*, **11**, 71.
Emmons T. P., Reeves J. M. & Fortson E. N. (1983) *Phys. Rev. Lett.*, **51**, 2089.
Emmons T. P., Reeves J. M. & Fortson E. N. (1984) *Phys. Rev. Lett.*, **52**, 86.

Engel M. H. & Nagy B. (1982) *Nature*, **296**, 837.
Engel M. H. & Nagy B. (1983) *Nature*, **301**, 496.
Engel M. H. & Macko S. A. (1997) *Nature*, **389**, 265.
Engel M. H., Macko S. A. & Silfer J. A. (1990) *Nature*, **348**, 47.
Fischer E. (1894) *Chem. Ber.*, **27**, 2985, 3189.
Florey J. J., Bonner W. A. & Massey G. A. (1977) *J. Am. Chem. Soc.*, **99**, 3622.
Fortson E. N. & Wilets L. (1980) *Adv. Atom. Molec. Phys.*, **16**, 319.
Frank F. C. (1953) *Biochim. Biophys. Acta*, **11**, 459.
Frondel C. (1978) *Am. Mineral.*, **63**, 17.
Gerike P. (1975) *Naturwissenschaften*, **62**, 38.
Gould S. J. (1989) *Wonderful Life*. W. W. Norton, New York.
Gould S. J. (1993) *Eight Little Piggies*. Jonathan Cape.
Greenberg J. M. (1997) *Proc SPIE*, **3111**, 226.
Greenberg J. M., Kouchi A., Niessen W., Irth H., van Paradijs J., de Groot M. & Hermsen W. (1994) *J. Biol. Phys.*, **20**, 61.
Guth A. (1989) In: *The New Physics* (Ed. by P. C. W. Davies). Cambridge University Press.
Hegstrom R. A. (1982) *Nature*, **297**, 643.
Hegstrom R. A. (1987) *BioSystems*, **20**, 49.
Hegstrom R. A. & Kondepudi D. K. (1990) *Sci. Am.*, **262**, 108.
Hegstrom R. A., Rein D. W. & Sandars P. G. H. (1979) *Phys. Lett. A*, **71**, 499.
Hegstrom R. A., Rein D. W. & Sandars P. G. H. (1980) *J. Chem. Phys.*, **73**, 2329.
Hoover R. B. (1997), *Proc. SPIE*, **3111**, 115.
Joyce G. F., Visser G. M., van Boeckel C. A. A., van Boom J., Orgel L. E. & van Westresen J. (1984) *Nature*, **310**, 602.
Kavasmaneck P. R. & Bonner W. A. (1977) *J. Am. Chem. Soc.*, **99**, 44.
Keszthelyi L. (1995) *Quart. Rev. Biophys.*, **28**, 473.
Kondepudi D. K. (1987) *BioSystems*, **20**, 75.
Kondepudi D. K. & Nelson G. W. (1983) *Phys. Rev. Lett.*, **50**, 1023.
Kondepudi D. K. & Nelson G. W. (1984) *Physica*, **125A**, 465.
Kondepudi D. K. & Nelson G. W. (1985) *Nature*, **314**, 438.
Kondepudi D. K., Kaufman R. J. & Singh N. (1990) *Science*, **250**, 975.
Kondepudi D. K., Bullock K. L, Digits J. A. & Yarborough P. D. (1995) *J. Am. Chem. Soc.*, **117**, 401.
Lazzeretti P. & Zanussi R. (1997) *Chem. Phys. Lett.*, **279**, 349.
Lazzeretti P. & Zanussi R. (1998) *Chem. Phys. Lett.*, **286**, 240.
Lee T. D. & Yang C. N. (1956) *Phys. Rev.*, **104**, 254.
Leger A., Mariotti J. M., Mennesson B., Ollivier M., Puget J. L., Rouan D. & Schneider J. (1996) *Icarus*, **114**, 139.
MacDermott A. J. (1995) *Orig. Life Evol. Biosphere*, **25**, 191.
MacDermott A. J. (1997) *Proc. SPIE*, **3111**, 272.
MacDermott A. J. & Tranter G. E. (1989) *Croatica Chemica Acta*, **62**, 165.
MacDermott A. J. & Tranter G. E. (1989a) *Chem. Phys. Lett.*, **163**, 1.
MacDermott A. J. & Tranter G. E. (1990) *Symmetries in Science IV* (Ed. by B. Gruber & J. H. Yopp) p. 67. Plenum New York.
MacDermott A. J., Tranter G. E. & Trainor S. J. (1992) *Chem. Phys. Lett.*, **194**, 152.
MacDermott A. J., Barron L. D., Brack A., Buhse T., Drake A. F., Emery R., Gottarelli G., Greenberg J. M., Haberle R., Hegstrom R. A., Hobbs K., Kondepudi

D. K., McKay C., Moorbath S., Raulin F., Sandford M., Schwartzman D. W., Thiemann W., Tranter G. E. & Zarnecki J. C. (1996) *Planet. Space Sci.*, **44**, 1441.

Mason S. F. (1982) *Molecular Optical Activity and the Chiral Discriminations*. Cambridge University Press.

Mason S. F. (1983) *Int. Rev. Phys. Chem.*, **3**, 217.

Mason S. F. (1986) *Nouv. J. Chim.*, **10**, 739.

Mason S. F. & Tranter G. E. (1984) *Mol. Phys.*, **53**, 1091.

Mason S. F. & Tranter G. E. (1985) *Proc. Roy. Soc. Lond. A*, **397**, 45.

Matthews C. N., Pesce-Rodriguez R. A. & Liebman S. A. (1997) *IAU Colloquium*, **161**, 179.

Meister A. (1965) *Biochemistry of the Amino Acids*. Academic Press, New York.

McKay D. S., Gibson E. K., Jr., Thomas-Keprta K. L., Vali H., Romanek C. S., Clemett S. J., Chillier X. D. F, Maechling C. R., Zare R. N. (1996) *Science*, **273**, 924.

Melcher G. (1974) *J. Mol. Evol.*, **3**, 121.

Mojzsis S. J., Arrhenius G., McKeegan K. D., Harrison T. M., Nutman A. P. & Friend C. R. L. (1996) *Nature*, **384**, 55.

Norden B. (1977) *Nature*, **266**, 567.

Noyes H. P., Bonner W. A. & Tomlin J. A. (1977) *Orig. Life*, **8**, 21.

Orgel L. E. (1996) *Orig. Life Evol. Biosphere*, **26**, 261.

Palache C., Erman G. B. & Frondel C. (1962) *Dana's System of Mineralogy*, 7th edn., Vol. III, p. 16. Wiley, New York.

Prieur B. E. (1994) *J. Biol. Phys.*, **20**, 301.

Rikken G. L. J. A. & Raupach E. (1997) *Nature*, **390**, 493.

Robert J. B. & Barra A. L. (1997) *Mol. Phys.*, in press.

Salam A. (1968) *Proc. 8th Nobel Symp., Elementary Particle Theory* (Ed. by N. Svartholm). Almquist and Wiksell, Stockholm.

Sandars P. G. H. (1980) *Fundamental Interactions and Structure of Matter* (Ed. by K. Crowse, J. Duclos, G. Fiorentini, & G. Torelli), p. 57. Plenum, New York.

Schopf J. W. (1993) *Science*, **260**, 640.

Schwartz A. W. & Orgel L. E. (1985) *Science*, **228**, 585.

Schwartzman D., McMenamin M. & Volk T. (1993) *BioScience*, **43**, 390.

Service R. F. (1999) *Science*, **286**, 1282.

Smith F. G, Jones D. H. P., Dick J. S. B. & Pike C. D. (1988) *Mon. Not. R. Ast. Soc.*, **233**, 305.

Soret C. (1899) *Arch. Sci. Phys. Nat.*, **7**, 80.

Spach G. & Brack A. (1983) *Structure, Dynamics and Evolution of Biological Macromolecules* (Ed. by C. Helene), p. 383. Reidel.

Thiemann W. & Teutsch H. (1990) *Orig. Life*, **20**, 121.

Teutsch H. (1988) Doctoral Thesis, University of Bremen.

Teutsch H. & Thiemann W. (1986) *Orig. Life*, **16**, 420.

Tranter G. E. (1985) *Nature*, **318**, 172.

Tranter G. E. (1986) *J. Chem. Soc. Chem. Commun.*, 60.

Tranter G. E., MacDermott A. J., Overill R. E. & Speers P. J. (1992) *Proc. R. Soc. Lond. A*, **436**, 603.

Ulbricht T. L. V. (1962) *Comp. Biochem.*, **4**, 1.

Ulbricht T. L. V. (1981) *Orig. Life*, **11**, 55.

Ulbricht T. L. V. & Vester F. (1962) *Tetrahedron*, **18**, 628.

van House J., Rich A. & Zitzewitz P. W. (1984) *Orig. Life*, **14**, 413.
Vistelius A. B. (1950) *Zapiski Vsyesoyuz. Mineral. Obsh.*, **79**, 191.
Weinberg S. (1967) *Phys. Rev. Lett.*, **19**, 1264.
Williams R. J. P. & Frausto da Silva J. J. R. (1996) *The Natural Selection of the Chemical Elements*. Oxford University Press.
Wolfrom M. L., Lemieux R. U. & Olin S. M. (1949) *J. Am. Chem. Soc.*, **71**, 2870.
Wolstencroft R. D. (1985) *IAU Symposium*, **112**, 171.
Wolstencroft R. D. (1996), poster at *Searching for Life in the Solar System and Beyond*, a Research Discussion Meeting held at the Royal Society of London, October 31st 1996.
Wu C. S., Ambler E., Hayward R. W., Hoppes D. D. & Hudson R. P. (1957) *Phys. Rev.*, **105**, 1413.
Wyrouboff G. (1896) *Bull. Soc. Mineral. Fr.*, **19**, 219.
Yamagata Y. (1966) *J. Theor. Biol.*, **11**, 495.

Chapter 3
Chirality at the Sub-Molecular Level: True and False Chirality

L. D. Barron

Department of Chemistry, University of Glasgow, University Avenue, Glasgow G12 8QQ, UK

INTRODUCTION

Scientists have been fascinated by handedness in the structure of matter ever since the concept first arose as a result of the discovery, in the early years of the last century, of natural optical activity in refracting media. This concept inspired major advances in physics, chemistry and the life sciences and continues to catalyse scientific and technological progress even today.

The subject can be considered to have started with the observation by Arago (1811) of colours in sunlight that had passed along the optic axis of a quartz crystal placed between crossed polarizers. Subsequent experiments by Biot (1812) established that the colours originated in the rotation of the plane of polarization of linearly polarized light (optical rotation)—rotation was different for light of different wavelengths (optical rotatory dispersion). The discovery of optical rotation in organic liquids such as turpentine (Biot 1815) indicated that optical activity could reside in individual molecules and could be observed even when the molecules were oriented randomly, unlike quartz where the optical activity is a property of the crystal structure, because molten quartz is not optically active.

After his discovery of circularly polarized light, Fresnel was able to understand optical rotation in terms of different refractive indices for the coherent right and left circularly polarized components of equal amplitude into which a linearly polarized light beam can be resolved. As the following statement shows (Fresnel 1824), this immediately gave him an important insight into the symmetry requirements for an optically active molecule or crystal:

> There are certain refracting media, such as quartz in the direction of its axis, turpentine, essence of lemon, *etc.*, which have the property of not transmitting with the same velocity circular vibrations from right to left and those from left to right. This may result from a peculiar constitution of the refracting medium or of

its molecules, which produces a difference between the directions right to left and left to right; such, for instance, would be a helicoidal arrangement of the molecules of the medium, which would present inverse properties accordingly as these helices were dextrogyrate or laevogyrate.

The celebrated resolution by Pasteur (1848) of crystals of the optically inactive (racemic) form of sodium ammonium tartrate into right- and left-handed hemihedral crystals which gave equal and opposite optical rotations in solution provided a dramatic confirmation of Fresnel's insight that optically active substances might be helicoidal at the microscopic level. This discovery emphasized that molecules must be pictured in three dimensions, and led ultimately to the concept of tetrahedral valences for the carbon atom and to the subject of stereochemistry.

Faraday (1846) discovered that optical activity could be induced in an otherwise inactive sample by a magnetic field. He observed optical rotation in a rod of lead borate glass placed between the poles of an electromagnet with holes bored through the pole pieces to enable a linearly polarized light beam to pass through. This effect is quite general—a Faraday rotation is found when linearly polarized light is transmitted through any crystal or fluid in the direction of a magnetic field, the sense of rotation being reversed on reversing the direction of either the light beam or the magnetic field.

At the time, the main significance of this discovery was to demonstrate conclusively the intimate connection between electromagnetism and light; but it also became a source of confusion to some physicists and chemists (including Pasteur) who failed to appreciate that there is a fundamental distinction between magnetic optical rotation and the natural optical rotation that is associated with handedness in the microstructure. That the two phenomena have fundamentally different symmetry characteristics is intimated by the fact that the magnetic rotation is additive when the light is reflected back though the medium whereas the natural rotation cancels.

Although he does not provide a formal definition, it can be inferred (Lowry 1964) from his original article which described in detail his experiments with salts of tartaric acid that Pasteur (1848) introduced the word *dissymmetric* to describe hemihedral crystals of a tartrate "which differ only as an image in a mirror differs in its symmetry of position from the object which produces it" and subsequently used this word to describe handed figures and handed molecules generally.

The two distinguishable mirror-image forms subsequently became known as enantiomers. A finite cylindrical helix provides a good example, because reflection reverses the screw sense. Dissymmetric figures are not necessarily asymmetric, meaning devoid of all symmetry elements, because they can have one or more proper rotation axes (the finite cylindrical helix has a twofold rotation axis through the midpoint of the coil, perpendicular to the long helix axis). Specifically, dissymmetry excludes improper rotation axes, *i.e.* a centre of inversion, reflection planes and rotation–reflection axes.

Pasteur attempted to extend the concept of dissymmetry to other aspects of the physical world in his search for universal forces which might be connected with handedness in molecules from the living world (Mason 1982). For example, he thought that a magnetic field, because it can induce optical rotation (the Faraday effect), generates the same type of dissymmetry as that possessed by an optically active molecule. As shall be seen, this idea is quite wrong and has been the source of much confusion. Pasteur was, however, correct in thinking that the combination of a linear motion with a rotation does generate the same type of dissymmetry as that of an optically active molecule.

The word dissymmetry was eventually replaced by *chirality*, meaning handedness, in the literature of stereochemistry. This word was first introduced into science by Lord Kelvin (1904), Professor of Natural Philosophy in the University of Glasgow, to describe a figure "if its image in a plane mirror, ideally realized, cannot be brought to coincide with itself".

Although Kelvin's definition of chiral is essentially the same as that used earlier by Pasteur for dissymmetric, the two words are not strictly synonymous in the broader context of modern chemistry and physics. Dissymmetry means the absence of certain symmetry elements, these being improper rotation axes in Pasteur's usage. Chirality has become a more positive concept in that it refers to the possession of the attribute of handedness, which has a physical content—in molecular physics this is the ability to support time-even pseudoscalar observables; in elementary particle physics chirality is defined as the eigenvalue of the Dirac matrix operator γ_5.

This chapter, to facilitate a proper understanding of the structure and properties of chiral molecules and of the factors involved in their synthesis and transformations, uses some principles of modern physics, especially fundamental symmetry arguments, to obtain a description of chirality deeper than that usually encountered in the literature of stereochemistry. A central result is that, although dissymmetry is sufficient to guarantee chirality in a stationary object such as a finite helix, dissymmetric systems are not necessarily chiral when motion is involved.

The words 'true' and 'false' chirality, corresponding to time-invariant and time-non-invariant enantiomorphism, respectively, were introduced by this author to draw attention to this distinction (Barron 1981a, b, 1982, 1986a), but it was not intended that this would become standard nomenclature. Rather, it was suggested that the word 'chiral' be reserved in future for systems that are truly chiral. The terminology of true and false chirality has, however, been taken up by others, especially in the area of absolute asymmetric synthesis, so for consistency it will be used in this article.

It shall be seen that, as intimated above, the combination of linear motion with a rotation does indeed generate true chirality, but that a magnetic field alone does not (in fact it is not even falsely chiral). Examples of systems with false chirality include a stationary rotating cone, and collinear electric and magnetic fields. The term 'false' should not be taken to be pejorative in any sense; indeed, false chirality can give rise to some fascinating new phenomena that are even more subtle and beautiful than those associated with true chirality.

The recent triumph of theoretical physics in unifying the weak and electromagnetic forces into a single 'electroweak' force has provided a new perspective on chirality. Because the weak and electromagnetic forces have turned out to be different aspects of the same, but more fundamental, unified force, the absolute parity violation associated with the weak force is now known to infiltrate to a tiny extent into all electromagnetic phenomena so that free atoms, for example, have very small optical rotations, and a tiny energy difference exists between the enantiomers of a chiral molecule (Hegstrom *et al.* 1980; Khriplovich 1991).

It shall be seen that the distinction between true and false chirality hinges on the symmetry operations that interconvert enantiomers, and that parity violation provides a cornerstone for the identification of true chirality (Barron 1986a). It is remarkable that parity violation provides a scientific basis for the general cosmic dissymmetry that Pasteur sensed over a century ago (Haldane 1960).

SYMMETRY PRINCIPLES

The symmetry arguments required for a proper understanding of chirality go beyond the purely spatial aspects with which most chemists are familiar and which usually employ point-group symmetry arguments to deduce qualitative information about molecules. In addition to conventional point group symmetry arguments, the fundamental symmetries of space inversion, time reversal and even charge conjugation have something to say about chirality at several levels including the experiments that show up optical activity observables, the objects generating these observables and the quantum states that these objects must be able to support. It is therefore appropriate to start by reviewing these fundamental symmetries.

NON-OBSERVABLES AND SYMMETRY OPERATIONS

Most symmetry arguments depend on the symmetries inherent in the laws which determine the operation of the physical world (Feynman *et al.* 1964). These symmetries can be associated with the impossibility of observing certain basic quantities (Lee 1981). Three such non-observables seem to be particularly fundamental—absolute chirality (meaning absolute right- or left-handedness), absolute direction of motion, and absolute sign of electric charge. A non-observable implies invariance of physical laws under an associated transformation and usually generates a conservation law or selection rule.

The transformation associated with absolute chirality is space inversion, represented by the operator P which inverts the positions of all the particles in a system through an arbitrary space-fixed origin. With the exception of those describing processes involving the weak force, all physical laws are unchanged under space inversion. If replacing the space coordinates (x,y,z) by $(-x,-y,-z)$ everywhere in equations describing physical laws (*e.g.* Newton's equations for mechanics or Maxwell's equations for electromagnetism) leaves those equations unchanged, all physical processes determined by those laws are said to conserve parity.

The transformation associated with absolute direction of motion is time reversal, represented classically by the operator T, which reverses the motions of all the particles in the system. If replacing the time co-ordinate (t) by $(-t)$ everywhere leaves equations describing physical laws unchanged, those laws are said to conserve time reversal invariance (or to have reversality). The name *time reversal* is rather unfortunate because it has mysterious connotations of travelling backwards in time, which is not the intended meaning. As Sachs (1987) has stressed, time reversal has nothing to do with the 'arrow of time' associated with irreversibility in thermodynamics. So it is best to think of time reversal as motion reversal.

The transformation associated with absolute sign of electric charge is charge conjugation, represented by the operator C which interconverts particles and antiparticles. Although this exotic operation from relativistic quantum field theory might seem to have no relevance outside elementary particle physics, we shall see that it has conceptual value in studies of molecular chirality.

An important consequence of the existence of symmetries in the laws which determine the operation of the physical world is that, if a complete experiment is subjected to an associated symmetry operation such as space inversion or time reversal, the resulting experiment should, in principle, be realizable (Feynman *et al.* 1964). A detailed consideration of the natural and magnetic optical rotation experiments shows that they do indeed conserve parity and reversality (Barron 1982); and such arguments can be used to predict or discount possible new effects (such as an electric analogue of the Faraday effect) without recourse to mathematical theories.

SYMMETRY OF PHYSICAL QUANTITIES

Physical quantities can be classified according to their behaviour under P, T and C. The physical quantities are first classified as *scalars*, *vectors* or *tensors* depending on their directional properties. A scalar such as temperature has magnitude but no associated direction; a vector such as velocity has magnitude and one associated direction; and a tensor such as electric polarizability has magnitudes associated with two or more directions.

Vectors such as position **r**, velocity **v** and linear momentum **p**, which change sign under parity P are called *polar* or true vectors. A vector such as angular momentum **L** = **r** × **p** whose sign is not changed by P is called an *axial* or pseudo vector—**L** is defined relative to the sense of rotation by a right-hand rule, and P does not change the sense of rotation. A *pseudoscalar* quantity is a number with no directional properties but which changes sign under P. A pseudoscalar is generated by taking the scalar product of a polar and an axial vector.

Physical quantities are also classified as *time-even* or *time-odd* depending on whether they are invariant or change sign under time reversal T. This behaviour is usually immediately obvious from a consideration of the motions of the constituent particles. Many physical quantities do not involve motion and so, of course, are time-even; examples are the energy scalar W, position vector **r** and electric dipole moment vector **μ**. Many others do involve motion and are time-odd, examples being the velocity vector **v**, linear momentum vector **p**, angular momentum vector **L** and magnetic dipole moment vector **m**. The elusive magnetic monopole, which has never been observed, transforms as a time-odd pseudoscalar.

The behaviour of the electric and magnetic field vectors **E** and **B** is particularly important here. The symmetry characteristics can be deduced by inspecting the physical systems which generate the fields. Thus **E** is generated by a pair of parallel plates carrying equal and opposite static charge distributions—because P swaps the charge distributions but T has no effect, **E** is a time-even polar vector. On the other hand **B** is generated by a cylindrical current sheet—because P does not affect the sense of circulation of the current but T reverses it, **B** is a time-odd axial vector.

Pseudoscalar quantities are of central importance in the discussion of molecular chirality because, as shown below, the natural optical activity phenomena supported by chiral molecules are characterized by time-even pseudoscalar observables such as optical rotation angle, rotational strength, Raman circular intensity difference, *etc.*

SYMMETRY IN QUANTUM MECHANICS

When symmetry operations are formulated in quantum mechanics, some important new features arise that are not present in the classical discussion.

Parity

The starting point for parity considerations is the invariance of the conventional (*i.e.* parity-conserving) Hamiltonian for a closed system of interacting particles to an inversion of the coordinates. P is now interpreted as an operator that changes the sign of the space coordinates in the Hamiltonian H and in the wavefunction $\psi(\mathbf{r})$.

Consider first the wavefunction:

$$P\psi(\mathbf{r}) = \psi(-\mathbf{r})$$

If $\psi(\mathbf{r})$ happens to be an eigenfunction of P we can write:

$$P\psi(\mathbf{r}) = p\psi(\mathbf{r}) \qquad (2)$$

The eigenvalues p are found by realizing that a double application amounts to the identity so that:

$$P^2\psi(\mathbf{r}) = p^2\psi(\mathbf{r}) = \psi(\mathbf{r}) \qquad (3)$$

from which is obtained:

$$p^2 = 1, p = \pm 1 \qquad (4)$$

Thus even (+) and odd (−) parity wavefunctions are defined according to whether they are invariant or simply change sign under P:

$$P\psi(+) = \psi(+), \ P\psi(-) = -\psi(-) \qquad (5)$$

Turning now to the Hamiltonian, its invariance under space inversion means that the following can be written:

$$PHP^{-1} = H, \text{ or } [P,H] = PH - HP = 0 \qquad (6)$$

Because P does not depend explicitly on time and commutes with the Hamiltonian, we can say that, if the state of a closed system has definite parity, that parity is conserved. This is called the law of conservation of parity (Landau & Lifshitz 1977).

If two eigenfunctions $\psi(+)$ and $\psi(-)$ of opposite parity have energy eigenfunctions which are degenerate, or nearly so, the system can exist in states of mixed parity with wavefunctions:

$$\psi_1 = \frac{1}{\sqrt{2}}[\psi(+)+\psi(-)] \qquad (7a)$$

$$\psi_2 = \frac{1}{\sqrt{2}}[\psi(+)-\psi(-)] \qquad (7b)$$

Clearly these two mixed-parity wavefunctions are interconverted by P:

$$P\psi_1 = \psi_2,\ P\psi_2 = \psi_1 \qquad (8)$$

It follows from eq. (6) that an important property of definite parity states is that they are true stationary states with constant energy $W(+)$ or $W(-)$, i.e.

$$\psi(\pm) = \psi^{(0)} e^{-iW(\pm)t/\hbar} \qquad (9)$$

but that mixed parity states are not. It shall be seen below that mixed parity states can become quasi-stationary when $W(+) \approx W(-)$, and can become true stationary states if H contains a parity-violating term.

All observables can be classified as having even or odd parity depending on whether they are invariant or change sign under space inversion. Even- and odd-parity operators $A(+)$ and $A(-)$ associated with these observables are therefore defined by:

$$PA(+)P^{-1} = A(+),\ PA(-)P^{-1} = -A(-) \qquad (10)$$

Because integrals taken over all space are only non-zero for totally symmetric integrands, the expectation values of these operators in a mixed state such as (7a) reduce to:

$$\langle\psi_1|A(+)|\psi_1\rangle = \tfrac{1}{2}[\langle\psi(+)|A(+)|\psi(+)\rangle + \langle\psi(-)|A(+)|\psi(-)\rangle] \qquad (11a)$$

$$\langle\psi_1|A(-)|\psi_1\rangle = \tfrac{1}{2}[\langle\psi(+)|A(-)|\psi(-)\rangle + \langle\psi(-)|A(-)|\psi(+)\rangle] \qquad (11b)$$

from which it follows that the expectation value of any odd-parity observable vanishes in any state of definite parity, i.e. a state for which either $\psi(+)$ or $\psi(-)$ is zero. This means that measurements on a system in a state of definite parity can reveal only observables with even parity, examples being electric charge, angular momentum and magnetic dipole moment. Measurements on a system in a state of mixed parity can reveal, in addition, observables with odd parity, examples being magnetic monopole, linear momentum and electric dipole moment (Barron 1982).

The optical rotatory parameter, being a pseudoscalar, has odd parity, which leads to the important conclusion that *resolved chiral molecules exist in mixed-parity quantum states*. The nature of these states, which correspond to the enantiomeric handed states ψ_L and ψ_R, is described below.

Time reversal

The classical time reversal operator T introduced above does not translate into a satisfactory quantum-mechanical operator. Instead, the operator:

$$\Theta = UK \tag{12}$$

where U is a unitary operator and K is the operator of complex conjugation, is taken as the time reversal operator in quantum mechanics (Sakurai 1985). It is possible to classify time-even and time-odd Hermitian operators $A(+)$ and $A(-)$ and their associated observables according to whether they are invariant or change sign under time reversal:

$$\Theta A(+)\Theta^{-1} = A(+)^\dagger, \; \Theta A(-)\Theta^{-1} = -A(-)^\dagger \tag{13}$$

Unlike the case of parity, however, it is not possible to classify a quantum state as being even or odd under time reversal because Θ, unlike P, does not have eigenvalues. (The operator Θ^2 does have eigenvalues, these being $+1$ for an even-electron system and -1 for an odd-electron system).

A useful illustration for our purposes is the effect of Θ on a general atomic state $|J,M\rangle$ where both orbital and spin angular momenta can contribute to the total electronic angular momentum specified by the usual quantum numbers J and M (Sakurai 1985):

$$\Theta|J,M\rangle = i^{2M}|J,M\rangle \tag{14}$$

Application of the time reversal operator has therefore generated a new quantum state, orthogonal to the original, corresponding to a reversal of the sense of the total angular momentum of the atom. Because they are interconverted by time reversal, such states can be loosely regarded as having 'mixed reversality' analogous to the mixed parity states, even though associated states of definite reversality do not exist. One important property of mixed reversality states is that they can support time-odd observables (Barron 1982). Notice, however, that the states $|J,M\rangle$ do still have definite parity since they are eigenstates of P:

$$P|J,M\rangle = (-1)^q |J,M\rangle \tag{15}$$

where q is the sum of the individual orbital angular momentum quantum numbers of all the electrons in the atom. This follows from the behaviour under space inversion of the spherical harmonics (Sakurai 1985).

Charge conjugation

A discussion of the effect of the charge conjugation operator C in quantum mechanics requires a formulation in terms of relativistic quantum field theory (Lee 1981; Weinberg 1995) and will not be given here. All that needs to be appreciated for the purposes of this article is that a charged particle is not in an eigenstate of C because charge conjugation generates a different quantum state corresponding to the associated antiparticle.

TRUE CHIRALITY

Optical activity is not necessarily the hallmark of chirality. Indeed, the failure to distinguish properly between natural and magnetic optical activity has often been a source of confusion in the literature of both chemistry and physics. As shown in this section, a symmetry classification of the observables associated with these two types of optical activity clearly shows that they are quite different and leads to a more precise definition of a chiral system.

NATURAL AND MAGNETIC OPTICAL ACTIVITY

Lord Kelvin (1904) was fully aware of the fundamental distinction between natural and magnetic optical rotation. This is evident, because his Baltimore Lectures contain the statement:

> The magnetic rotation has neither left-handed nor right-handed quality (that is to say, no chirality). This was perfectly understood by Faraday, and made clear in his writings, yet even to the present day we frequently find the chiral rotation and the magnetic rotation of the plane of polarized light classed together in a manner against which Faraday's original description of his discovery of the magnetic polarization contains ample warning.

He might have had Pasteur in mind who, judging from his writings, was a persistent offender. For example, because a magnetic field induces optical rotation, Pasteur thought that by growing crystals, normally holohedral, in a magnetic field a magnetically-induced dissymmetry would be manifest in hemihedral crystal forms; the resulting crystals, however, retained their usual holohedral forms (Mason 1982). Lord Kelvin's viewpoint was reinforced much later by Zocher & Török (1953), who discussed the space–time symmetry aspects of natural and magnetic optical activity from a general classical viewpoint and recognized that quite different asymmetries are involved.

The required symmetry classifications can be obtained by comparing the results of optical rotation measurements before and after subjecting the sample plus any applied field to space inversion and time reversal (Barron 1982).

Consider first natural optical rotation. Under space inversion P an isotropic collection of chiral molecules is replaced by a collection of the corresponding enantiomeric molecules, and an observer with a linearly polarized probe light beam will measure equal and opposite optical rotation angles before and after the inversion. This indicates that the observable has odd parity, and it is easy to see that it is a pseudoscalar (rather than, say, a polar vector) because it is invariant with respect to any proper rotation in space of the complete sample. Under time reversal T an isotropic collection of chiral molecules is unchanged, so the optical rotation is unchanged. Hence *the natural optical rotation observable is a time-even pseudoscalar*.

Conder next the Faraday effect, where optical rotation is induced in an isotropic collection of achiral molecules by a static uniform magnetic field collinear with the light beam. Under P the molecules and the magnetic field direction are unchanged, so the same magnetic optical rotation will be observed. This indicates that the observable has even parity, and it can be further deduced that it is an axial vector (rather than, say, a scalar) by noting that a proper rotation of the sample plus magnetic field through 180° about any axis perpendicular to the field reverses the relative directions of the magnetic field and the probe light beam and so changes the sign of the observable. Under T the collection of molecules (even if they are individually paramagnetic) can be regarded as unchanged provided it is isotropic in the absence of the field; again the relative directions of the magnetic field and the probe light beam are, however, reversed so the optical rotation changes sign. Hence *the magnetic optical rotation observable is a time-odd axial vector.*

The same conclusions are obtained from a more fundamental approach in which operators are defined whose expectation values generate the optical activity observables (Barron 1981a, 1982). It is found that the natural optical rotation observable is generated by a time-even odd-parity operator, and the magnetic optical rotation observable by a time-odd even-parity operator. Another approach is to look at the associated molecular property tensors—all the contributions to natural optical rotation are generated by time-even tensors, and all those to magnetic optical rotation by time-odd tensors (Buckingham *et al.* 1971).

This analysis reveals that the nature of molecular quantum states supporting natural optical rotation is quite different from that of quantum states supporting magnetic optical rotation. From the discussion above it is clear that the former must have, among other things, mixed parity and the latter mixed reversability. The former is associated with spatial dissymmetry; whereas the latter originates in a different type of dissymmetry associated with a lack of time reversal invariance.

A NEW DEFINITION OF CHIRALITY

From the foregoing, it should now be clear that the hallmark of a chiral system is that it can support time-even pseudoscalar observables. This leads to the following definition which enables chirality to be distinguished from other types of dissymmetry (Barron 1986a,b):

> *True chirality is exhibited by systems that exist in two distinct enantiomeric states that are interconverted by space inversion, but not by time reversal combined with any proper spatial rotation.*

This means that the spatial enantiomorphism shown by truly chiral systems is time-invariant. Spatial enantiomorphism that is time-non-invariant has different characteristics that this author has called false chirality to emphasize the distinction. Notice that a magnetic field on its own is not even falsely chiral because there is no associated spatial enantiomorphism.

A stationary object such as a finite helix that is chiral according to the traditional stereochemical definition is accommodated by the first part of this definition—space inversion is a more fundamental operation than the mirror reflection traditionally invoked, but provides an equivalent result. Time reversal is irrelevant for a stationary object, but the full definition is required to identify more subtle sources of chirality in which motion is an essential ingredient. A few examples will make this clear.

TRANSLATING SPINNING ELECTRONS, PHOTONS AND CONES

Consider an electron, which has a spin quantum number $s = ½$ with $m_s = ±½$ corresponding to the two opposite projections of the spin angular momentum on to a space-fixed axis. A stationary spinning electron is not a chiral object because space inversion P does not generate a distinguishable P-enantiomer (Fig. 3.1a).

An electron translating with its spin projection parallel or antiparallel to the direction of propagation has true chirality, however, because P interconverts distinguishable left- and right-spin polarized versions propagating in opposite directions, whereas time reversal T does not (Fig. 3.1b). (The projection of the spin angular momentum **s** of a particle along its direction of motion is called the helicity, λ, = **s.p**/|**p**|. Spin-½ particles can have $\lambda = ±\hbar/2$, the positive and negative states being called right- and left-handed; this, however, corresponds to the opposite sense of handedness to that used in the usual definition of right- and left-circularly polarized light in classical optics.)

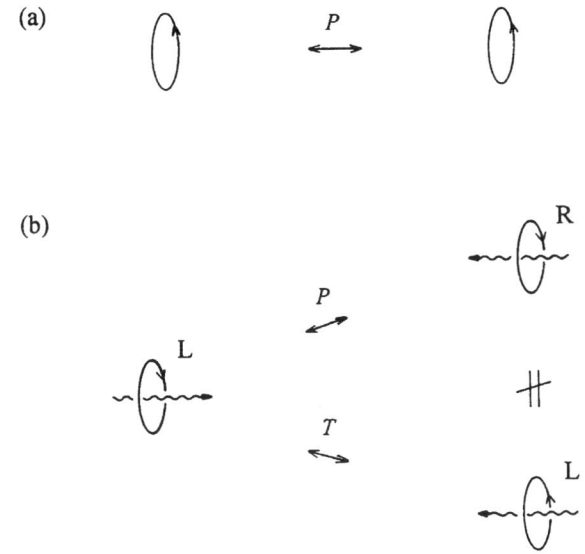

Fig. 3.1 The effect of parity P and time reversal T on the motions of (a) a stationary spinning particle and (b) a translating spinning particle. L and R refer to left- and right-handed helicities according to the usual definition in particle physics.

The photons in a circularly polarized light beam propagating as a plane wave are in spin angular momentum eigenstates characterized by $s = 1$ with $m_s = ±1$ corresponding to projections of the spin angular momentum vector parallel or antiparallel, respectively, to the propagation direction. The absence of states with $m_s = 0$ is connected with the fact that photons, being massless, have no rest frame and so always move with the velocity of light. Considerations the same as those in Fig. 3.1b show that a circularly polarized photon has true chirality.

Now consider a cone spinning about its symmetry axis. Because P generates a version that is not superposable on the original (Fig. 3.2a), it might be thought that this is a chiral system. The chirality is, however, false because T followed by a rotation R_π through 180° about an axis perpendicular to the symmetry axis generates the same system as space inversion (Fig. 3.2a). If, however, the spinning cone is also translating along the axis of spin, T followed by R_π now generates a system different from that generated by P alone (Fig. 3.2b). Hence a *translating* spinning cone has true chirality.

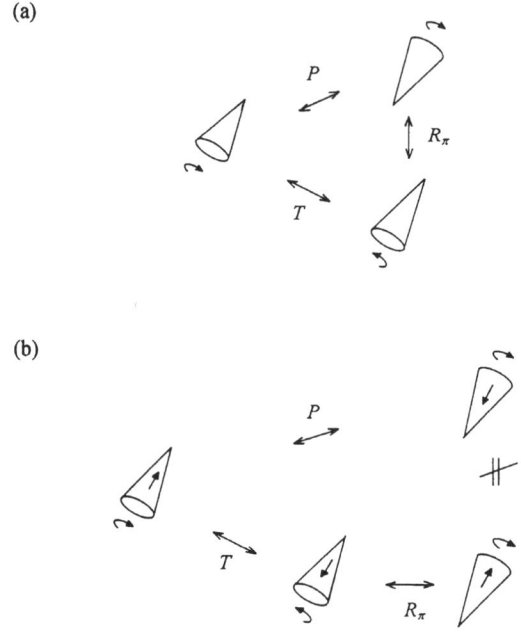

Fig. 3.2 The effect of P, T and the 180° rotation R_π on (a) a stationary spinning cone and (b) a translating spinning cone.

These considerations expose a link between chirality and special relativity because the chirality that an observer perceives in a spinning electron, for example, depends on the relative motions of the observer and the electron (see below).

ELECTRIC, MAGNETIC AND GRAVITATIONAL FIELDS

It is clear that neither a static uniform electric field **E** (a time-even polar vector) nor a static uniform magnetic field **B** (a time-odd axial vector) constitutes a chiral system. Likewise for time-dependent uniform electric and magnetic fields. No combination of a static uniform electric and a static uniform magnetic field can, furthermore, constitute a chiral system—collinear electric and magnetic fields do indeed generate spatial enantiomorphism (Curie 1894), but it is time-non-invariant and so corresponds to false chirality. Thus, as shown in Fig. 3.3a, parallel and antiparallel arrangements are interconverted by space inversion and are not superposable. As shown in Fig. 3.3b, however, they are also interconverted by time reversal combined with a rotation through 180°. Zocher & Török (1953) also recognized the flaw in Curie's suggestion. They called the collinear arrangement of electric and magnetic fields a time-asymmetric enantiomorphism and said that it does not enable time-symmetric optical activity.

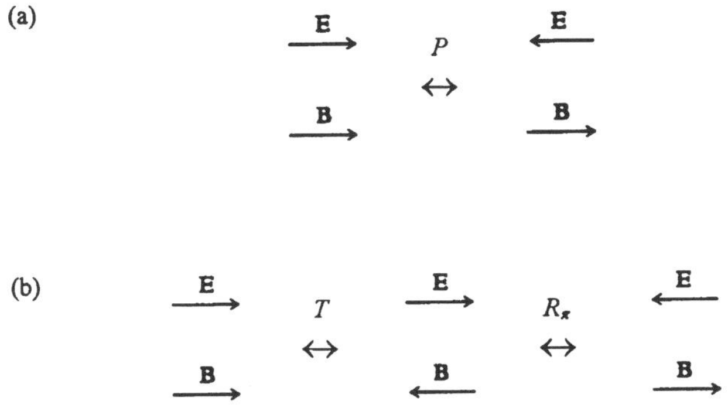

(a)   **E** →   P   ← **E**
             ↔
    B →   **B** →

(b)   **E** →   T   **E** → R_π ← **E**
          ↔   ↔
    B →   **B** ←   **B** →

Fig. 3.3 The effect of P and T on collinear static uniform electric and magnetic fields showing (a) that the distinguishable parallel and antiparallel arrangements of **E**,**B** are interconverted by P, and (b) that they are also interconverted by T combined with R_π and so exhibit time-non-invariant enantiomorphism (false chirality).

In fact, the basic requirement for two collinear vectorial influences to generate chirality is that one transforms as a polar vector and the other as an axial vector, with both either time-even or time-odd. The second case is exemplified by *magneto-chiral* phenomena such as a birefringence and a dichroism induced in a chiral sample by a uniform magnetic field collinear with the propagation vector **k** of a light beam of arbitrary polarization (Wagnière & Meier 1982; Barron & Vrbancich 1984; Rikken & Raupach 1997; Kleindienst & Wagnière 1998). Thus parallel and antiparallel arrangements of **B** and **k** are true chiral enantiomers because they cannot be interconverted by time reversal, because **k**, like **B**, is time-odd.

Analogous to collinear electric and magnetic fields is the case of a rapidly rotating vessel with the axis of rotation perpendicular to the earth's surface (Dougherty 1980). Here we have the time-odd axial angular momentum vector of the spinning vessel either parallel or antiparallel to the earth's gravitational field, itself a time-even polar vector. The physical influence in this instance, therefore, has false chirality.

ABSOLUTE ASYMMETRIC SYNTHESIS

The use of an external physical influence to produce an enantiomeric excess in what would otherwise be a racemic product of a prochiral chemical reaction is known as an absolute asymmetric synthesis. The subject still attracts much interest and controversy (Bonner 1988, 1990; Keszthelyi 1995; Avalos *et al.* 1998; Feringa & van Delden 1999), not least because it is an important ingredient in considerations of the pre-biotic origins of biological molecules (Cline 1996). Indeed the subject can be considered to have started with Pasteur's search for a universal dissymmetric influence which might be responsible for handedness in the chemistry of life (Haldane 1960; Mason 1982).

TRULY CHIRAL INFLUENCES

If an influence can be classified as truly chiral we can be confident that it has the correct symmetry characteristics to induce absolute asymmetric synthesis, or some associated process such as preferential asymmetric decomposition, in any conceivable situation, although, of course, the influence might be too weak to produce an observable effect. In this respect it is important to remember Jaeger's dictum (Jaeger 1930; Bonner 1990): "The necessary conditions will be that the externally applied forces are a *conditio sine qua non* for the initiation of the reaction which would be impossible without them."

Consider a unimolecular process in which an achiral molecule R generates a chiral molecule M or its enantiomer M*:

$$M \underset{k_b}{\overset{k_f}{\rightleftharpoons}} R \underset{k_f^*}{\overset{k_b^*}{\rightleftharpoons}} M^* \tag{16}$$

In the absence of a chiral influence, M and M* have the same energy, of course, so no enantiomeric excess can exist if the reaction is left to reach thermodynamic equilibrium. In the presence of such an influence, however, M and M* will have different energies.

This is easily seen from a simple symmetry argument. Consider a collection of single enantiomers M in the presence of a right-handed chiral influence $(Ch)_R$, say. Under *P*, the collection of enantiomers M becomes an equivalent collection of mirror-image enantiomers M* and the right-handed chiral influence $(Ch)_R$ becomes the equivalent left-handed chiral influence $(Ch)_L$. Assuming parity is conserved, this indicates that the energy of M in the presence of $(Ch)_R$ is equal to that of M* in the presence of $(Ch)_L$; because parity (or any other symmetry operation) does not provide a relation between the energy of M and M* in the presence of the same influence, be it $(Ch)_R$ or $(Ch)_L$, they will in general have different energies. Hence an enantiomeric excess can now exist at equilibrium. There will also be kinetic effects because the enantiomeric transition states will also have different energies.

Circularly polarized photons or longitudinal spin-polarized electrons are the obvious choice, and several examples in asymmetric synthesis or preferential asymmetric decomposition are known (Bonner 1988, 1990; Keszthelyi 1995; Avalos *et al.* 1998; Feringa & van Delden 1999). Less obvious is the use of an *unpolarized* light beam collinear with a magnetic field. As discussed in above, this system has true chirality so there can be confidence that it can induce absolute asymmetric synthesis. The most favourable mechanism with this type of influence would seem to be one based on magneto-chiral dichroism (Wagnière & Meier 1983).

FALSELY CHIRAL INFLUENCES

It is important to appreciate that, unlike the case of a truly chiral influence, enantiomers M and M* remain strictly isoenergetic (neglecting the very small difference as a result of parity violation) in the presence of a falsely chiral influence such as collinear electric and magnetic fields.

Again this can be seen from a simple symmetry argument. Under P, the collection of enantiomers M becomes the collection M* and the parallel arrangement, say, of **E** and **B** becomes antiparallel (Fig. 3.3a). The antiparallel arrangement of **E** and **B**, however, becomes parallel again under time reversal T and, furthermore, will regain its original orientation under a rotation through 180° (Fig. 3.3b); but these last two operations will have no effect on an isotropic collection of chiral molecules, even if paramagnetic. Hence the energy of the collection M is the same as that of the collection M* in parallel (or antiparallel) electric and magnetic fields.

When considering the possibility or otherwise of absolute asymmetric synthesis being induced by a falsely chiral influence, a distinction must be made between reactions that have been left to reach thermodynamic equilibrium (*thermodynamic control*) and reactions that have not attained equilibrium (*kinetic control*). The case of thermodynamic control is quite clear—because M and M* remain strictly isoenergetic in the presence of a falsely chiral influence, such an influence cannot induce absolute asymmetric synthesis in a reaction mixture which is isotropic in the absence of the influence and which has been allowed to reach thermodynamic equilibrium (Mead *et al.* 1977). The situation is, however, less straightforward for reactions under kinetic control for, as discussed in the next section, if microscopic reversibility and the associated detailed balancing were to break down an enantiomeric excess could develop.

The breakdown of microscopic reversibility induced by a falsely chiral influence: enantiomeric detailed balancing

It is possible that conventional detailed balancing, and the associated kinetic principles, might not be valid for reactions involving chiral molecules under a falsely chiral influence (Barron, 1987). This suggestion was inspired by a remark of Lifshitz & Pitaevskii (1981) that, for a system comprising chiral molecules of just one enantiomer, detailed balancing in the literal sense does not obtain because space inversion and well as time reversal is applied to each microscopic process, so that a completely different system is generated which cannot be compared with the original to deduce new information about its properties. This can be seen from a quantum-mechanical description of the microscopic reaction event (Sakurai 1985). The amplitude for a transition from some initial linear momentum state **p** to some final state **p**′ is written $\langle \mathbf{p}'|\mathfrak{I}|\mathbf{p}\rangle$ where $\mathfrak{I}$ is the operator responsible for the transition. If $\mathfrak{I}$ involves purely electromagnetic interactions it will be invariant under both parity and time reversal, which enables us to write:

$$\langle \mathbf{p}'|\mathfrak{I}|\mathbf{p}\rangle \stackrel{\text{under}\,T}{=} \langle -\mathbf{p}|\mathfrak{I}|-\mathbf{p}'\rangle \stackrel{\text{under}\,P}{=} \langle \mathbf{p}*|\mathfrak{I}|\mathbf{p}*'\rangle \tag{17}$$

where the particles have been allowed to be chiral, the star denoting the P enantiomer. The first equality in eq. (17), obtained from time reversal alone, is the basis of the conventional principle of microscopic reversibility and, when averaged over the complete system of reacting particles at equilibrium, of the principle of detailed balancing (Tolman 1938). The second equality, obtained by applying space inversion to the time-reversed transition amplitude, describes the inverse process involving the *enantiomeric* particles.

Conventional detailed balancing is usually adequate for the kinetic analysis of chemical reactions, even those involving chiral molecules, because conventional microscopic reversibility based on the assumption of T invariance is usually valid. As illustrated in Fig. 3.4, this might be conceptualized in terms of a potential energy profile that is the same in the forward and backward directions for a given reaction.

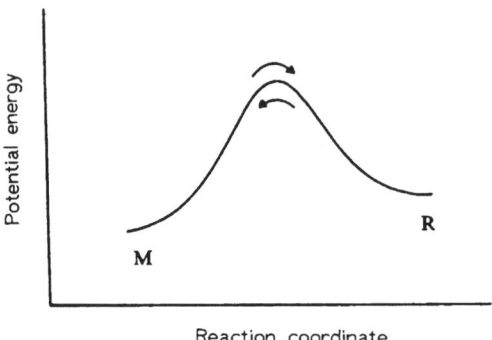

Fig. 3.4 Microscopic reversibility conceptualized as the same potential energy profile for the forward and backward processes.

In the presence of a falsely chiral influence such as collinear electric and magnetic fields, however, time reversal alone is not a symmetry operation because a different influence would be generated—space inversion must also be applied to recover the original relative orientations of **E** and **B**. The first equality in eq. (17) is therefore no longer valid, and any kinetic analysis must be based on the relationship:

$$\langle \mathbf{p}'|\mathfrak{I}|\mathbf{p}\rangle \stackrel{\text{under }TP}{=} \langle \mathbf{p}*|\mathfrak{I}|\mathbf{p}*'\rangle \qquad (18)$$

which implies *enantiomeric* microscopic reversibility.

Hence, as illustrated in Fig. 3.5, for reactions of chiral molecules in situations where only the combined *TP* invariance holds, the potential energy profiles for the forward and backward *enantiomeric* reactions are the same, but the forward and backward profiles for the reaction of a given enantiomer are in general different. This might be modelled in terms of different velocity-dependent contributions to the potential energy profiles for the forward and reverse reactions involving a particular enantiomer induced by the falsely chiral influence.

Although conventional chemical kinetics is founded on the assumption of microscopic reversibility, the possibility of a breakdown in the presence of a time-odd influence such as a magnetic field does not conflict with any fundamental principles. Indeed, in his classic paper on irreversible processes, Onsager (1931) recognized that microscopic reversibility does not apply when external magnetic fields are present.

Onsager's prescription of reversing **B** along with the motions of the interacting particles does not however, restore microscopic reversibility when **B** is a component of a falsely chiral influence; even then there will only be observable consequences if the particles are chiral (Barron 1994a). This is because, if the particles are achiral, the *P* enantiomers M and M* are indistinguishable so that the barriers to the left and right of R in Fig. 3.5 must coalesce, which is only possible if the forward and reverse barriers shown for production of a particular enantiomer become identical; but if M is not identical to M* they can, in general, remain distinct.

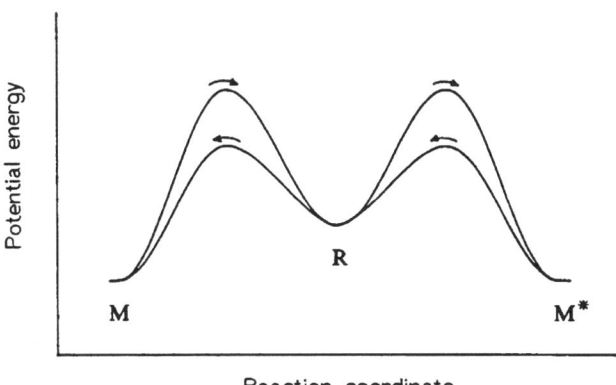

Fig. 3.5 Potential energy profiles for enantiomeric reactions in the presence of a falsely chiral influence illustrating the principle of enantiomeric microscopic reversibility.

Unitarity and Thermodynamic Equilibrium

From the foregoing, it seems that enantiomeric microscopic reversibility and the associated enantiomeric detailed balancing allows $k_f \neq k_f^*$ and $k_b \neq k_b^*$ if the reaction depicted by eq. (16) takes place in the presence of a falsely chiral influence such as collinear **E**, **B**. This would, however, generate an enantiomeric excess at thermodynamic equilibrium, which conflicts with the requirement that the concentrations of M and M* must be the same.

This conflict between kinetic and thermodynamic requirements can be resolved by including other pathways by which M and M* can be interconverted. Then at *true* thermodynamic equilibrium (*i.e.* when all possible interconversion pathways are in equilibrium), equal concentrations of M and M* obtain because the different enantiomeric excesses associated with each separate pathway sum to zero. The proof of this assertion (Barron, 1987) follows from the unitarity of the scattering matrix, which corresponds to the fact that the sum of the transition probabilities from a given initial state to all possible final states is unity.

The argument is similar to that used to demonstrate the existence of equal numbers of particles and antiparticles at thermodynamic equilibrium, even when *CP* violation (see below) destroys the equality between the rates for specific particle→antiparticle and antiparticle→particle transitions in the big bang model of the early universe (Kolb & Wolfram 1980). All this is consistent with the demonstration that it is *unitarity*, rather than microscopic, reversibility that lies behind the validity of Boltzmann's H-theorem and hence the second law (Aharoney 1973; Weinberg 1995).

CHIRALITY AND SYMMETRY VIOLATION

A symmetry violation (often called symmetry non-conservation) arises when what was thought to be a non-observable is actually observed (Lee 1981). As discussed above, the relevant non-observables here, namely absolute chirality, absolute sense of motion and absolute sign of electric charge, are associated with the symmetry operations P, T and C, respectively. Even if one or more of these three symmetries is violated, however, the combined symmetry of CPT is always thought to be conserved (Lee 1981; Weinberg 1995). A consideration of symmetry violation, and how it differs from spontaneous symmetry breaking, provides considerable insight into the phenomenon of molecular chirality.

THE FALL OF PARITY

Before 1957 it had been accepted as self-evident that handedness is not built into the laws of nature. If two objects exist as non-superposable mirror images of each other, such as the two enantiomers of a chiral molecule, it did not seem reasonable that nature should prefer one over the other. Any difference between enantiomeric systems was thought to be confined to the sign of pseudoscalar observables—the mirror image of any complete experiment involving one enantiomer should be realizable, with any pseudoscalar observable (such as optical rotation angle) changing sign but retaining *precisely* the same magnitude. Then Lee & Yang (1956) pointed out that, unlike the strong and electromagnetic interactions, there was no evidence for parity conservation in processes involving the weak interaction. Of the experiments they suggested, that performed by Wu *et al.* (1957) is the most famous.

The Wu experiment studied the β-decay process:

$$^{60}Co \rightarrow {}^{60}Ni + e^- + \bar{\nu}_e$$

in which, essentially, a neutron decays *via* the weak interaction into a proton, an electron e^- and an electron antineutrino $\bar{\nu}_e$. The nuclear spin magnetic moment **I** of each ^{60}Co nucleus was aligned with an external magnetic field **B**, and the angular distribution of the emitted electrons measured. It was found that the electrons were emitted preferentially in the direction *antiparallel* to that of **B** (Fig. 3.6a).

As discussed above, **B** and **I** are axial vectors and so do not change sign under space inversion, whereas the electron propagation vector **k** does because it is a polar vector. Hence in the corresponding space-inverted experiment the electrons should be emitted *parallel* to the magnetic field (Fig. 3.6b). It is only possible to reconcile Figs 3.6a and b with parity conservation if there is no preferred direction for electron emission (an isotropic distribution), or if the electrons are emitted in the plane perpendicular to **B**.

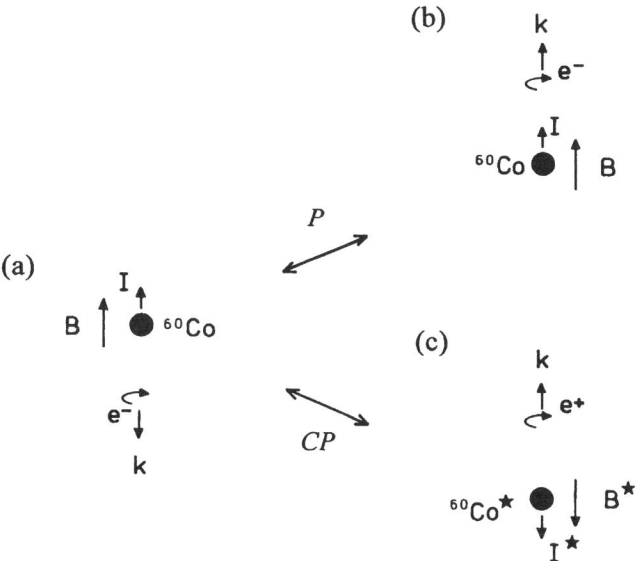

Fig. 3.6 Parity violation in β-decay. Only experiment (a) is found; the space-inverted version (b) cannot be realized. Symmetry is recovered in experiment (c) which is obtained from (a) by invoking *CP* (charge conjugation simultaneously with space inversion). Anti-Co is represented by Co*, and **B*** and **I*** are reversed relative to **B** and **I** because the charges of the moving source particles change sign under *C*.

The observation of Fig. 3.6a alone provides unequivocal evidence for parity violation. Another important aspect of parity violation in β-decay is that the emitted electrons have a 'left-handed' longitudinal spin polarization, being accompanied by 'right-handed' antineutrinos. The corresponding antiparticles emitted in other β-decay processes, namely positrons and neutrinos, have the opposite handedness.

In fact symmetry is recovered by invoking charge conjugation simultaneously with space inversion: the missing experiment is to be found in the antiworld! Thus it can be seen from Fig. 3.6c that the combined operation of *CP* interconverts the two equivalent experiments for which nature seems to have no preference. This result implies that *P* violation is accompanied here by *C* violation—absolute charge is distinguished because the charge that we call negative is carried by electrons, which are emitted with a left-handed spin polarization.

It should be noted that the Wu experiment provides a good example of a truly chiral system, as defined in above. The two experiments Fig. 3.6a and Fig. 3.6b are enantiomeric with respect to space inversion, but cannot be interconverted by time reversal combined with any proper spatial rotation. Nature's preference for one experiment over the other is an example of the discrimination between the enantiomers of a truly chiral system shown by the parity-violating weak interaction, which transforms as a time-even pseudoscalar (next section).

PARITY VIOLATION IN ATOMS AND MOLECULES: TRUE ENANTIOMERS

After the Wu experiment described above, the original Fermi theory of the weak interaction was upgraded to take account of parity violation. This was achieved by reformulating the theory in such a way that the interaction takes the form of a left-handed pseudoscalar (Aitcheson & Hey 1989). Several technical problems remained, however; these were finally overcome in the 1960s in the celebrated work of Weinberg, Salam and Glashow which unified the theory of the weak and electromagnetic interactions into a single *electroweak* interaction theory.

The conceptual basis of the theory rests on two pillars—gauge invariance and spontaneous symmetry breaking (Aitchison & Hey 1989), the details of which are beyond the scope of this article. In addition to accommodating the massless photon and the two massive charged W^+ and W^- particles which mediate the charge-changing weak interactions, a new massive particle, the neutral intermediate vector boson Z^0, was predicted which can generate a new range of *neutral current* phenomena including parity-violating effects in atoms and molecules (Khriplovich 1991). In one of the most important experiments of all time, these three particles were detected in 1983 at CERN in proton-antiproton scattering experiments (Rubbia 1985).

Manifestations of parity violation in atoms are now observed routinely in the form of optical activity phenomena such as tiny optical rotations in vapours of heavy metals, and provide tests of the standard model of elementary particle physics at low energies (Commins 1993). Hegstrom *et al.* (1988) have provided an appealing pictorial representation of the associated atomic chirality in terms of a helical electron probability current density.

Chiral molecules support a unique manifestation of parity violation in the form of a lifting of the exact degeneracy of the energy levels of mirror-image enantiomers (Hegstrom *et al.* 1980; Khriplovich 1991). Being pseudoscalars, the parity-violating weak neutral current terms V_{pv} in the molecular Hamiltonian are odd under space inversion:

$$PV_{pv}P^{-1} = -V_{pv} \qquad (19)$$

As discussed below, the enantiomeric handed quantum states ψ_L and ψ_R of a chiral molecule are examples of the mixed parity states (eq. (7)) and so are interconverted by P. It then follows that V_{pv} shifts the energies of the enantiomeric states in opposite directions:

$$\langle\psi_L|V_{pv}|\psi_L\rangle = \langle P\psi_R|V_{pv}|P\psi_R\rangle = \langle\psi_R|P^\dagger V_{pv}P|\psi_L\rangle = -\langle\psi_R|V_{pv}|\psi_R\rangle = \varepsilon \qquad (20)$$

It is possible to compute the parity-violating energy difference $2|\varepsilon|$ by *ab initio* methods (Mason & Tranter 1985); the values obtained are of the order of 10^{-20} to 10^{-17} hartree, roughly 10^{-17} to 10^{-14} kT where k is Boltzmann's constant. It is intriguing that of all the cases treated so far the L amino acids and the D sugars, which dominate the chemistry of living organisms, are found to be the more stable enantiomers (Mason & Tranter 1985; Tranter *et al.* 1992).

Being a time-even pseudoscalar, V_{ev} is the archetypal chiral interaction potential and hence serves as a fundamental test to distinguish between true and false chirality

(Barron 1986a). Thus the development above (eq. (20)) demonstrates that parity violation lifts the degeneracy of the P enantiomers of a truly chiral system. Parity violation does not, however, lift the degeneracy of the P enantiomers of a falsely chiral system such as a rotating cone. This follows from the fact that, although P_{pv} is odd under space inversion, it is invariant under both time reversal and any proper spatial rotation—because the last two operations together interconvert the two P enantiomers of a falsely chiral system, it follows from a development analogous to eq. (20) that the energy difference is zero.

Because, on account of the weak neutral current, the space-inverted enantiomers of a truly chiral object are not strictly degenerate, they are not true enantiomers (because the concept of enantiomers implies the *exact* opposites). So where is the true enantiomer of a chiral object to be found? In the antiworld, of course! Just as symmetry is recovered in the Wu experiment above by invoking CP rather than P alone, one might expect true enantiomers to be interconverted by CP; in other words, the molecule with the opposite absolute configuration but composed of antiparticles should have exactly the same energy as the original (Barron 1981a, b, 1982; Jungwirth *et al.* 1989), which means that a chiral molecule is associated with two distinct pairs of true enantiomers (Fig. 3.7).

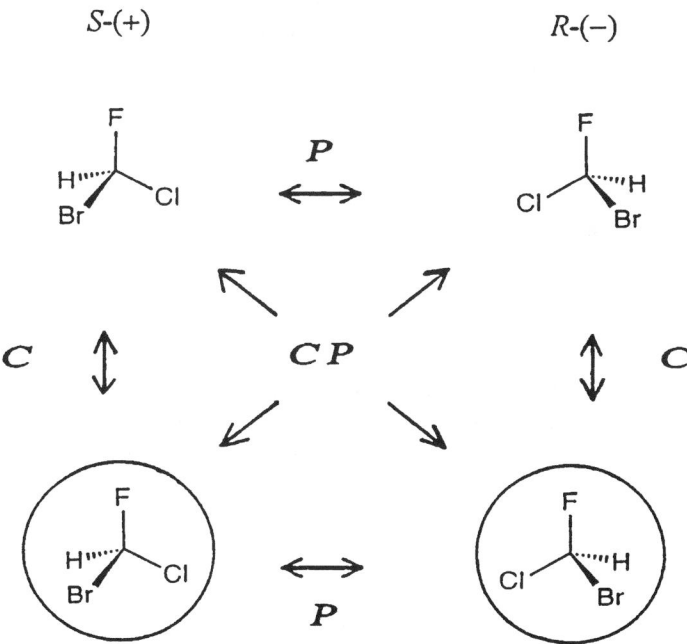

Fig. 3.7 The two pairs of true enantiomers (strictly degenerate) of a chiral molecule that are interconverted by CP. The structures within the circles are antimolecules built from the antiparticle versions of the constituents of the original molecules. The strict degeneracy remains even if CP is violated (Barron, 1994b).

Because *P* violation automatically implies *C* violation here, it also follows that there is a small energy difference between a chiral molecule in the real world and the corresponding chiral molecule with the same absolute configuration in the antiworld.

The original proof was based on the *CPT* theorem with the assumption that *T* is not violated; it has since been shown, from an extension of the proof that a particle and its antiparticle have identical rest mass, that the *CP* enantiomers of a chiral molecule remain strictly degenerate even in the presence of *CP* violation provided *CPT* invariance holds (Barron 1994b).

This more general definition of the enantiomer of a chiral system is consistent with the chirality that free atoms display on account of parity violation. The weak neutral current generates only one type of chiral atom in the real world: the conventional enantiomer of a chiral atom obtained by space inversion alone does not exist. Clearly the enantiomer of a chiral atom is generated by the combined *CP* operation. Thus the corresponding atom composed of antiparticles will of necessity have the opposite 'absolute configuration' and, like the true *CP* enantiomer of a chiral molecule, will show an exactly equal but opposite sense of optical rotation (Barron 1981a, 1982).

The space-inverted enantiomers of objects such as translating spinning electrons or cones that are chiral only on account of their motion also show parity-violating energy differences. One manifestation is that, as discussed below, left-handed and right-handed particles (or antiparticles) have different weak interactions. Again, true enantiomers are interconverted by *CP*—for example, a left-handed electron and a right-handed positron. Notice that, because a photon is its own antiparticle (Weinberg 1995), right- and left-handed circularly polarized photons are automatically true enantiomers.

VIOLATION OF TIME REVERSAL AND CP VIOLATION

Violation of time reversal was observed in the famous experiment of Christenson *et al.* (1964) involving measurements of rates for different decay modes of the neutral *K*-meson, the K^0 (Gottfried & Weisskopf, 1984; Weinberg, 1995). Despite intensive efforts since then, no other system has shown the effect. As remarked by Cronin (1981), nature has provided us with just one extraordinarily sensitive system to convey a cryptic message that has still to be deciphered. Although unequivocal, the effects are very small; certainly nothing like the parity-violating effects in weak processes which can sometimes be absolute. In fact *T* violation itself is not observed directly—rather, the observations show *CP* violation from which *T* violation is implied *via* the *CPT* theorem.

One manifestation of *CP* violation is the following decay rate asymmetry of the long-lived neutral *K*-meson, the K_L:

$$\Delta = \frac{\text{rate}\left(K_L \to \pi^- e_r^+ \nu_l\right)}{\text{rate}\left(K_L \to \pi^+ e_l^- \widetilde{\nu}_r\right)} = 1.00648 \qquad (21)$$

As the formula indicates, K_L can decay into either positive pions π^+ plus left-helical electrons e_l^- plus right-helical antineutrinos $\widetilde{\nu}_r$; or into negative antipions π^- plus right-helical positrons e_r^+ plus left-helical neutrinos ν_l. Because these two sets of decay products are interconverted by *CP*, this decay-rate asymmetry indicates that *CP* is violated. If we naively represent these decay processes in the form of the following 'chemical equilibrium':

$$\pi^+ + e_l^- + \tilde{v}_r \underset{k_b}{\overset{k_f}{\rightleftharpoons}} K_L \underset{k_f^*}{\overset{k_b^*}{\rightleftharpoons}} \pi^- + e_r^+ + v_l \qquad (22)$$

a parallel is established with absolute asymmetric synthesis associated with a breakdown in microscopic reversibility as discussed in above, because in both cases $k_f \neq k_f^*$. Thus the K_L and the two sets of decay products in eq. (22) are the equivalents, with respect to *CP*, of R, M and M* in eq. (16) with respect to *P*. Because the *CPT* theorem guarantees that the two distinct *CP*-enantiomorphous influences are interconverted by *T*, we can picture the decay rate asymmetry here as arising from a breakdown in microscopic reversibility because of a time-non-invariant *CP*-enantiomorphism in the forces of nature (Barron 1987), which places *CP* violation within the conceptual framework of false chirality. The analogy is completed by the fact that, as mentioned above, the asymmetries cancel out when summed over all possible channels at true thermodynamic equilibrium.

CP violation is analogous to chemical catalysis!

In chemistry, a catalyst is defined as a substance that is not consumed in a chemical reaction and which increases the reaction rate at a given temperature by reducing the activation energy but without affecting the free energy change for the reaction. It follows that a falsely chiral influence acts as a special type of catalyst because it modifies potential energy barriers to change relative rates of formation of enantiomeric products without affecting the relative energies of reactants and products (remember that a falsely chiral influence does not lift the degeneracy of *P*-enantiomeric chiral molecules). Because the *CP*-violating interaction responsible for the decay rate asymmetry of the K_L does not lift the degeneracy of the two sets of *CP*-enantiomeric products (a particle and its antiparticle have identical rest mass if *CPT* invariance holds), its action is analogous to that of a special type of chemical catalyst, because it affects the kinetics but not the thermodynamics of the reaction (Barron 1994b).

CPT VIOLATION

The *CPT* theorem of relativistic quantum field theory which states that, even if one or more of the symmetries *C*, *P* and *T* is violated, the combined operation of *CPT* is always conserved (Lee 1981; Sachs 1987; Weinberg 1995), has featured prominently in this article. This theorem can be proved in great generality with only minor assumptions. There are three important consequences of *CPT* invariance: the rest mass of a particle and its antiparticle are equal; the particle and antiparticle lifetimes are the same (even though decay rates for individual channels may not be equal); and the electromagnetic properties such as charge, magnetic moment, *etc.* are opposite. This last consequence immediately reveals a fatal flaw in the recent suggestion that a circularly polarized photon supports a static axial magnetic field (Evans 1993) because, on account of the fact that a photon is its own antiparticle (Weinberg 1995), it requires that any such field must be zero.

Despite being the cornerstone of modern elementary particle physics, the possibility that even *CPT* symmetry might be violated to a very small extent should nonetheless be contemplated. The simplest tests focus on the measurement of the rest mass of a particle and that of its associated antiparticle, because any difference would reveal a violation of *CPT* (Lee 1981). Also, the photon's magnetic field mentioned above could serve as a 'non-observable' to test *CPT* invariance. The world of atoms and molecules might ultimately provide the best testing ground, because one of the motivations for current plans to manufacture antihydrogen in significant quantities is to use ultra-high resolution spectroscopy to compare the $1S$–$2S$ energy intervals in atomic hydrogen and antihydrogen to one part in 10^{18} as a test of *CPT* invariance to much higher precision than any previous measurements (Eades *et al.* 1993; Eades 1996).

As mentioned above, the *CP*-enantiomers of a chiral molecule remain strictly degenerate even in the presence of *CP* violation, provided *CPT* is conserved. Hence any attempt to measure an energy difference between a chiral molecule and its mirror-image made of antiparticles (assuming that antimolecules can be manufactured routinely sometime in the distant future!) would in fact constitute a search for both *CP* and *CPT* violation together (Barron 1994b) and not just *CP* violation as previously supposed (Quack 1993).

THE MIXED-PARITY STATES OF A CHIRAL MOLECULE

It has been shown that, because a chiral molecule can support pseudoscalar observables, it must exist in mixed-parity internal quantum states. The nature of these quantum states in relation to the chiral molecular framework is now outlined, together with the intriguing consequences of including a small parity-violating term in the Hamiltonian.

THE DOUBLE WELL MODEL

The origin of these mixed-parity states can be appreciated by considering vibrational wavefunctions associated with the 'inversion' mode of a molecule such as ammonia which interconverts the two equivalent configurations shown in Fig. 3.8 (Townes and Schawlow 1955). If the planar configuration were the most stable, the adiabatic potential energy function would have the parabolic form shown on the left with simple harmonic vibrational levels equally spaced. If a potential energy barrier is raised gradually in the middle, the two pyramidal configurations become the most stable and the energy levels approach each other in pairs.

For an infinitely high potential barrier, the pairs of levels are exactly degenerate, as shown on the right. The rise of the central potential barrier modifies the vibrational wavefunctions as shown, but does not destroy their parity. The even- and odd-parity wavefunctions $\psi(+)$ and $\psi(-)$ describe stationary states in all circumstances. On the other hand, the wavefunctions ψ_L and ψ_R corresponding to the system in its lowest state of oscillation and localized completely in the left and right wells, respectively, are not true stationary states.

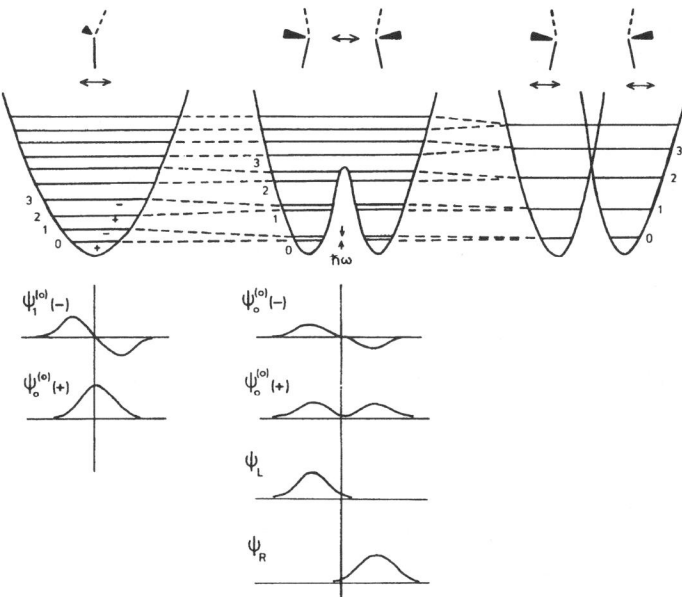

Fig. 3.8 The vibrational states of a molecule that can invert between two equivalent configurations. $\psi^{(0)}(+)$ and $\psi^{(0)}(-)$ are the amplitudes of the definite parity stationary states with energies $W(+)$ and $W(-)$ and ψ_L and ψ_R are the two mixed-parity handed non-stationary states at $t = 0$ and $t = \pi/\omega$ with $\hbar\omega$ the tunnelling splitting.

The wavefunctions ψ_L and ψ_R are obtained from the following combinations of the even- and odd-parity wavefunctions:

$$\psi_L = \frac{1}{\sqrt{2}}\left[\psi^{(0)}(+) + \psi^{(0)}(-)\right] \tag{23a}$$

$$\psi_R = \frac{1}{\sqrt{2}}\left[\psi^{(0)}(+) - \psi^{(0)}(-)\right] \tag{23b}$$

which are explicit examples of the mixed-parity wavefunctions given in eqs (7).

The wavefunctions given in eqs (23) are in fact specializations of the general time-dependent wavefunction of a degenerate two-state system (Cohen-Tannoudji *et al.*, 1977). To be precise, it is assumed that at $t = 0$ the system is in the left well. Then at a later time:

$$\begin{aligned}\psi(t) &= \frac{1}{\sqrt{2}}\left[\psi^{(0)}(+)e^{-iW(+)t/\hbar} + \psi^{(0)}(-)e^{-iW(-)t/\hbar}\right] \\ &= \frac{1}{\sqrt{2}}\left[\psi^{(0)}(+) + \psi^{(0)}(-)e^{-i\omega t}\right]e^{-iW(+)t/\hbar}\end{aligned} \tag{24}$$

where $\hbar\omega = W(-) - W(+)$ is the energy separation of the two opposite parity states, which in this context is interpreted as a splitting arising from tunnelling through the potential energy barrier separating the two wells. Thus at $t = 0$ the part of eq. (24) in square brackets reduces to that of eq. (23a) corresponding to the molecule being found in the left well, as required; and at $t = \pi/\omega$ it reduces to that of eq. (23b) corresponding to the molecule being found in the right well. The angular frequency ω is interpreted as the frequency of a complete inversion cycle. The tunnelling splitting $\hbar\omega$ is determined by the height and width of the barrier, and is zero if the barrier is infinite.

Because a similar potential energy diagram can be drawn for any chiral molecule with a high barrier separating left and right wells associated with the two enantiomeric states, we now have a model for the source of the mixed-parity internal (vibrational–electronic) states of a resolved enantiomer. The horizontal axis might represent the position of an atom above a plane containing three different atoms, the torsion co-ordinate of a chiral biphenyl, or some more complicated collective co-ordinate of the molecule.

If such a state is prepared, but the tunnelling splitting is finite, its energy will not be sharply defined because it is a superposition of two opposite parity states of different energy. The splitting of the two definite parity states, and hence the uncertainty in the energy of an enantiomer, is inversely proportional to the left–right conversion time π/ω. This is an example of the general result that the width of an energy level corresponding to a quasi-stationary state with average lifetime t is $\Delta W = \hbar/t$ (Cohen-Tannoudji, 1977).

A central point is, therefore, the relationship between the time-scale of the optical activity (or any time-even pseudoscalar observable) measurement and the lifetime of the resolved enantiomer. A new aspect of the uncertainty principle seems to arise here which may be stated loosely as follows (Barron, 1982):

If, for the duration of the measurement, there is complete certainty about the enantiomer, there is complete uncertainty about the parity of the quantum state of the molecule; whereas if there is complete certainty about the parity of the quantum state, there is complete uncertainty about the enantiomer.

Thus experimental resolution of the definite parity states of tartaric acid, for example, an enantiomer of which has a lifetime probably greater than the age of the universe, is impossible unless the duration of the experiment is virtually infinite; whereas for a non-resolvable chiral molecule such as H_2O_2 spectroscopic transitions between definite parity states are observed routinely.

TWO-STATE SYSTEMS AND PARITY VIOLATION

The nature of the mixed parity states of a resolved chiral molecule can be explored further by considering the quantum mechanics of a two-state system. This provides insight into the apparent paradox of the stability of optical enantiomers, which was recognized at the beginning of the quantum era when it was found that the existence of optical enantiomers was difficult to reconcile with basic quantum mechanics (Hund 1927, Quack 1989).

The essence of the 'paradox' is that, because the potential energy term in the Hamiltonian of a molecule originates in Coulombic interactions between point charges, the complete molecular Hamiltonian is always invariant under space inversion and so the energy eigenstates (the stationary states) must be parity eigenstates. A resolved chiral molecule cannot, however, be in such an eigenstate because the parity operation generates the mirror-image enantiomer, which is a different system. Yet typical chiral molecules such as alanine seem to be no less stable than typical achiral molecules. Hund (1927) suggested a resolution of the paradox based on the argument that typical chiral molecules have such large barriers to inversion that the lifetime of a prepared enantiomer is virtually infinite.

Hund's approach has been brought up to date by adding a small parity-violating term to the potential energy in the Hamiltonian which can result in the two enantiomeric states becoming the true stationary states (Harris & Stodolsky 1978; Harris 1980). Details of the quantum-mechanical development can be found elsewhere (Harris 1980; Barron 1982, 1991). The essence of the results are neatly summarized by expressions for the optical rotation angle $\alpha(t)$, or indeed any time-even pseudoscalar observable characteristic of a truly chiral system, as a function of time. Thus if the molecule is prepared initially in one of the handed states, say ψ_L:

$$\alpha(t) = \alpha_L \left\{ \varepsilon^2 + \delta^2 \cos\left[2(\delta^2 + \varepsilon^2)^{\frac{1}{2}} t / \hbar \right] \right\} / (\delta^2 + \varepsilon^2) \qquad (25)$$

where α_L is the optical rotation angle of the left-handed enantiomer, $2|\varepsilon|$ is the parity-violating energy difference between the two handed states ψ_L and ψ_R, and $2|\delta|$ is the energy difference between the two definite parity states being interpreted as the tunnelling splitting. If $\delta \gg \varepsilon$ (tunnelling splitting much greater than parity-violating splitting), eq. (25) reduces to:

$$\alpha(t) = \alpha_L \cos(2\delta t / \hbar) \qquad (26a)$$

which describes the optical rotation oscillating between the equal and opposite values associated with the two enantiomers in accord with eq. (24). If, on the other hand, $\varepsilon \gg \delta$, eq. (25) reduces to:

$$\alpha(t) = \alpha_L \qquad (26b)$$

showing that the optical rotation does not change with time.

Eq. (26b) is a manifestation of the fact that, as $\delta/\varepsilon \to 0$, the handed states ψ_L and ψ_R become the true stationary states. This limit actually obtains for typical chiral molecules: δ is extremely small because typical tunnelling times are of the order of millions of years, and although ε is also very small it is many orders of magnitude larger, because the associated times are of the order of seconds to days. Hence at low temperature and in a vacuum, a prepared enantiomer will retain its handedness essentially for ever (Harris & Stodolsky 1978). These considerations therefore suggest that the ultimate answer to the paradox of the stability of optical enantiomers might lie in the weak interactions. The situation is actually more complicated, however, because the influence of the environment must also be considered (Quack 1989; Vager 1997).

PARITY VIOLATION AND SPONTANEOUS PARITY BREAKING

The appearance of parity-violating phenomena is interpreted in quantum mechanics by saying that, contrary to what had been previously supposed, the Hamiltonian lacks inversion symmetry because of the presence of the pseudoscalar weak interaction potential. This means that P and H no longer commute, so the associated law of conservation of parity no longer holds. Such *symmetry violation* must be clearly distinguished from *spontaneous symmetry breaking*: current usage in the physics literature applies the latter term to describe the situation that arises when a system displays a lower symmetry than that of its Hamiltonian (Michel, 1980; Ludwig and Falter, 1988).

Natural optical activity is therefore a phenomenon arising from spontaneous parity breaking because, as we have seen, a resolved chiral molecule has lower symmetry than does its associated Hamiltonian—if the small parity-violating term in the Hamiltonian is neglected, the symmetry operation that the Hamiltonian has but the chiral molecule lacks is parity, and it is this parity operation which interconverts the two enantiomeric parity-broken states.

It should be mentioned that the term 'spontaneous symmetry breaking' is often reserved for quantum-mechanical systems with an infinite number of degrees of freedom for which there can be special dynamic instabilities which inevitably generate an unsymmetric state (Strocchi 1985). Quack (1989) adheres to this usage but, to accommodate the unsymmetric states that occur in finite systems such as molecules and which are so important in molecular chirality, he has introduced the term 'symmetry breaking *de facto*' to describe the situation where the unsymmetric states arise through the choice of initial conditions. Quack (1989) also introduced the term 'symmetry breaking *de lege*' to describe what is called here symmetry violation for the case where the asymmetry resides in the Hamiltonian.

The conventional view, formulated in terms of the double-well model above, is that parity violation plays no part in the stabilization of chiral molecules. The optical activity remains observable only as long as the observation time is short compared with the interconversion time between enantiomers, which is proportional to the inverse of the tunnelling splitting. Such spontaneous parity-breaking optical activity therefore averages to zero over a sufficiently long observation time.

Hence statements such as "…processes involving pseudoscalar quantities will not obey the law of parity…" (Ulbricht 1959) are indicative of a common misconception—the law of parity is saved in systems having spontaneous parity breaking because their pseudoscalar properties average to zero over a sufficiently long observation period on account of tunnelling or, equivalently, it is possible to perform (at least in principle) the space-inverted experiment. In either interpretation absolute chirality is not observable.

These considerations lead us to an important criterion for distinguishing between optical activity generated through spontaneous parity breaking from that generated through parity violation. The former are time-dependent and average to zero; the latter are constant in time (recall from the previous section that the handed states become the stationary states when $\varepsilon \gg \delta$). Hence if a small chiral molecule could be isolated sufficiently from the environment, a parity-violating contribution is indicated if the optical activity remains observable for longer than the expected interconversion time between the enantiomers (Harris & Stodolsky 1981). Notice that, because it arises entirely as a result of parity violation, the tiny optical rotation shown by a free atomic vapour will be constant in time.

There is considerable interest in the development of quantitative measures of the 'chirality content' of chiral objects, including molecules (Buda *et al.* 1992; Katzenelson *et al.* 1996). Although such measures are of mathematical interest in the context of static geometry and topology and might have practical applications in chemistry, it should be clear from the discussion above that the chirality content of molecular structures in the form of some fundamental time-even pseudoscalar quantity analogous to, say, energy (a time-even scalar) is in fact a will-o'-the-wisp (Barron 1996). Thus chirality content evaporates under close quantum-mechanical scrutiny because, neglecting parity violation, chiral molecules are not in stationary states of the Hamiltonian and so any pseudoscalar quantity will average to zero on an appropriate time-scale.

CHIRALITY AND RELATIVITY

It was demonstrated above that a spinning sphere or cone translating along the axis of spin has true chirality. This is an interesting concept because it exposes a link between chirality and special relativity. This is evident if a particle with right-handed helicity moving away from an observer is considered. If the observer accelerates to a sufficiently high velocity that he starts to catch up with the particle, it will then appear to be moving towards the observer and so take on a left-handed helicity. In its rest frame, the helicity of the particle is undefined so its chirality vanishes. Only for massless particles such as photons and neutrinos is the chirality conserved since they always move at the velocity of light in any reference frame.

This relativistic aspect of chirality is, in fact, a central feature of modern elementary particle theory, especially in connection with the weak interaction where the parity-violating aspects are velocity-dependent. A good example is provided by the interaction of electrons with neutrinos. Neutrinos are quintessential chiral objects because only left-helical neutrinos and right-helical antineutrinos exist (Gottfried & Weisskopf 1984).

Consider first the extreme case of electrons moving close to the velocity of light. Only left-helical relativistic electrons interact with left-helical neutrinos *via* the weak force; right-helical relativistic electrons do not interact at all with neutrinos. But right-helical relativistic positrons interact with right-helical antineutrinos. For non-relativistic electron velocities, the weak interaction still violates parity but the amplitude of the violation is reduced to order v/c (Gottfried & Weisskopf 1984). This is used to explain the interesting fact that the $\pi^- \ne e^- \tilde{v}_e$ decay is a factor of 10^4 smaller than the $\pi^- \ne \mu^- \tilde{v}_\mu$ decay, even though the available energy is much larger in the first decay.

Thus in the rest frame of the pion, the lepton (electron or muon) and the antineutrino are emitted in opposite directions so that their linear momenta cancel. Also, because the pion is spinless, the lepton must have a right-handed helicity to cancel the right-handed helicity of the antineutrino. Thus both decays would be forbidden if e and μ had the velocity c because the associated maximum parity violation dictates that both be pure left-handed. Because of its much greater mass, however, the muon is emitted much more slowly than the electron so there is a much greater amplitude for it to be emitted with a right-handed helicity.

It should be mentioned that the discussion in the previous paragraph applies only to charge-changing weak processes, mediated by W^+ or W^- particles. Weak neutral current processes, mediated by Z^0 particles, are rather different because, even in the relativistic limit, both left- and right-handed electrons participate but with slightly different amplitudes (Aitchison & Hey 1989).

Up to this point in the article, the word *chirality* has been used in its qualitative chemical sense. In elementary particle physics, however, chirality is given a precise quantitative meaning (Itzykson & Zuber 1980): it is the eigenvalue of the Dirac matrix operator γ_5 taking values $+1$ and -1 for massless fermions with right- and left-handed helicities, respectively, and having the opposite signs for the corresponding antiparticles (*i.e.* chirality, but not helicity, changes sign under C: chirality has the opposite sign for particles and antiparticles with the same helicity). Hence massless fermions, such as neutrinos, which always move at the velocity of light are in eigenstates of γ_5 and so have a precise chirality. Fermions with mass, such as electrons, always move more slowly than c and so do not have well-defined chirality. On the other hand helicity, as defined in above, can be defined for both massless and massive particles; but only for the former is it invariant to the frame of the observer.

There is another, rather different, connection between chirality and relativity. It was shown above that spontaneous parity-breaking optical activity is distinguished from parity-violating optical activity because the first is time-dependent whereas the second does not depend on time. Because a clock on a moving object slows down relative to a stationary observer, a molecule with spontaneous parity-breaking optical activity will become increasingly stable with increasing velocity relative to a stationary observer, and were it able actually to achieve the speed of light it would become infinitely stable. This means that spontaneous parity-breaking optical activity in a chiral object moving at the speed of light is indistinguishable from parity-violating optical activity.

CHIRALITY IN TWO DIMENSIONS

Chirality in two dimensions arises when there are two distinct enantiomers interconverted by space inversion but not by any proper rotation within the plane (rotation out of the plane requires a third dimension, which is not accessible). In two dimensions, however, the parity operation is no longer equivalent to an inversion through the co-ordinate origin as it is in three dimensions because this does not change the handedness of the two co-ordinate axes: instead an inversion of just one of the two axes is required (Halperin et al. 1989; Wilczek 1990).

For example, if the axes (x,y) are in the plane with z perpendicular, then the parity operation could be taken as producing $(-x,y)$ which is equivalent to a mirror reflection across the line defined by the y-axis. Hence an object such as a scalene triangle (one with three sides of different length) which is achiral in three dimensions becomes chiral in the two dimensions defined by the plane of the triangle because mirror reflection across any line within the plane generates a triangle which cannot be superposed on the original through any rotation about z, the axis perpendicular to the plane. Notice that a reflection across a second line, perpendicular to the first, generates a triangle superposable on the original, which demonstrates why an inversion of both axes, $(x,y) \rightarrow (-x,-y)$ is not acceptable as the parity operation in two dimensions.

Consider a surface covered with an isotropic layer (meaning no preferred orientations in the plane) of molecules. If the molecules are achiral, there will be an infinite number of mirror reflection operations possible across lines within the plane which generate an indistinguishable isotropic layer. But if the surface molecules are chiral such mirror reflection operations do not exist because any such reflection would generate the distinct isotropic surface composed of enantiomeric molecules.

Such considerations are by no means purely academic. For example, chiral molecules on an isotropic surface can show new chiroptical phenomena such as huge circular intensity differences in second harmonic scattering which are generated through pure electric dipole processes (Hicks *et al.* 1994; Hecht & Barron 1996): the equivalent time-even pseudoscalar observables in light scattered from chiral molecules in bulk three-dimensional samples are three orders of magnitude smaller because electric dipole–magnetic dipole and electric dipole–electric quadrupole processes are required.

The concept of false chirality arises in two dimensions also. For example, the sense of a spinning electron on a surface with its axis of spin perpendicular to the surface is reversed under the two-dimensional parity operation (unlike in three dimensions). Because electrons with opposite spin sense are non-superposable in the plane, a spinning electron on a surface would seem to be chiral. However, the apparent chirality is false because the sense of spin is also reversed by time reversal (like in three dimensions). The enantiomorphism is therefore time-non-invariant, the system being invariant under the combined operation PT but not under P and T separately.

Although not referred to as such, the concept of false chirality arises in the anyon theory of high-temperature superconductivity. Anyons are spinning particles (or quasiparticle excitations) obeying a type of quantum statistics which can vary continuously between those of fermions and bosons and which have been suggested as inhabiting the two-dimensional world of copper oxide planes (Wilczek 1990). Attempts to detect polarized light-scattering observables with symmetry signatures that break P and T separately but which are PT-invariant overall have featured prominently in efforts to find experimental support for the anyon theory of high-temperature superconductivity (Wen & Zee 1991).

CONCLUDING REMARKS

This account of chirality at the sub-molecular level has drawn on several concepts from modern physics, especially the fundamental symmetry arguments usually employed in the context of elementary particles. A deeper perspective of molecular chirality from that usually encountered in chemistry is thereby obtained, which reinforces Heisenberg's perception of a kinship between the theory of elementary particles and quantum chemistry (Heisenberg 1966).

It is remarkable that Pasteur's glimpse of a cosmic dissymmetric force, which he thought might somehow be connected with biomolecular handedness, anticipated by a century the discovery of parity violation in the weak interactions, and also in all electromagnetic processes *via* the weak neutral current. Despite some tantalizing hints, however, it has yet to be established that parity violation has any connection with handedness in general, or homochirality in particular, in the chemistry of life.

The concept of false chirality (more properly called time-non-invariant enantiomorphism) has emerged as something even more subtle than true chirality (time-invariant enantiomorphism). For example, whereas true chirality in the weak interactions is responsible for parity violation, the deeper form of cosmic dissymmetry responsible for that most enigmatic of all natural phenomena discovered to date, namely *CP* violation and the concomitant *T* violation in certain elementary particle processes, falls within the conceptual framework of false chirality. This realization enables *CP* violation to be conceptualized in terms of chemical concepts such as catalysis.

Clearly Pasteur stumbled on a very rich seam when he started mining the subject of molecular handedness; this seam extends far beyond the realm of stereochemistry. Indeed, one could say that the study of chiral systems both true and false, and the special phenomena which they support, provides a peephole into the very fabric of the universe!

ACKNOWLEDGEMENTS

The author thanks the Engineering and Physical Sciences Research Council for the award of a Senior Fellowship.

REFERENCES

Aharoney A. (1973) In: *Modern Development in Thermodynamics* (Ed. by B. Gal-Or), p. 95. Wiley, New York.
Aitchison I. J. R. & Hey A. J. (1989) *Gauge Theories in Particle Physics*. Adam Hilger, Bristol.
Arago D. F. J. (1811) *Mém. Inst. France*, Part 1, 93.
Avalos M., Babiano R., Cintas P., Jiménez J. L., Palacios J. C. & Barron L. D. (1998) *Chem. Rev.*, **98**, 2391
Barron L. D. (1981a) *Chem. Phys. Lett.*, **79**, 392.
Barron L. D. (1981b) *Mol. Phys.*, **43**, 1395.
Barron L. D. (1982) *Molecular Light Scattering and Optical Activity*. Cambridge University Press.
Barron L. D. (1986a) *Chem. Phys. Lett.*, **123**, 423.
Barron L. D. (1986b) *J. Am. Chem. Soc.*, **108**, 5539.
Barron L. D. (1987) *Chem. Phys. Lett.*, **135**, 1.
Barron L. D. (1991) In: *New Developments in Molecular Chirality* (Ed. by P. G. Mezey), p. 1. Kluwer, Dordrecht..
Barron L. D. (1994a) *Science*, **266**, 1491.
Barron L. D. (1994b) *Chem. Phys. Lett.*, **221**, 311.
Barron L. D. (1996) *Chem. Eur. J.*, **2**, 743.
Barron L. D. & Vrbancich J. (1984) *Mol. Phys.*, **51**, 715.
Biot J. B. (1812) *Mem. Inst. France*, **1**, 1.
Biot J. B. (1815) *Bull. Soc. Philomat.*, 190.
Bonner W. A. (1988) *Topics in Stereochemistry*, **18**, 1.
Bonner W. A. (1990) *Origins Life*, **20**, 1.

Buckingham A. D., Graham C. & Raab R. E. (1971) *Chem. Phys. Lett.*, **8**, 622.
Buda A. B., Auf der Hyde T. & Mislow K. (1992) *Angew. Chem. Int. Ed. Engl.*, **31**, 989.
Christenson J. H., Cronin J. W., Fitch V. L. & Turlay, R. (1964) *Phys. Rev. Lett.*, **13**, 138.
Cline D. B. (Ed.) (1996) *Physical Origin of Homochirality in Life*, AIP Conference Proceedings 379. American Institute of Physics, Woodbury, New York.
Cohen-Tannoudji C., Diu B & Laloë F. (1977) *Quantum Mechanics*, Vol. 1. Wiley, New York.
Commins E. D. (1993) *Physica Scripta*, **T46**, 92.
Cronin J. W. (1981) *Rev. Mod. Phys.*, **53**, 373.
Curie P. (1894) *J. Phys. (Paris)* (3) **3**, 393.
Dougherty R. C. (1980) *J. Am. Chem. Soc.*, **102**, 380.
Eades J. (1996) *Nature*, **379**, 674.
Eades J., Hughes R. J. & Zimmermann C. (1993) *Physics World*, **6**, July, 44.
Evans M. W. (1993) *Adv. Chem. Phys.*, **85**, Part 2, 51, 97.
Faraday M. (1846) *Phil. Mag.*, **28**, 294.
Feringa B. L. & van Delden R. A. (1999) *Ang. Chem. Int. Ed.*, **38**, 3418.
Feynman R. P., Leighton R. B. & Sands M. (1964) *The Feynman Lectures on Physics*. Addison-Wesley, Reading, Massachusetts.
Fresnel A. (1824) *Bull. Soc. Philomat.*, 147.
Gottfried K. & Weisskopf V. F. (1984) *Concepts of Particle Physics*, Vol. 1. Clarendon Press, Oxford.
Haldane J. B. S. (1960) *Nature*, **185**, 87.
Halperin B. I., March-Russell J. & Wilczek F. (1989) *Phys. Rev.*, **B40**, 8726.
Harris R. A. & Stodolsky L. (1978) *Phys. Lett.*, **78B**, 313.
Harris R. A. (1980) In: *Quantum Dynamics of Molecules* (Ed. by R. G. Woolley), p. 357. Plenum Press, New York.
Harris R. A. & Stodolsky L. (1981) *J. Chem. Phys.*, **74**, 2145.
Hecht L. & Barron L. D. (1996) *Mol. Phys.*, **89**, 61.
Hegstrom R. A., Rein D. W. & Sandars P. G. H. (1980) *J. Chem. Phys.*, **73**, 2329.
Hegstrom R. A., Chamberlain J. P., Seto K. & Watson R. G. (1988) *Am. J. Phys.*, **56**, 1086.
Heisenberg W. (1966) *Introduction to the Unified Field Theory of Elementary Particles*. Wiley, New York.
Hicks J. M., Petralli-Mallow T. & Byers J. D. (1994) *Faraday Discuss.*, **99**, 341.
Hund F. (1927) *Z. Phys.*, **43**, 805.
Itzykson C. & Zuber J. B. (1980) *Quantum Field Theory*. McGraw–Hill, New York.
Jaeger F. M. (1930) *Optical Activity and High-Temperature Measurements*. McGraw–Hill, New York.
Jungwirth P., Skála L. & Zahradník R. (1989) *Chem. Phys. Lett.*, **161**, 502.
Katzenelson O., Zabrodsky Hel-Or H. & Avnir D. (1996) *Chem. Eur. J.*, **2**, 174.
Lord Kelvin (1904) *Baltimore Lectures*. C. J. Clay and Sons, London.
Keszthelyi L. (1995) *Quart. Rev. Biophys.*, **28**, 473.
Khriplovich I. B. (1991) *Parity Nonconservation in Atomic Phenomena*. Gordon & Breach, Philadelphia.
Kleindienst P. & Wagnière G. H. (1998) *Chem. Phys. Lett.*, **288**, 89.
Kolb E. W. & Wolfram S. (1980) *Nucl. Phys.*, **B172**, 224.

Landau L. D. & Lifshitz L. D. (1977) *Quantum Mechanics*. Pergamon Press, Oxford.
Lee T. D. (1981) *Particle Physics and Introduction to Field Theory*. Harwood, Chur.
Lee T. D. & Yang C. N. (1956) *Phys. Rev.*, **104**, 254.
Lifshitz E. M. & Pitaevskii L. P. (1981) *Physical Kinetics*. Pergamon Press, Oxford.
Lowry T. M. (1935) *Optical Rotatory Power*. Longman, Green and Co., London. Reprinted in 1964 by Dover, New York.
Ludwig W. & Falter C (1988) *Symmetries in Physics*. Springer, Berlin.
Mason S. F. (1982) *Molecular Optical Activity and the Chiral Discriminations*. Cambridge University Press.
Mason S. F. & Tranter G. E. (1985) *Proc. Roy. Soc.* **A397**, 45.
Mead C. A., Moscowitz A., Wynberg H. & Meuwese F. (1977) *Tetrahedron Lett.*, 1063.
Michel L. (1980). *Rev. Mod. Phys.*, **52**, 617.
Onsager L. (1931) *Phys. Rev.*, **37**, 405.
Pasteur L. (1848) *Ann. Chim.*, **24**, 442.
Quack M. (1989) *Ang. Chem. Int. Ed. Engl.*, **28**, 571.
Quack M. (1993) *J. Mol. Struct.*, **292**, 171.
Rikken G. L. J. A. & Raupach E. (1997) *Nature*, **390**, 493
Rubbia C. (1985) *Rev. Mod. Phys.*, **57**, 699.
Sachs R. G. (1987) *The Physics of Time Reversal*. University of Chicago Press.
Sakurai J. J. (1985) *Modern Quantum Mechanics*. Benjamin/Cummings, Menlo Park, California.
Strocchi F. (1985) *Elements of Quantum Mechanics of Infinite Systems*. World Scientific, Singapore.
Tolman R. C. (1938) *The Principles of Statistical Mechanics*. Oxford University Press.
Townes C. H. & Schawlow A. L. (1955) *Microwave Spectroscopy*. McGraw–Hill, New York. Reprinted in 1975 by Dover, New York.
Tranter G. E., MacDermott A. J., Overill R. E. & Speers P. J. (1992) *Proc. Roy. Soc.* **A436**, 603.
Ulbricht T. L. V. (1959) *Quart. Rev. Chem. Soc.*, **13**, 48.
Vager Z. (1997) *Chem. Phys. Lett.*, in press.
Wagnière G. & Meier A. (1982) *Chem. Phys. Lett.*, **93**, 78.
Wagnière G. & Meier A. (1983) *Experientia*, **39**, 1090.
Weinberg S. (1995) *The Quantum Theory of Fields*, Vol. 1. Cambridge University Press.
Wen X. G. & Zee A. (1991) *Phys. Rev.* **B43**, 5595.
Wilczek F. (1990) *Fractional Statistics and Anyon Superconductivity*. World Scientific, Singapore.
Wu C. S., Ambler E., Hayward R. W., Hoppes D. D. & Hudson R. P. (1957) *Phys. Rev.*, **105**, 1413.
Zocher H. & Török C. (1953) *Proc. Natl. Acad. Sci. USA*, **39**, 681.

Chapter 4
The Molecular Basis of Chiral Recognition

J. Greer
Pharmaceutical Discovery, Abbott Laboratories, Illinois, USA
I. W. Wainer
Gerontology Research Center, National Institute on Aging, National Institutes of Health, Baltimore, MD, USA

INTRODUCTION

In 1848, Louis Pasteur used a hand lens and a pair of tweezers to separate the sodium ammonium salt of paratartaric acid into two piles, one of left-handed crystals and the other of right-handed crystals, thereby accomplishing the enantioselective resolution of racemic tartaric acid (Pasteur 1948). Thus, the first reported mechanism for the recognition and differentiation of enantiomeric molecules was visual recognition of crystal structures.

Pasteur also accomplished the first chemical resolution of a racemic compound through the formation of diasteromeric salts (Pasteur 1948; Drayer 1993). In 1853 he neutralized a solution of the optically pure alkaloid L-cinchonine with racemic tartaric acid. The solution was left to crystallize and the first crop of crystals comprised entirely the L-tartrate- L-cinchonine salt.

With this experiment, Pasteur demonstrated that the transformation of enantiomers into diastereomers converted molecules with the same physicochemical properties (enantiomers) into compounds with different physicochemical properties (diastereomers). Thus, while the solubilities of D- and L-tartaric acid were identical, those of the diastereomeric D- and L-tartaric acid-L-cinchonine salts were not. The resolution of enantiomers through their conversion into diastereomeric salts or covalent diastereomeric derivatives is now a routine chemical procedure.

In biological systems, the differentiation between enantiomeric compounds is conceptually similar to the chemical process, *i.e.* based upon conversion of enantiomers into diastereomers. In this case, however, instead of diastereomeric compounds, the key step is the formation of transient diastereomeric complexes. This mechanism is based upon the capacity of one chiral molecule (the selector) to interact with the enantiomers of a second (the selectand). Differentiation or molecular chiral recognition is the result of energy differences between the diastereomeric selector–selectand complexes.

The chiral selectors in biological systems are most often biopolymers which derive their chirality from L-amino acid backbones and the resulting secondary and tertiary structures. The most prominent are enzymes, receptors and carrier proteins such as serum albumin. In addition to amino acid-based biopolymers, carbohydrates such as amylose and cellulose (Yashima & Okamoto 1997) and cyclodextrins (Francotte 1997) also have the capacity for chiral recognition. Native and derivatized forms of these biopolymers have been utilized in *in vitro* systems for the analytical and preparative chromatographic separations of chiral compounds (Wainer 1993; Francotte 1997; Yashima & Okamoto 1997).

The recognition of three-dimensional structure during enzyme–substrate or ligand–receptor interactions is a complicated but basic aspect of biological and pharmacological processes. Yet within the very complexity of this process lies the simplicity of nature and a key to understanding many basic pharmacological processes. This realization can also be attributed to Pasteur who, in 1858, reported that the *dextro* form of ammonium tartrate was more rapidly destroyed by the mould *Penicillium glaucum* than the *laevo* isomer (Pasteur 1901; Drayer 1993). This was the first report of molecular chiral recognition.

This chapter is designed to explore the mechanisms behind molecular chiral recognition. Because this is a broad topic, the scope has been limited to the interactions of small molecules and peptides with enzymes. It is our hope that this brief examination will shed some light on this process and lay the basis for another 150 years of exploration of this intricate interaction.

KINETIC AND THERMODYNAMIC ASPECTS OF ENZYMATIC ENANTIOSELECTIVITY

KINETICS OF ENZYMATIC REACTIONS

The basic mechanisms of enzyme catalysis have been extensively studied and have been the subject of several volumes, c.f. (Dixon & Webb 1979). The usual approach to this topic utilizes the Michaelis–Menten theory of enzyme kinetics which assumes that enzymatic reactions are multiple-step processes (Dixon & Webb 1979; Halgas 1992). The simplest form of this mechanism is outlined in Scheme 1; where E is the enzyme, S is the substrate, [ES] is the enzyme–substrate complex and P is the product.

$$\mathrm{E} + \mathrm{S} \underset{k_{-1}}{\overset{k_1}{\rightleftharpoons}} [\mathrm{ES}] \overset{k_2}{\to} \mathrm{E} + \mathrm{P}$$

Scheme 1 General process of enzymatic transformations

In this process, k_1 and k_{-1} are the rate constants for the forward and reverse reactions, respectively and k_2 is the rate that the [ES] dissociates to E and P. In the Michaelis–Menten approach, it is assumed that $k_2 \ll k_{-1}$; the Michaelis constant, K_m is defined as:

$$K_m = k_{-1}/k_1 = [\mathrm{S}] \cdot [\mathrm{E}]/[\mathrm{ES}] \tag{1}$$

THERMODYNAMICS OF ENZYME REACTIONS

Although enzymatic conversions can be described using only kinetic terms, there are also interrelated thermodynamic processes. This is illustrated in Fig. 4.1, in terms of a free energy reaction co-ordinate diagram. In this illustration, the free energy of the system, ΔG, is initially at the level of the substrate, S, increases as the activated enzyme–substrate complex is formed, reaches a maximum at the transition state, $[ES]^{\#}$, and then decreases until it reaches the free energy of the product, P.

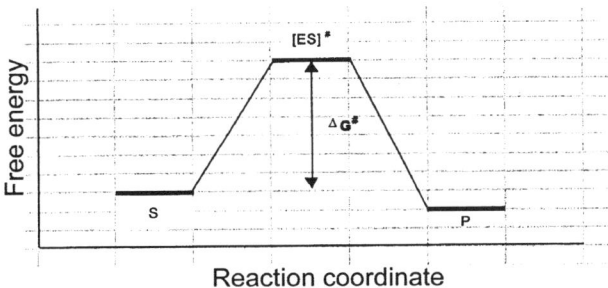

Fig. 4.1 Free energy reaction co-ordinate diagram for the enzymatic conversion of S to P.

The key aspect of this process is the formation of the activated enzyme–substrate complex, $[ES]^{\#}$. The $[ES]^{\#}$ is not the [ES] described by Michaelis–Menten kinetics and cannot be described in the terms of K_m (Dixon & Webb 1979). The description of the transformation of the system from initial ground state to transition state requires the use of the theory of absolute reaction rates described by Eyring (Dixon & Webb 1979). The resulting relationship is:

$$\Delta G^{\#} = RT \ln K^{\#} \qquad (2)$$

where $\Delta G^{\#}$ is the free energy of the transition state and $K^{\#}$ is the equilibrium constant reflecting the distribution of the substrate between the ground state and the transition state.

ENZYMATIC ENANTIOSELECTIVITY

The ability of an enzyme to discriminate between the enantiomers of a chiral substrate can be a result of kinetic and/or thermodynamic factors. In the first instance, molecular chiral recognition occurs at the binding step producing differences in affinities between the substrate and the enzyme. Thus, Michaelis–Menten kinetics might be different for each enantiomer (Testa & Mayer 1988).

Kinetic-based enantioselectivity is exemplified by the *in vitro* oxidation of *R*-felodipine, which proceeds more rapidly than that of *S*-felodipine (Eriksson *et al.* 1991). The difference between the rates of metabolism for *R*- and *S*-felodipine is associated with a lower K_m value for the *R* isomer. Thus, the enantioselectivity (defined as α) of the oxidation of *R*- and *S*-felodipine can be described in terms of their relative K_m values as:

$$\alpha = K_{mR}/K_{mS} \tag{3}$$

where K_{mR} is the K_m for *R*-felodipine and K_{mS} is the K_m for *S*-felodipine.

Enzymatic enantioselectivity can also occur at the catalytic step because of differences between the energies of the activated transition states, $\Delta G^{\#}$. For example, in characterized human liver microsomes, the rate of formation (V_{max}) of *S*-norfluoxetine from *S*-fluoxetine was greater than for the formation of *R*-norfluoxetine from *R*-fluoxetine, while the K_m values for *R*- and *S*-fluoxetine were equivalent (Stevens & Wrighton 1993). In this instance, the enantioselectivity of the process arose from differences between the activated diastereomeric enzyme–substrate complexes, $[ES]^{\#}_R$ and $[ES]^{\#}_S$. This process is depicted graphically in Fig. 4.2.

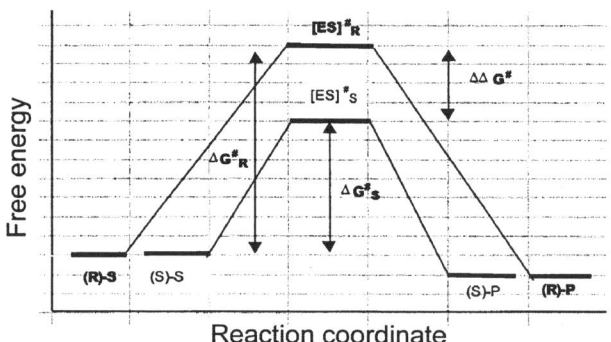

Fig. 4.2 Free energy reaction co-ordinate diagram for the enantioselective enzymatic conversion of *R*-S and *S*-S.

In this example, the two enantiomers form activated enzyme–substrate complexes, $[ES]^{\#}_R$ and $[ES]^{\#}_S$, with different free energies. The free energy changes for each enantiomer from ground state to activated complex can be expressed, by use of eq. (2) as:

$$\Delta G^{\#}_R = RT \ln K^{\#}_R \tag{4}$$

and

$$\Delta G^{\#}_S = RT \ln K^{\#}_S \tag{5}$$

The difference between the activation energies for $[ES]^{\#}_R$ and $[ES]^{\#}_S$ is the source of the observed enantioselectivity (α) and can be expressed as:

$$\Delta\Delta G^{\#} = \Delta G^{\#}_R - \Delta G^{\#}_S = RT \ln K^{\#}_R / K^{\#}_S = RT \ln \alpha \tag{6}$$

MOLECULAR CHIRAL RECOGNITION MECHANISMS

THE 'THREE-POINT' MODEL

The enantioselective differences between the kinetic and/or thermodynamic parameters observed for enzymes and chiral substrates arise from the specific molecular interactions between the selector (enzyme) and selectand (substrate). These interactions include hydrogen bonding, electrostatic, hydrophobic and steric interactions. To result in molecular chiral recognition, the 3-dimensional spatial arrangement of the selectand requires a complementary 3-dimensional structure on the selector with which to form a sufficient and necessary number of interactions.

To specify the origin of molecular chiral recognition, one must specify the nature of the various interactions between the species involved. It is also necessary to define a model with which to characterize the requirements of enantioselective recognition. The first model was proposed by Easson & Stedman (1933). In their mechanism, enantioselective receptor binding was the result of the differential binding of two enantiomers to a common site produced by a 'three-point contact' model between ligand and receptor (Fig. 4.3).

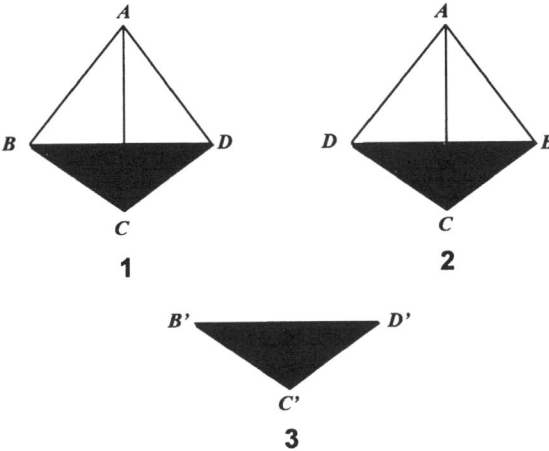

Fig. 4.3 Easson and Stedman's three-point interaction model of molecular chiral recognition.

In this model, enantiomer **1** was more active than its enantiomorph **2** because **1** was more tightly bound to the receptor (**3**). The differential binding is a result of the sequence of the substituents, BCD, around the chirally substituted carbon atom which form a triangular face of the tetrahedral bond array. For **1**, the sequence matches the complementary triad of binding sites on **3**, B'C'D', leading to a '3-point' interaction. The enantiomorph **2** has a mirror-image sequence, DCB, and its interaction with **3** occurs at only two of the three sites on the receptor surface producing a relatively weaker ligand–receptor interaction.

The 'three-point' interaction model was overlooked for 15 years until Ogston (1948) resurrected it to explain the enzymatic decarboxylation of L-serine to glycine.

The pivotal step in this conversion was the stereoselective decarboxylation of the prochiral intermediate metabolite aminomalonic acid. In Ogston's model, the carboxylic moieties on the aminomalonic acid become inequivalent because of the existence of three non-equivalent binding sites on the enzyme, one of which was responsible for the decarboxylation (Fig. 4.4A).

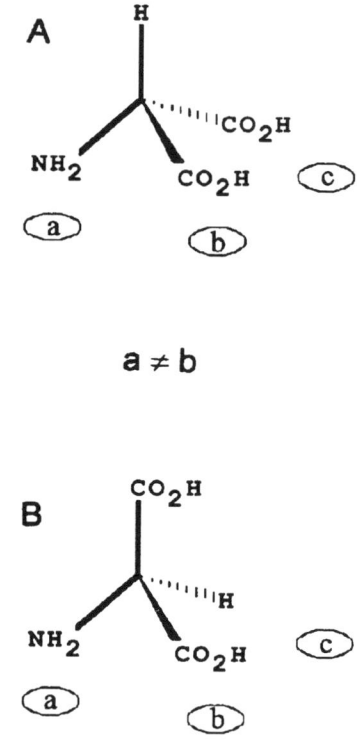

Fig. 4.4 A. The 'three-point' interaction model for the enzymatic transformation of aminomalonic acid as proposed by Ogston. B. A second possible interaction between substrate and enzyme. a, b and c are three non-equivalent areas on the enzyme and c is designated the catalytic site.

A RE-EXAMINATION OF THE 'THREE-POINT' MODEL

Ogston published his work in 1948 and by the mid-1950s, the 'three-point' interaction model was generally accepted as the source of biological enantioselectivity. Although it is clear the 'three-point' interaction model works in several situations, is it an accurate description of how the dissymmetry of one molecule is perceived by a second?

In principle, the model is a static picture of bimolecular process, essentially the 'lock and key' model of enzymatic activity (Fisher 1894). The 'lock and key' model has, however, been superseded by the understanding that enzymatic conversions involve mutually induced conformational adjustments of the substrate and enzyme (Pauling 1948; Dixon & Webb 1979). Thus, it is necessary to develop a dynamic description of molecular chiral recognition.

The best place to begin is with a re-examination of Ogston's model. The first assumption which must be addressed is the assignment of three distinct 'binding sites' on the enzyme. Surely, the catalytic area of an enzyme is best described as a spatial environment or a cavity, not as a point or single interaction site. Binding sites which position and constrain the substrate relative to the enzyme do not necessarily have to lie within the catalytic site; for example, the hydrophobic binding site of chymotrypsin. If this is so, Ogston's model contains only two interaction sites which produce binding interactions between the substrate and enzyme. Because the binding sites are not equivalent, the bound substrate is presented to the enzyme in two distinct spatial configurations, Figs 4.4A, B. The chirality of the enzyme places the catalytic site in a position such that only one of the configurations can be decarboxylated (Fig. 4.4A), leading to the production of a single enantiomeric product.

In this example enantioselectivity results from a two-point directional interaction and from the chirality of the enzyme. This model is similar to that previously proposed by Sokolov and Zefirov (1991) which is described by the authors as the 'rocking tetrahedron', Fig. 4.5. In this approach to molecular chiral recognition, the substrate is secured to the enzyme by two binding interactions which must be non-equivalent and directional so that only one orientation is possible. The tethered substrate still has conformational mobility and the two hydrogen atoms sweep out overlapping, but not identical, steric volumes. Where and to what extent the active site of the enzyme interacts with these steric volumes determines the enantioselectivity of the process. If the chirality of the enzymes places the interaction perpendicular to the plane of the substrate, no enantioselectivity is observed. As the deviation from the perpendicular increases, so does the enantioselectivity.

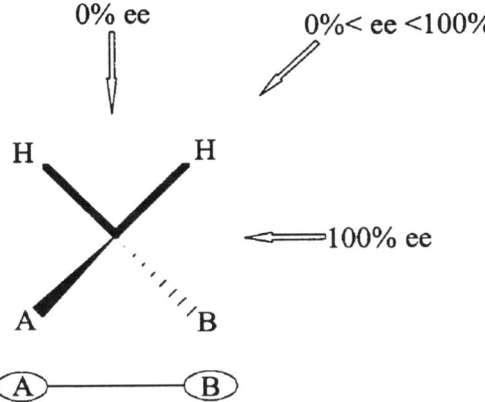

Fig. 4.5 The 'rocking tetrahedron' model of molecular chiral recognition. Where the arrows indicate the direction of enzymatic interaction with the substrate and *ee* represents the enantiomeric excess produced by the enzymatic conversion.

CONFORMATIONALLY-DRIVEN MOLECULAR CHIRAL RECOGNITION

In the molecular chiral recognition model presented in Fig. 4.4 the enantiomeric substrates undergo conformational modifications before binding to the enzyme. The enantioselectivity of the process results from the energy differences between the binding conformations assumed by the substrate enantiomers and/or energy differences arising from conformational adjustments of the enzyme to each of the enantiomeric solutes. These differences should be reflected in different K_m values for each enantiomer and the enantioselectivity of the process will be described by eq. (3).

In the molecular chiral recognition model depicted in Fig. 4.5 conformational adjustments of the substrate enantiomers and/or the enzyme occur after the formation of the diastereomeric enzyme–substrate complexes and enantioselective differentiation occurs during the transformation of these enzyme–substrate complexes into activated complexes, i.e. $[ES] \rightarrow [ES]^{\#}$. The enantioselectivity of this process will be described by eq. (6).

Both of these approaches describe 'conformationally driven' molecular chiral recognition mechanisms. Within an enantioselective enzyme-catalysed reaction, each of these mechanisms will contribute to any observed enantioselectivity. The distinction between kinetic-driven and thermodynamic-driven molecular chiral recognition is merely a reflection of the relative contributions of these constituents to a single process.

THE N-DECHLOROETHYLATION OF IFOSFAMIDE ENANTIOMERS

Ifosfamide, IFF (Fig. 4.6), an effective oxazaphorine alkylating agent widely used in cancer chemotherapy, is also a chiral molecule containing an asymmetrically substituted phosphorus atom and exists in two enantiomeric forms, R-IFF and S-IFF. IFF is clinically administered as a racemate

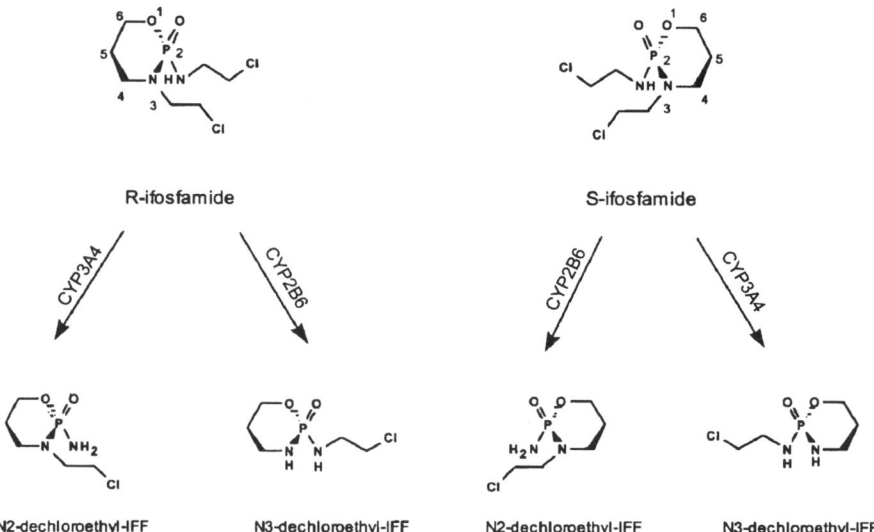

Fig. 4.6 N-Dechloroethylation of R- and S-ifosfamide.

IFF is a prodrug with pharmacological activity resulting from enzymatic transformation into active metabolites. The antitumour efficacy of the agent is associated with initial hydroxylation at the C4 position on the oxazaphosphorine ring followed by intracellular decomposition which produces isophosphoramide mustard, a DNA cross-linking agent (Allen & Creaven 1972). In man this transformation is predominately mediated by a microsomal cytochrome P450 (CYP) isoenzyme identified as CYP3A (Bullock *et al.* 1997).

In addition to the 4-hydroxylation pathway, there is a second IFF metabolic route which involves β-oxidation of one the molecule's two chloroethyl moieties (Allen & Creaven 1972; Wainer *et al.* 1996). This transformation takes place at either the 2-N- or 3-N-position on the oxazaphosphorine ring producing the chiral N-dechloroethylated (DCE) metabolites *R*- and *S*-2-DCE-IFF and *R*- and *S*-3-DCE-IFF, Fig. 4.6. It is important to note that *R*-IFF is metabolized into *R*-2-DCE-IFF and *S*-3-DCE-IFF whereas *S*-2-DCE-IFF and *R*-3-DCE-IFF arise from *S*-IFF. The apparent inversion in stereochemical configuration is an artefact introduced by the Cahn–Ingold–Prelog system of chemical nomenclature.

Clinical studies of IFF metabolism have identified two distinct urinary excretion patterns (Wainer *et al.* 1996). In patients, the cumulative excretion of *R*-3-DCE-IFF and *R*-2-DCE-IFF correlated significantly as did the cumulative excretion of *S*-2-DCE-IFF and *S*-3-DCE-IFF. These results suggest that IFF N-dechloroethylations are mediated by two or more CYPs

Recent studies have identified the microsomal enzymes responsible for the majority of the N-dechloroethylation of IFF (Granvil *et al.* 1996; Bullock *et al.* 1997) and the metabolic scheme for IFF N-dechloroethylation is outlined in Fig. 4.6. These studies demonstrate that CYP3A mediates the N-dechloroethylation of *R*-IFF at the N-2 position and the N-dechloroethylation of *S*-IFF at the N-3 position whereas CYP 2B6 mediates the N-dechloroethylation of *R*-IFF at the N-3 position and the N-dechloroethylation of *S*-IFF at the N-2 position. Thus, the two IFF enantiomers are metabolized by the same enzymes but at different positions and to different extents. The enantioselectivity of this process is reflected in the cumulative urinary excretion pattern which is *R*-3-DCE-IFF >> *S*-3-DCE-IFF = *S*-2-DCE-IFF > *R*-2-DCE-IFF [22].

A molecular chiral recognition mechanism has recently been proposed to explain positional (N-2 rather than N-3) and enantioselective aspects of IFF metabolism (Wainer *et al.* 1996). In this mechanism, *R*-IFF and *S*-IFF are bound at the active sites of the enzymes through hydrogen bonds involving the phosphoramide oxygens of the IFF molecules. Because of the chirality of the IFF enantiomers, *R*-IFF and *S*-IFF are positioned as mirror images relative to the enzymes as illustrated in Fig. 4.7. This results in a situation where the N-2 position of *R*-IFF and the N-3 position of *S*-IFF are positioned close to each other while the opposite situation occurs for the N-3 position of *R*-IFF and the N-2 position of *S*-IFF.

By use of this model, the source of the positional selectivity can be readily discerned. The enantioselectivity of the processes is not, however, as clear. The CYP3A-mediated N-dechloroethylation of *S*-IFF proceeds to a much greater extent than that of *R*-IFF whereas the CYP2B6-mediated N-dechloroethylation of *S*-IFF and *R*-IFF are equivalent. This indicates that the molecular chiral recognition mechanism is similar to that proposed by Sokolov & Zefirov (1991), *i.e.* the 'rocking tetrahedron', Fig. 4.5.

With IFF enantioselectivity is the result of a two-point directional interaction and the chirality of the CYP3A and CYP2B6 enzymes. Once the IFF enantiomers are secured to the enzymes, the N-dechloroethyl moieties sweep out overlapping but not identical steric volumes. With CYP3A, the chirality of the enzyme places the oxidative interaction closer to the N-2-chloroethyl group on S-IFF than the N-3-chloroethyl group on R-IFF, thus producing the observed enantioselectivity.

When the enantiomers are immobilized in the active site of CYP2B6, the interactive moieties of the enzyme are equidistant from both N-dechloroethyl groups, and no enantioselectivity is observed. This molecular chiral recognition model not only explains the observed metabolic transformations, but also raises some interesting questions about the structures of CYP3A and CYP2B6. The active sites of these enzymes seem to exist in approximately mirror image forms. Clearly, this could be accomplished by changes as small as exchange of groups 'a' and 'c' of Figs 4.3 or 4.4. The genetic and evolutionary aspects of this relationship are clearly interesting areas for future research.

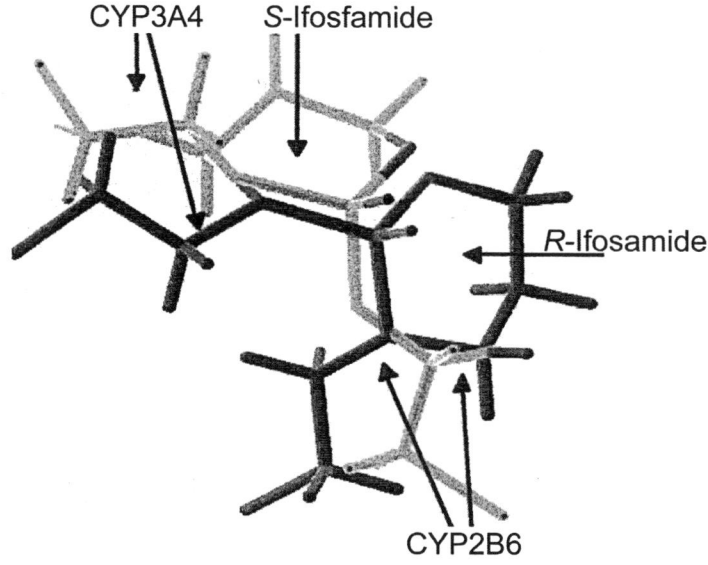

Fig. 4.7 Molecular chiral recognition model for the N-dechloroethylation of ifosfamide. The light blue molecule is S-Ifosfamide; dark blue is R-Ifosfamide.

INTERACTION OF PEPTIDES AND LARGER MOLECULES WITH CHIRAL BIOPOLYMERS

BINDING OF PEPTIDES AND PEPTIDE-LIKE MOIETIES TO A PROTEIN

The mode of binding of peptides to proteins usually involves an extensive set of interactions. The peptide itself is a relatively complex molecule usually comprising three or more amino acid residues, for example in TRH, to over twenty, as in endothelin or atrial naturietic factor (Kaumaya & Hodges 1996). These peptides usually consist of natural L amino acids, each residue of which has at least one asymmetric centre. Modified peptides can be formed from quite an extensive list of synthetic amino acids. Some peptides, both natural and synthetic are cyclic. The rings can be formed from disulphide bridges, other side-chain–side-chain covalent links, or even main chain–main chain cyclizations.

The interaction of peptides with proteins serves several different roles in biology. Peptides can be substrates for protein enzymes and in some instances act as inhibitors. In other instances the peptide might merely bind to the protein receptor as an effector to initiate a response, where it is called an agonist, or to inhibit a response giving an antagonist. Interestingly, in a number of peptides it has been observed that conversion of one or more specific residues from L to D configuration, or even cyclization, can result in changing the peptide from an agonist to an antagonist as occurs with luteinizing hormone-releasing hormone (LH-RH), for example (Rees *et al.* 1974; Beattie *et al.* 1975). Unfortunately, the structural basis or mechanism of this transition is not yet understood.

Structural information about how peptides bind to proteins is available for several systems. Most of these data come from X-ray crystallographic studies of protein/peptide complexes (Bernstein *et al.* 1977). Some are from high-resolution solution NMR structures of such complexes. The example that will be used here is from studies of the human immunodeficiency virus (HIV) aspartic protease which is target for the treatment of AIDS. In this case, the peptide that binds is a substrate and cannot be studied by X-ray crystallography or NMR spectroscopy because of rapid cleavage by the protease enzyme. The binding has, therefore, been probed by use of peptidic inhibitors.

The initial crystal structures determined for HIV protease were of the native protein without any inhibitor bound (Navia *et al.* 1989; Wlodawer *et al.* 1989). The first inhibitor/protein complex examined (Miller *et al.* 1989) was of a peptide analogue called MVT-101 (Fig. 4.8).

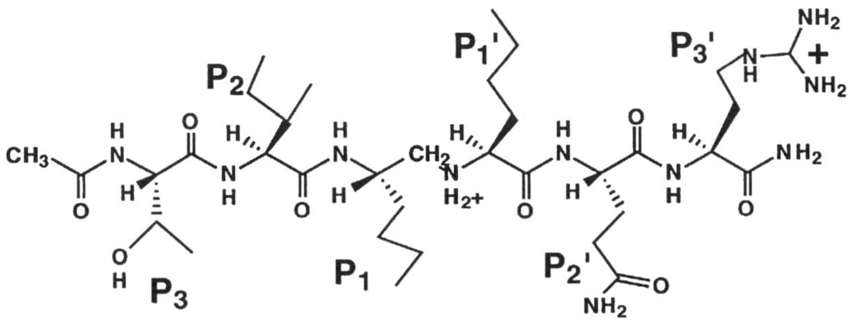

Ac-Thr-Ile-Nle^{R}Nle-Gln-Arg

Fig. 4.8 Structure of the pseudo-peptide MVT-101, an inhibitor of HIV protease. The peptide bond at the cleavage site has been modified to form a reduced amide (symbolized by the 'R') to prevent cleavage of this peptide by the enzyme thereby producing an inhibitor.

Note that by convention (Schechter & Berger 1967) N-terminal residues are labelled outwards from the scissile bond as P_1, P_2, P_3, *etc.*, whereas those C-terminal from the scissile bond are labelled P_1', P_2', P_3', *etc.* (Fig. 4.8). The resulting crystal structure of HIV protease with MVT-101 bound was significantly different from that of the unbound form of the enzyme (Fig. 4.9). Thus, there is clearly an induced fit of this protein to bind ligand. Details of the binding of MVT-101 to the active site can be seen in Fig. 4.10. Each of the peptide side chains fits into specific subsites. There are several specific hydrogen bond interactions between the main chain polar groups of the peptide and the active site residues of the protein. Details of these specific interactions have been used to aid the design (Greer *et al.* 1994) of effective inhibitors of HIV protease to be used in anti-AIDS therapy (Deeks & Volberding 1997; Hoetelmans *et al.* 1997.

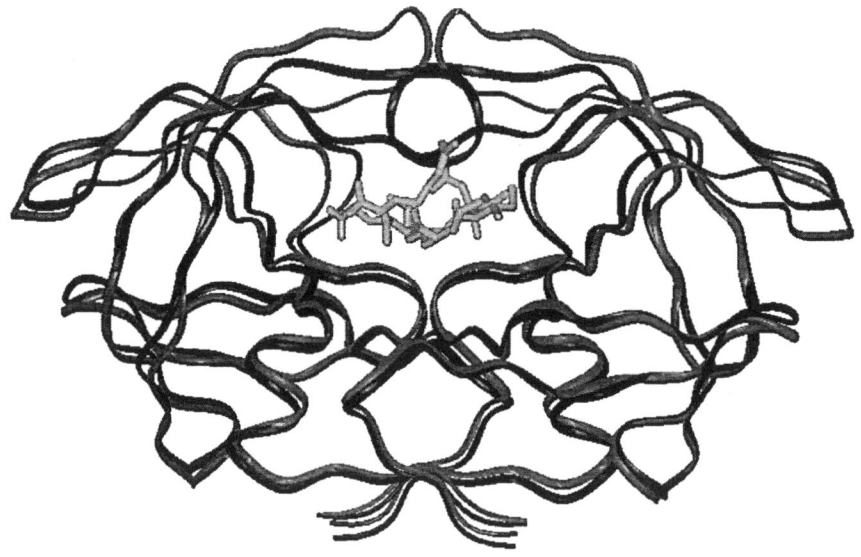

Fig. 4.9 Ribbon representation of the unbound form of HIV protease (green) with the inhibitor-bound form (blue). The inhibitor (shown in orange) fits into a groove in the molecule. Two loops in the protein are open or perhaps disordered in the absence of inhibitor but change their conformation to envelop the inhibitor when it binds in the active site.

It can be seen from Fig. 4.9 that HIV protease is composed of two monomers that are related to each other by a twofold rotation axis, *i.e.*, one monomer can be generated from the other by a 180° rotation. Thus, the HIV protease molecule has C2 symmetry. It is, however, clear that a peptide, natural substrate or inhibitor such as MVT-101 does not have C2 symmetry. Accordingly, on the basis of the three-dimensional structure of HIV protease it was decided to design C2 symmetric peptide-like inhibitors (Erickson *et al.* 1990; Kempf *et al.* 1990).

The design concept is shown in Fig. 4.11. Two series of compounds were generated—mono-ols and diols. From the very earliest compounds synthesized, the diols were, in general, more potent than the mono-ols (Table 4.1). Consequently, the chemists started with the C2 symmetric diols and modified them to improve inhibitory potency, specificity, pharmacokinetics, and oral bioavailability. The first compound which was taken to the clinic was ABT-003 (Table 4.2) (Kempf *et al.* 1991). This compound was extremely potent against HIV protease with an inhibitory potency of 0.084 nM. The structure of ABT-003 on the protein was determined by X-ray crystallography; this showed that it interacted in much the same way as the peptides (Fig. 4.12) (Hosur *et al.* 1994). Even though the hydrogen bonding groups are positioned slightly differently, the C2 symmetric enzyme active site easily accommodates the differences and binds the C2 symmetric inhibitor very well.

Table 4.1 Potency of mono-ol and diol inhibitors. Data are taken from Hosur *et al.* (1994).

Mono-ol K_I (nM)

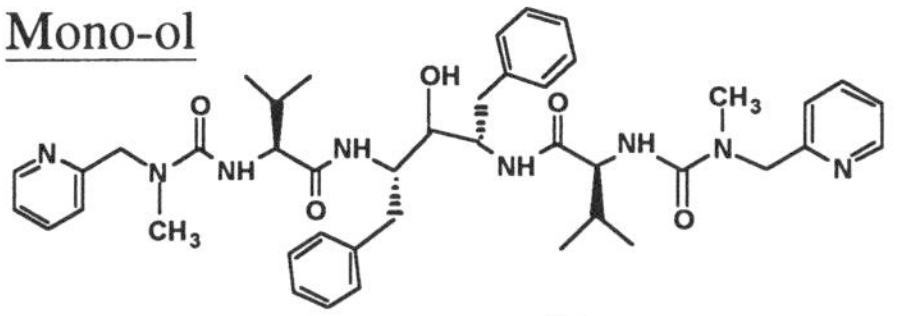

A-77272   4.3

Diol

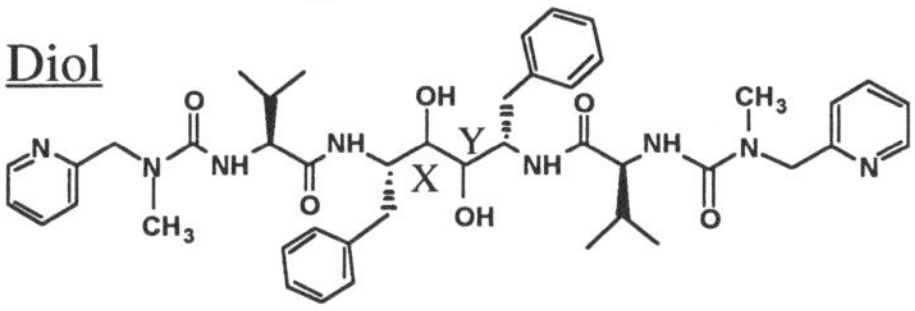

	X	Y	
A-77003	R-OH	S-OH	0.084
A-76889	R-OH	R-OH	1.00
A-76928	S-OH	S-OH	0.077

Table 4.2 Potency of enantiomers of ABT-003 inverting four centres outside the diol region of the molecule.

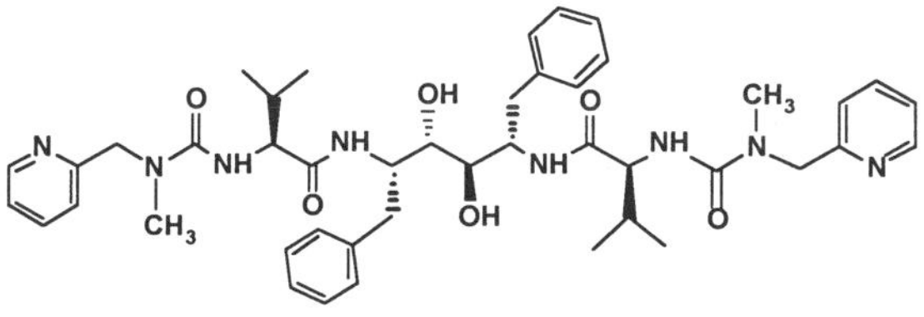

K_I (nM)

A-77003   0.084

A-85271   8100

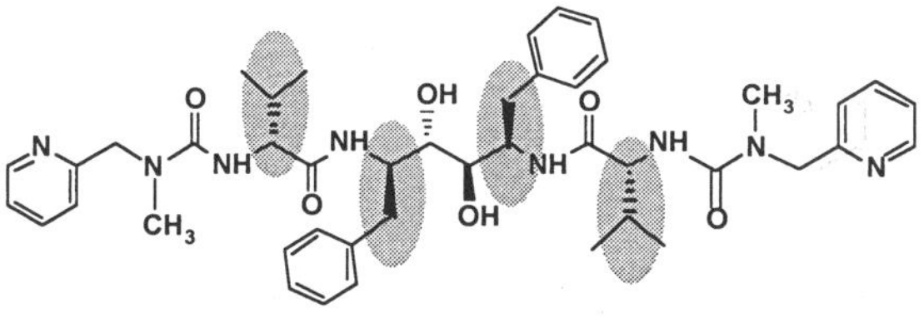

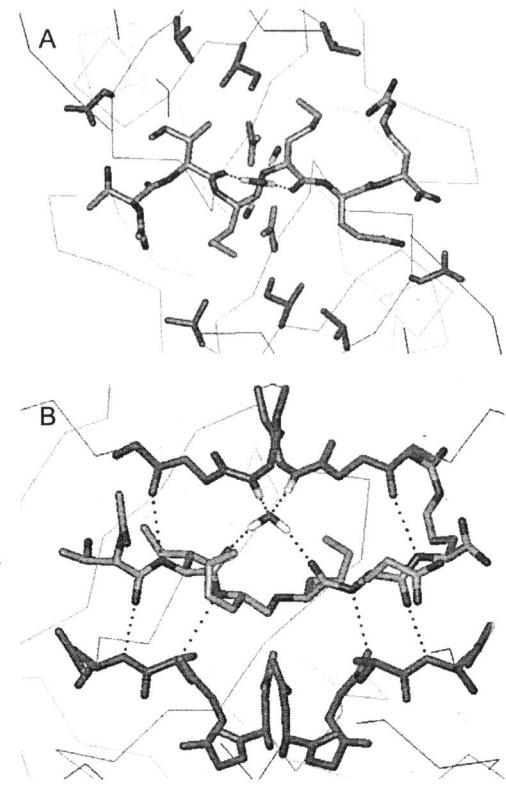

Fig. 4.10. Details of binding of MVT-101 (orange carbons) to the active site of HIV protease (green carbons and purple thin bonds). (A) Note the significant number of binding pockets that exist for this interaction. (B) Several specific hydrogen bonds are also made in this view of the active site which is orthogonal to that shown in part A.

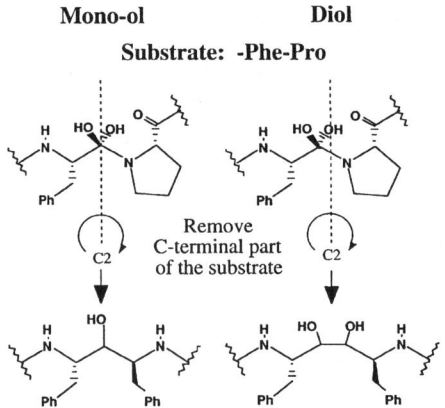

Fig. 4.11 The C-terminal part of the substrate is removed and the N-terminal rotated by 180° to generate a new C-terminus to produce a molecule that has C2 symmetry.

This operation can be performed in two ways. In the first instance, the axis is placed on the carbon of the tetrahedral intermediate and the resulting molecule is a mono-ol. If the axis is placed halfway along the bond that will be cleaved by the enzyme then the rotation operation will produce a diol.

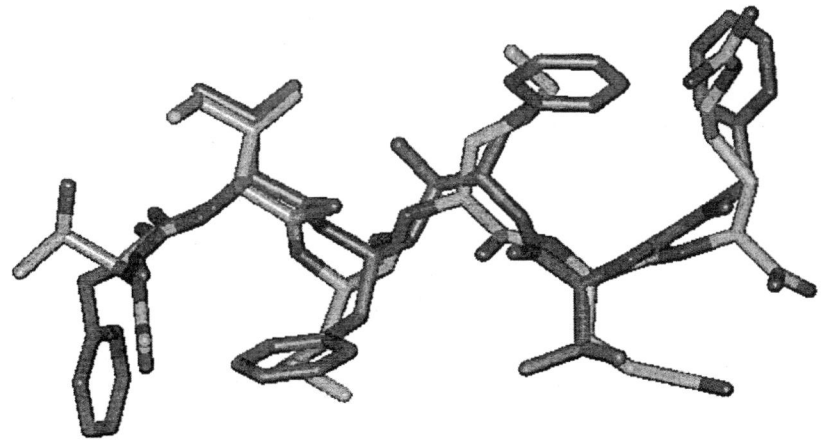

Fig. 4.12 Structure of a C2 symmetric diol inhibitor, ABT-003 (shown in purple), superimposed on the MVT-101 inhibitor (orange bonds) in the same view as Fig 4.10A. Note that all the side chains of ABT-003 fit properly on top of the respective side chains of MVT-101 even though the latter is a straight peptide and the former has C2 symmetry.

HOW CRITICAL ARE THE CHIRAL CENTRES FOR BINDING TO THE ENZYME?

Having seen how a chiral peptidic molecule makes specific interactions with a chiral active site in the protein, how critical is the chirality of the peptide to the binding interactions? How does it affect the binding affinity and what are the structural consequences in terms of binding mode to the protein? The importance of the handedness of the inhibitor was examined for several analogues in the HIV protease inhibitor series. Inhibitors of the diol series are very complex molecules. Altogether, there are six asymmetric centres in this molecule—at the two hydroxyl positions in the centre of the molecule and at P_1, P_2, P_1', and P_2'. Note that because this molecule is C2 symmetric $P_1' = P_1'$ and $P_2 = P_2'$. How are the potency and binding mode of ABT-003 affected by modifying the asymmetric centres in this molecule?

Modification of Multiple Centres

In the first study, the stereochemistry at the diol core was examined while retaining the same epimers as in ABT-003 at the remaining four centres. Considering just the hydroxyls, there are three possible forms: the *R,R*, *S,S* and *R,S* (= *S,R*). (By contrast, notice that the mono-ol series has only one possible form at the hydroxyl because of the C2 symmetry.) These are different molecules and might be expected to have different potencies as inhibitors and maybe even different binding modes on the enzyme.

The binding of this set of molecules has been reported (Hosur *et al.* 1994). The potencies are shown in Table 4.1. Whereas the *S,S* form is highly potent and equivalent to ABT-003, the *R,S* form, the *R,R* form is approximately tenfold less potent. How do these three compounds bind to the active site? It should be noted that of the three compounds, only two, the *R,R* and *S,S* forms, have true C2 symmetry. This twofold symmetry is violated by the hydroxyls in the *R,S* form. How do these compounds bind to the active site?

Fig. 4.13 shows a superposition of the three inhibitors as they bind in the active site of HIV protease. The close overlap indicates that they bind very similarly and that it is the subtle, detailed interactions of the inhibitor with the protein that are responsible for the differences in binding affinity (Hosur *et al.* 1994). In any case, tenfold differences in potency, which correspond to a little over 1 kcal mol^{-1} of energy are at or beyond the limit of the current ability to rationalize binding data from structural information.

In a second study, the compound inverted at all four remaining asymmetric centres, P_1, P_2, P_1', and P_2', was prepared to see what its inhibitory potency would be. This compound was one-thousandfold weaker in inhibitory potency than ABT-003 (Table 4.2). Indeed, this compound was so weak, its mode of binding could not be determined crystallographically to see how it differed from that of the parent molecule, ABT-003 (unpublished observations).

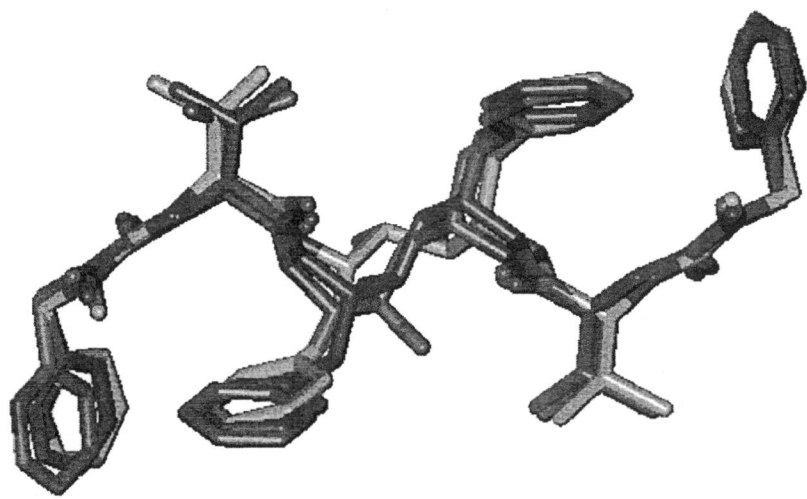

Fig. 4.13 ABT-003 (purple bonds, the *R,S* form) and the two related inhibitors (A-76889, green bonds, the *R,R* form and A-76928, orange bonds, the *S,S* form) produced by varying the stereocentres of the diol core. The inhibitors were overlapped by superimposing the protein structures from their respective crystal structures. The differences between the binding modes of these inhibitors are very small except in the region of the diols. One is forced to conclude that the differences in potency are a result of subtle interactions between the inhibitor and particular active site residues (Hosur *et al.* 1994).

Modification of One Centre at a Time

Further optimization of the HIV protease inhibitors was necessary to discover a compound with optimum potency, pharmacokinetics, and oral bioavailability for successful clinical treatment of AIDS. The efforts of the chemists produced the compound ABT-538 also known as ritonavir (Table 4.3) (Kempf *et al.* 1995). This compound had an IC_{50} of 0.015 nM against the enzyme. Its mode of binding to the protein is shown in Fig. 4.14A.

Table 4.3 Modification of enantiomeric sites in ABT-538 (ritonavir) at P_1 and P_2, one at a time.

Structure	Compound	K_I (nM)
	ABT-538 Ritonavir	0.015
	A-98658	1.6
	A-117673	0.0

The compound fits into the various active site pockets remarkably well, which is consistent with the very potent inhibition properties of this compound. Note that this compound no longer has exact C2 symmetry because there is no P_2' residue on the C-terminal half of the inhibitor. Violating the exact C2 symmetry was necessary to optimize all the appropriate properties of this molecule.

The epimeric centres of this compound were inverted systematically. The first compound prepared was A-98658 which inverts the centre at the P_1 benzyl group. Although P_1' is related to this group by the twofold axis, it was not inverted in this molecule. The resulting compound loses significant potency relative to ritonavir—the IC_{50} is 1.6 nM, one-hundredth that of the parent (Table 4.3). This is a dramatic loss for the modification of only one asymmetric centre and shows how critically these groups fit into the respective binding subsites in the enzyme active site.

A crystal structure was determined for A-98658 to see how the molecule actually fits into the active site. This result is shown in Fig. 4.14B. Note that there is no large change in the binding mode of A-98658 relative to ABT-538. The only significant difference occurs in the immediate region of the inversion of P_1 where the change in handedness causes the benzyl side chain to move and overlap with the enzyme active site, in particular with the isopropyl side chain of residue Val82. It is this overlap which is apparently responsible for the large loss in inhibitory affinity of A-98658.

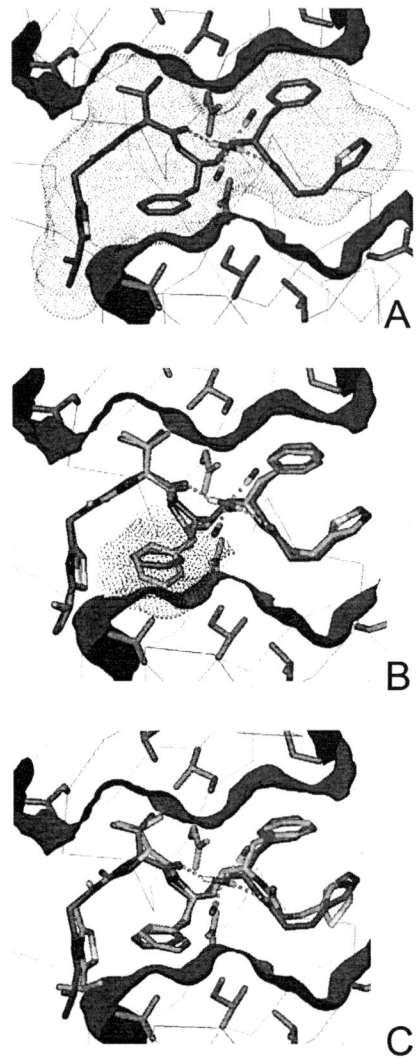

Fig. 4.14 (A) Structure of ritonavir (ABT-538) (blue structure) in the active site of HIV protease (green atoms). The blue surface shows the close complementary fit of the inhibitor to the enzyme active site. (B) The structure of A-98658 (orange) with the centre at P_1 inverted shown together with the structure of ritonavir (blue). The red patch on the green protein surface shows the area of the interaction where close contact occurs between the inhibitor and the enzyme, in particular with the side-chain of residue Val82 (shown in white) explaining the 100-fold loss in activity. (C) The structure of A-117673 (orange), inverted at P_2, together with ritonavir (blue). Minor conformational changes throughout the structure enable A-117673 to place the Val P_2 side-chain into the P_2 pocket, as does ritonavir, without perturbing the interaction with the enzyme.

The next compound prepared inverted the centre at the Val P_2 site of ABT-538 to form A-117673 (Table 4.3) (Kempf *et al.* 1997). It was fully expected that this compound would suffer a similar loss in potency such as when P_1 was inverted in A-98658. Instead, quite surprisingly, this compound was found to be equipotent with the parent (Table 4.3). Once again, the crystal structure of the complex of A-117673 with HIV protease was determined to enable understanding of the structural basis for this preservation of binding.

What was seen is shown in Fig. 4.14C. The binding of A-117673 is almost identical to that of ritonavir. The inversion of the centre at P_2 is accommodated by very minor movements throughout the molecule which enable insertion of the P_2 side chain into the same pocket in A-117673 as in ritonavir, even though the takeoff direction of the side chain from the main chain is very different in the epimer. No clashes are observed to occur with the protein which is completely consistent with the retention of full potency by this analog.

CONCLUDING REMARKS

The preceding illustrative examples serve to emphasize the importance of conformational flexibility in chiral recognition. It follows, therefore, that the three-point interaction model first introduced by Easson and Stedman does not wholly describe the chiral recognition process. It has had its successes, for example in the rational design of relatively rigid chiral selectors for the resolution of relatively rigid small drug molecules and in explaining the interactions between relatively rigid small drug molecules and biomolecules. Modelling the more complex chiral selectors and drug molecules now being encountered will, however, require the more dynamic approaches described here.

REFERENCES

Allen L. M. & Creaven P. J. (1972), *In vitro* activation of isophosphamide (NSC-109724), a new oxazaphosphorine, by rat liver microsomes. *Cancer Chemother. Rep.* **56**, 603–610.

Beattie C. W., Corbin A., Foell T. J., Garsky V., McKinley W. A., Rees R. A. W., Sarantakis D. & Yardley J. P. (1975), Luteinizing hormone-releasing hormone. Antiovulatory activity of analogues substituted in positions 2 and 6. *J. Med. Chem.* **18**, 1247.

Bernstein F. C., Koetzle T. F., Williams G. J., Meyer E. J., Brice M. D., Rodgers J. R., Kennard O., Shimanouchi T. & Tasumi M. (1977) The Protein Data Bank: a computer-based archival file for macromolecular structures. *J. Mol. Biol.* **112**, 535–542.

Bullock P., Lin S., Lake K., Reimer M., Flarakos T., Daly E. & Wainer I. W. (1997) *ISSC Proceedings (8th North American ISSx Meeting)* **12**, 107.

Deeks S. G. & Volberding P. A. (1997) HIV-1 protease inhibitors. *Aids Clin. Rev.*, 145–185.

Dixon M. & Webb E. C. (1979) In: *Enzymes*, 3rd edn. pp. 55–75, 169–172, 272–274. Academic Press, New York.

Drayer D. (1993) The early history of stereochemistry. In: *Drug Stereochemistry, Analytical Methods and Pharmacology*, 2nd edn. (Ed. by I. W. Wainer), pp. 5–14. Marcel Dekker, New York.

Easson E. H. & Stedman E. (1933) CLXX—Studies on the relationship between chemical constitution and physiological action. Molecular dissymmetry and physiological activity. *Biochem. J.* **27**, 1257.

Erickson J., Neidhart D. J., VanDrie J., Kempf D. J., Wang X. C., Norbeck D. W., Plattner J. J., Rittenhouse J. W., Turn M., Wideburg N., Kohlbrenner W. E., Simmer R., Helfrich R., Paul D. A. & Knigge M. (1990) Design, activity, and 2.8Å crystal structure of a C2 symmetric inhibitor complexed to HIV-1 protease. *Science* **249**, 527–533.

Eriksson V. G., Lundahl J., Baarnhiel M. C. & Regardh C. G. (1991) Stereo-selective metabolism of felodipine in human liver microsomes from rat, dog and human. *Drug Metab. Disp.* **19**, 889–894.

Fisher E. (1894) Einfluss der configuration auf die wirkung der enzyme. *Ber. Dtsch Chem. Ges.* **27**, 2985–2993.

Francotte E. (1997) Preparation of drug enantiomers by chromatographic resolution on chiral stationary phases. In: *The Impact of Stereochemistry on Drug Development and Use*, (Ed. by H. Y. Aboul-Enien and I. W. Wainer), pp. 633–683. John Wiley & Sons, New York.

Granvil C. P., Sharkawi M., Ducharme J., Madan A., Sanzghiri U., Parkinson A. & Wainer I. W. (1996) *ISSX Proceedings (7th North American ISSx Meeting)* **10**, 360.

Greer J., Erickson J. W., Baldwin J. J. & Varney M. D. (1994) Application of the three-dimensional structures of protein target molecules to structure-based drug design. *J. Med. Chem.* **37**, 1035–1054.

Halgas J. (1992) In: *Biocatalysts in Organic Synthesis*, (Studies in Organic Chemistry, Vol. 46), pp. 3–6. Elsevier, Amsterdam.

Hoetelmans R. M., Meenhorst P. L., Mulder J. W., Burger D. M., Koks C. H. & Beijnen J. H. (1997) Clinical pharmacology of HIV protease inhibitors: focus on saquinavir, indinavir, and ritonavir. *Pharm. World Sci.* **19**, 159–175.

Hosur M. V., Bhat N. T., Kempf D. J., Baldwin E. T., Liu B., Gulnik S., Wideburg N. E., Norbeck D. W., Appelt K. & Erickson J. W. (1994), Influence of stereochemistry on activity and binding modes for C2 symmetry-based diol inhibitors of HIV-1 protease. *J. Am. Chem. Soc.* **116**, 847–855.

Kaumaya P. T. P. & Hodges R. S. (Eds) (1996) *Peptides: Chemistry, Structure and Biology*, Proceedings of the Fourteenth American Peptide Symposium, June 18–23, 1995. Mayflower Scientific, Kingswinford, UK.

Kempf D. J., Norbeck D. W., Codacovi L., Wang X. C., Kohlbrenner W. E., Wideburg N., Paul D. A., Knigge M. F., Vasavanonda S., Craig-Kennard A., Saldivar A., Rosenbrook J. W., Clement J. J., Plattner J. J. & Erickson J. (1990) Structure-based C2 symmetric inhibitors of HIV protease. *J. Med. Chem.* **33**, 2687–2689.

Kempf D. A., Marsh K., Paul D. A., Knigge M. F., Norbeck D. W., Kohlbrenner W. E., Codacovi L., Vasavanonda S., Bryant P., Wang X. C., Wideburg N. E., Clement J. J., Plattner J. J. & Erickson J. (1991) Antiviral and pharmacokinetic properties of C2 symmetric inhibitors of human immunodeficiency virus type 1 protease. *Antimicrob. Agents Chemother.* **35**, 2209–2214.

Kempf D. J., Marsh K. C., Denissen J. F., McDonald E., Vasavanonda S., Flentge C. A., Green B. E., Fino L., Park C. H., Kong X. P., Wideburg N. E., Saldivar A., Ruiz L., Kati W. M., Sham H. L., Tobins T., Stewart K. D., Hsu A., Plattner J. J., Leonard J. M. & Norbeck D. W. (1995) ABT-538 is a potent inhibitor of human immunodeficiency virus protease and has high oral bioavailability in humans. *Proc. Natl. Acad. Sci. USA* **92**, 2484–2488.

Kempf D. J., Molla A., Marsh K. C., Park C., Rodrigues A. D., Korneyeva M., Vasavanonda S., McDonald E., Flentge C. A., Muchmore S., Wideburg N. E., Saldivar A., Cooper A., Kati W. M., Stewart K. D. & Norbeck D. W. (1997) Lack of stereospecificity in the binding of the P2 amino acid of ritonavir to HIV protease. *Bioorg. Med. Chem. Lett.* **7**, 699–704.

Miller M., Schneider J., Sathyanarayana B. K., Toth M. V., Marshall G. R., Clawson L., Selk L., Kent S. B. & Wlodawer A. (1989) Structure of complex of synthetic HIV-1 protease with a substrate-based inhibitor at 2.3 A resolution. *Science* **246**, 1149–1152.

Navia M. A., Fitzgerald P. M. D., McKeever B. M., Leu C.-T., Heimbach J. C., Herber W. K., Sigal I. S., Darke P. L. & Springer J. P. (1989) Three-dimensional structure of the aspartyl protease from human immunodeficiency virus HIV-1. *Nature* **337**, 615–620.

Ogston A. G. (1948) Interpretation of experiments on metabolic processes using isotopic tracer elements. *Nature* **29**, 963.

Pasteur L. (1948) *Researches on the Molecular Asymmetry of Natural Organic Products*, Alembic Club Reprints, No. 14, reissue edition. F. and S. Livingstone, Edinburgh.

Pasteur L. (1901) On the asymmetry of naturally occurring organic compounds. In: *The Foundation of Stereochemistry: Memoirs by Pasteur, Van't Hoff, Le Bel and Wislicenus* (Ed. by G. M. Richardson), pp. 1–33. American Book Co., New York.

Pauling L. (1948) Molecular structure and biological specificity. *Chem. Ind. (Suppl)*, 1.

Rees R. W. A., Foell T. J., Chai S. Y. & Grant N. (1974) Synthesis and biological activities of analogues of the luteinizing hormone-releasing hormone (LH-RH) modified at position 2. *J. Med. Chem.* **17**, 1016.

Schechter I. & Berger A. (1967) On the size of the active site in proteases. I. Papain. *Biochem Biophys Res Commun.* **27**(2), 157–162.

Sokolov, V. I. & Zefirov N. S. (1991) Enantioselectivity in two-point binding: Rocking tetrahedron model. *Doklady Akademii Nauk SSSR* **319**, 1382–1383.

Stevens J. C. & Wrighton S. A. (1993) Interaction of the enantiomers of fluoxetine and norfluoxetine with human liver cytochrome P450. *J. Pharmacol. Exp. Ther.* **262**, 964–971.

Testa B. & Mayer J. M. (1988) Stereoselective drug metabolism and its significance in drug research. *Progr. Drug Res.* **32**, 249–303.

Wainer I. W. (1993) HPLC chiral stationary phases for the stereochemical resolution of enantiomeric compounds. In: *Drug Stereochemistry, Analytical Methods and Pharmacology,* 2nd edn. (Ed. by I. W. Wainer), pp. 139–182. Marcel Dekker, New York.

Wainer I. W., Booth T. D. & Colvin M. (1996) *ISSX Proceedings (7th North American ISSx Meeting)* **10**, 233.

Wainer I. W., Ducharme J. & Granvil C. P. (1996) The N-dechloroethylation of ifosfamide: Using stereochemistry to obtain an accurate picture of a clinically relevant metabolic pathway. *Cancer Chemother. Pharmacol.* **37**, 332–336.

Wlodawer A., Miller M., Jaskolski M., Sathyanarayana B. K., Baldwin E., Weber I. T., Seld L. M., Clawson L., Schneider J. & Kent S. B. H. (1989) Conserved folding in retroviral proteases: crystal structure of a synthetic HIV-1 protease. *Science* **245**, 616–621.

Yashima E. & Okamoto Y. (1997) Chiral discrimination mechanism of polysaccharides chiral stationary phases. In: *The Impact of Stereochemistry on Drug Development and Use*, (Ed. by H. Y. Aboul-Enien and I. W. Wainer), pp. 345–376. John Wiley & Sons, New York.

Chapter 5
Chirality in Drug Design and Development

David J. Triggle
The Graduate School, State University of New York at Buffalo, Buffalo NY 14260-1200, USA

INTRODUCTION

We live in an asymmetric world where life is neither fair nor even handed—sociologically or physically. This asymmetry plays out in our understanding and definition of biologically important interactions—including those at pharmacological receptors. Paradoxically, this world of asymmetric molecular interactions results in considerable symmetry of organic anatomy and our considerations of this anatomy in terms of 'beauty' are linked closely to an appearance of physical symmetry (Enquist & Arak 1994).

Stereoselective interactions are so common that they are frequently considered to be a defining component of the specificity of drug–receptor interactions. This stereoselectivity of interaction represents a challenge to the definition of drug actions, long recognized chemically, but now increasingly faced at the clinical, development and regulatory levels (Triggle 1997).

Pasteur first explicitly recognized the issue of biologically important stereochemistry in 1860 when he stated that the differences between tartaric acid molecules, which he was able to separate by differential crystallization, must be because of the asymmetric non-superimposition of these molecules. Pasteur was able to demonstrate, moreover, that the laevorotatory and dextrorotatory forms of tartaric acid were differentially destroyed by moulds and yeasts. He wrote very explicitly (Pasteur 1860; Holmstedt 1990):

> There cannot be the slightest doubt that the only and exclusive cause of this difference in the fermentation of the two tartaric acids is caused by the opposite molecular arrangements of the tartaric acids. In this way, the idea of the influence of the molecular asymmetry of natural organic products is introduced into physiological studies, this important characteristic being perhaps the only distinct line of demarcation which we can draw today between dead and living matter. I have in fact set up a theory of molecular asymmetry, one of the most

important and wholly surprising chapters of the science, which opens up a new, distant but definite horizon for physiology.

The stereochemical basis of drug action was investigated by Arthur Cushny as early as the beginning of this century (Cushny 1926). He was able to show the stereochemistry of the action of atropine and related compounds including the hyoscines and homatropines. Although the enantiomers behaved in a *qualitatively* similar manner, they were *quantitatively* distinct. Similar distinctions were made between the enantiomers of catecholamines. This quantitative distinction between the enantiomers of a drug is a common, although not singular, consequence of stereoselective drug action. These stereoselective interactions have been reviewed on several occasions (Ariëns *et al.* 1983; Smith 1983, 1989; Dalgaard 1990; Holmstedt *et al.* 1990; Casy 1993a; Wainer 1993). Additionally, Cushny showed a clinical enantiomeric distinction whereby the (–) enantiomer, but not the (+) enantiomer, of hyoscine was effective in producing 'twilight sleep' (Holmstedt 1990; Cushny 1926).

The issues of the stereoselectivity of drug action go beyond simple academic and scientific considerations, however. Clinical significance derives from considerations of the efficacy of a single enantiomer versus a racemate, from considerations of stereoselective metabolism and disposition, and from the impact of the route of administration and of patient variability. Regulatory issues are derived from considerations that racemic drugs can represent separate agents in fixed combinations; development issues are derived from considerations of the costs, including those for chemical synthesis, of pursuing a single enantiomer or a racemic mixture.

THE STEREOSELECTIVITY OF DRUG–RECEPTOR INTERACTIONS

The chirality of drug–receptor interactions originates from the chirality of the fundamental building blocks of receptors—the L amino acid of proteins. This chirality is conserved and regenerated through the dominant or exclusive stereochemistry of amino acid and protein synthesis. The ratio of D/L amino acids in proteins is increased by spontaneous racemization, a process facilitated by heat, water and other life-damaging processes. Consequently, this ratio can be used as an index of biological origin in areas as diverse as amino acids in meteorites (Cronin & Pizzarello 1997) and the patentness of mitochondrial DNA from Neanderthal man (Krings 1997).

Receptors derived from D amino acids, although not of terrestrial occurrence, would be expected to have the same physical properties as the L receptor, but the opposite stereochemistry of ligand recognition. The synthesis of the D enzyme HIV-1 protease revealed it to have the expected reciprocal chirality to the natural L enzyme with respect to both substrates and inhibitors (deL Milton *et al.* 1992; Petsko 1992). In fact, D amino acid-containing peptides are relatively common in nature, being found in antibiotic families such as the gramicidins and lantibiotics (Jung 1992; Kreil 1997), in the dermorphin and other peptides from the South American tree *Phyllomedusa sauvagei* which contains D-alanine (Montecucchi *et al.* 1981; Broccardo *et al.* 1981) and the Ca^{2+} channel toxins from the funnel-web spider, *Agelenopsis aperta*, which contain a post-translationally introduced D-serine that is important both to biological activity and to channel subtype selectivity (Heck *et al.* 1994; Kreil 1994; Kuwada *et al.* 1994).

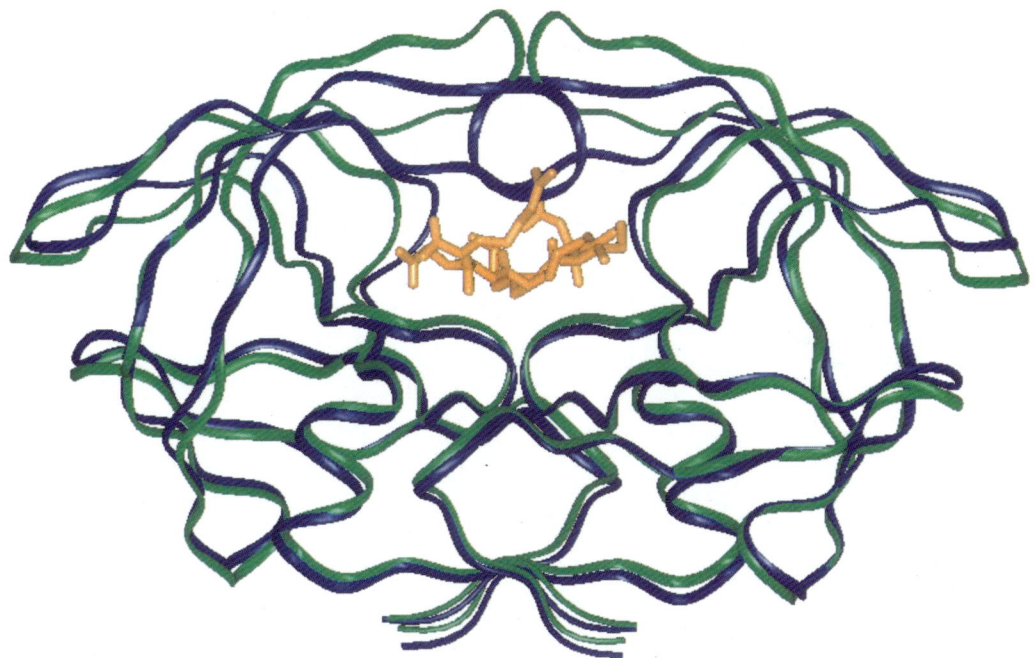

Fig. 4.9 Ribbon representation of the unbound form of HIV protease (green) with the inhibitor-bound form (blue). The inhibitor (shown in orange) fits into a groove in the molecule. Two loops in the protein are open or perhaps disordered in the absence of inhibitor but change their conformation to envelop the inhibitor when it binds in the active site.

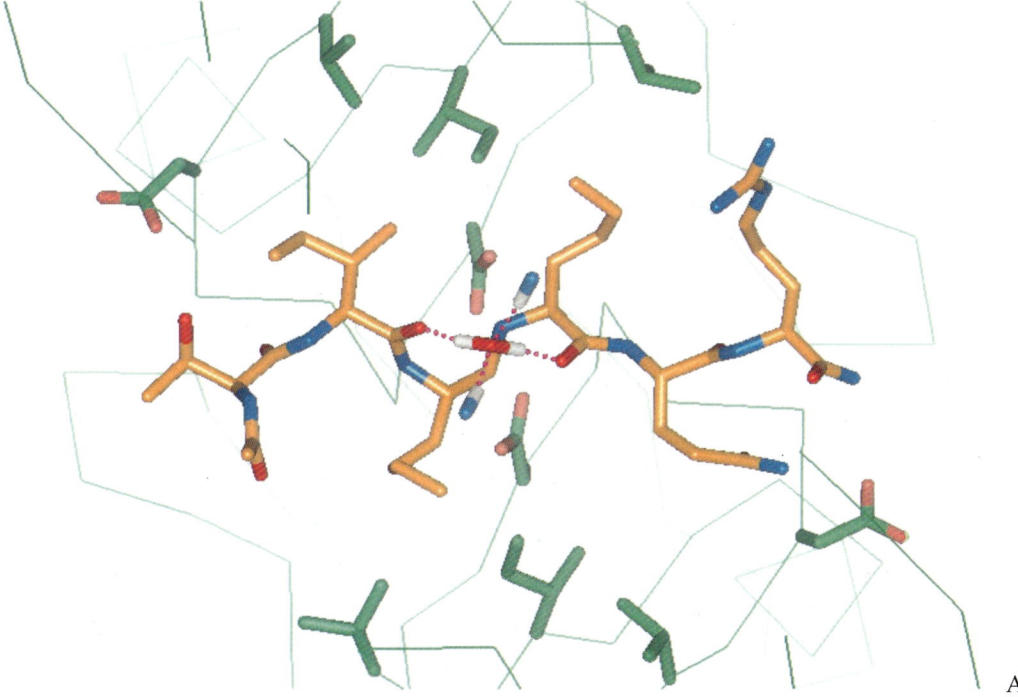

Fig. 4.10 (A)
Details of binding of MVT-101 (orange carbons) to the active site of HIV protease (green carbons and purple thin bounds). (A) Note the significant number of binding pockets that exist for this interaction. (B) Several specific hydrogen bonds are also made in this view of the active site which is orthogonal to that shown in part A. (See part B over page.)

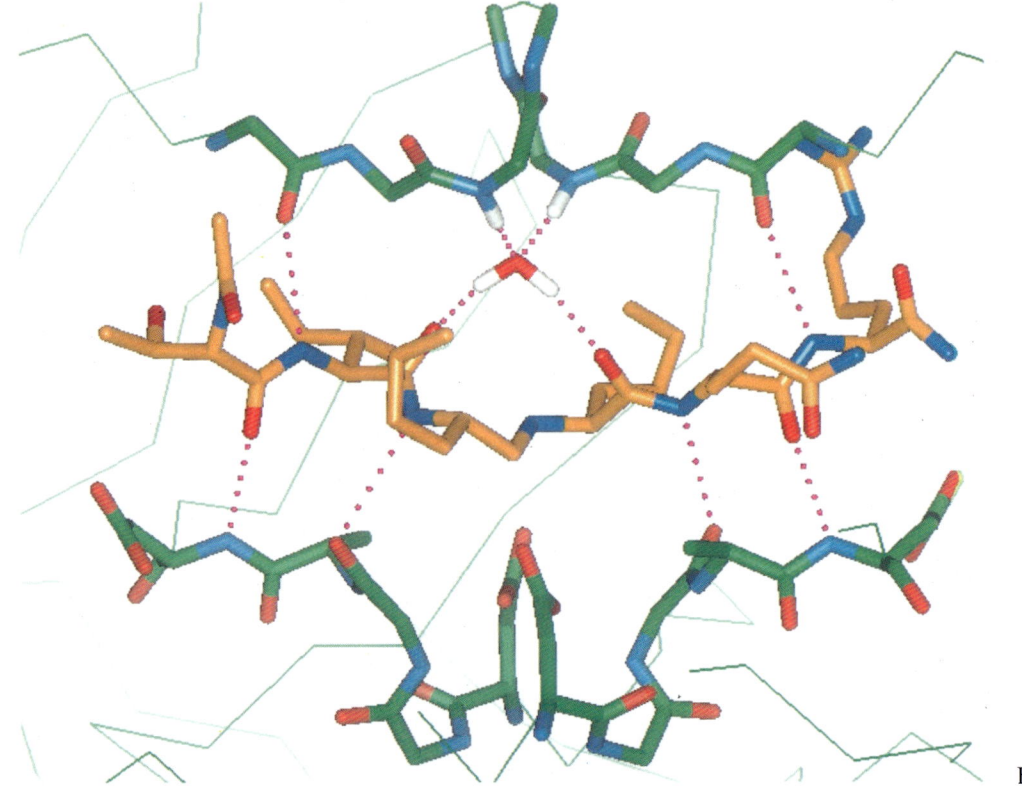

Fig. 4.10 (B) See previous page for caption.

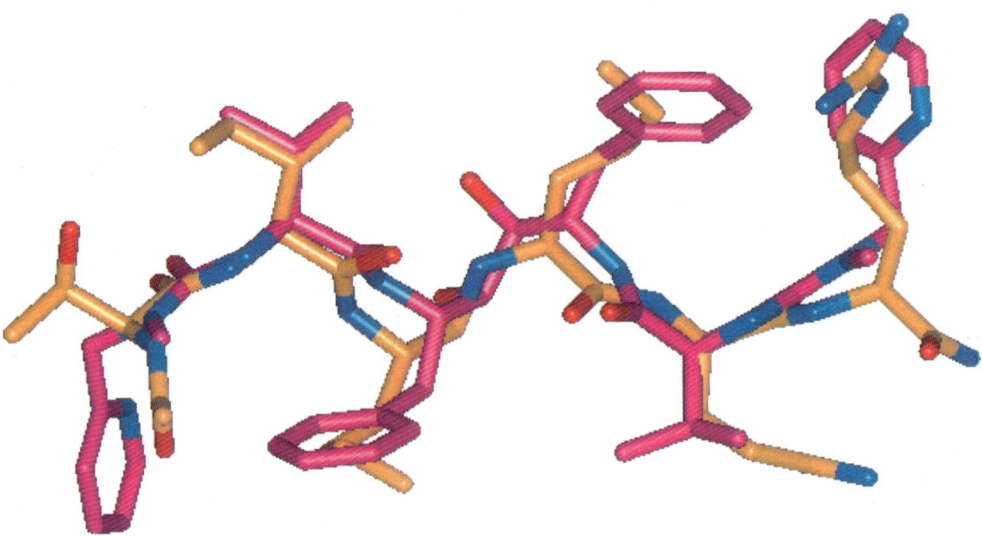

Fig. 4.12 Structure of a C2 symmetric diol inhibitor, ABT-003 (shown in purple), superimposed on the MVT-101 inhibitor (orange bonds) in the same view as Fig 4.10A. Note that all the side chains of ABT-003 fit properly on top of the respective side chains of MVT-101 even though the latter is a straight peptide and the former has C2 symmetry.

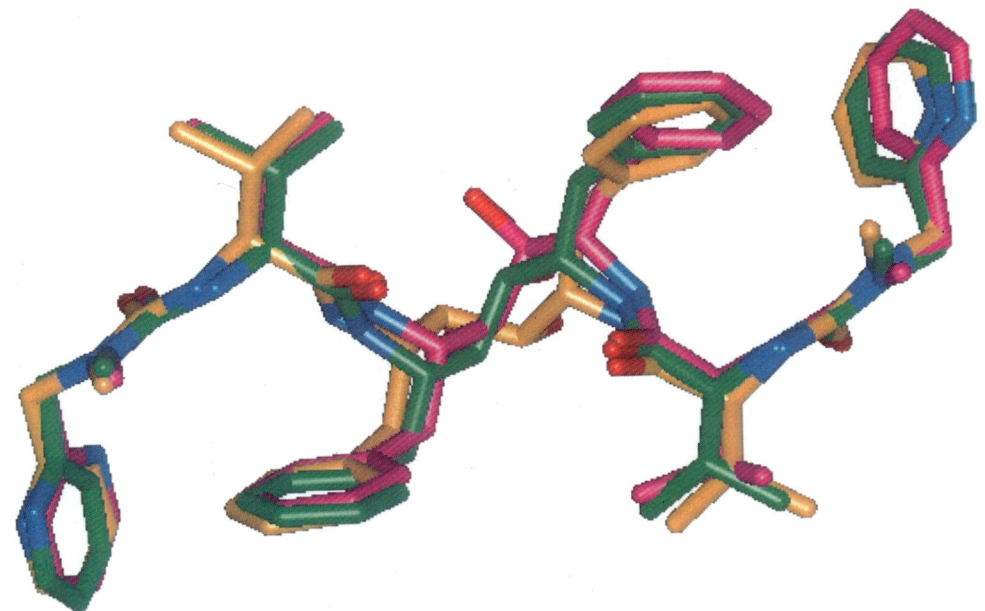

Fig. 4.13 ABT-003 (purple bonds, the *R,S* form) and the two related inhibitors (A-76889, green bonds, the *R,R* form and A-76928, orange bonds, the *S,S* form) produced by varying the stereocentres of the diol core. The inhibitors were overlapped by superimposing the protein structures from their respective crystal structures. The differences between the binding modes of these inhibitors are very small except in the region of the diols. One is forced to conclude that the differences in potency are a result of subtle interactions between the inhibitor and particular active site residues (Hosur *et al.* 1994).

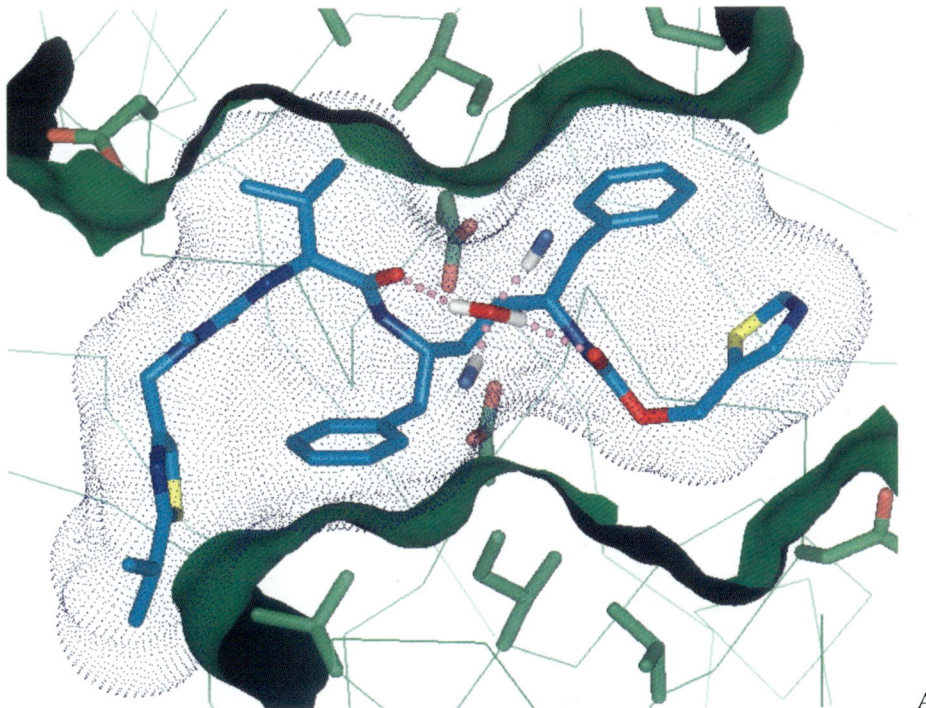

Fig. 4.14 (A) Structure of ritonavir (ABT-538) (blue structure) in the active site of HIV protease (green atoms). The blue surface shows the close complementary fit of the inhibitor to the enzyme active site.

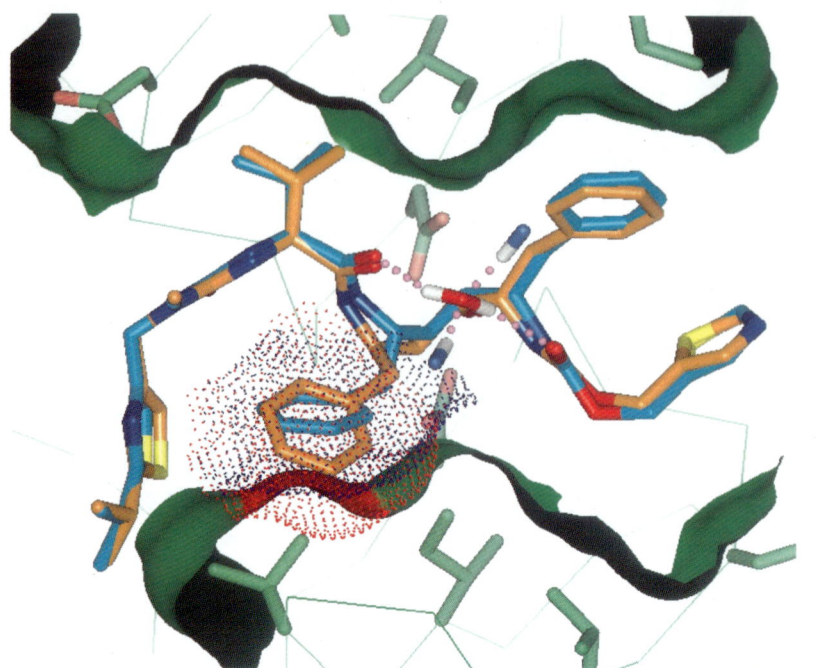

Fig. 4.14 (B) The structure of A-98658 (orange) with the centre at P_1 inverted shown together with the structure of ritonavir (blue). The red patch on the green protein surface shows the area of the interaction where close contact occurs between the inhibitor and the enzyme, in particular with the side-chain of residue Val82 (shown in white) explaining the 100-fold loss in activity.

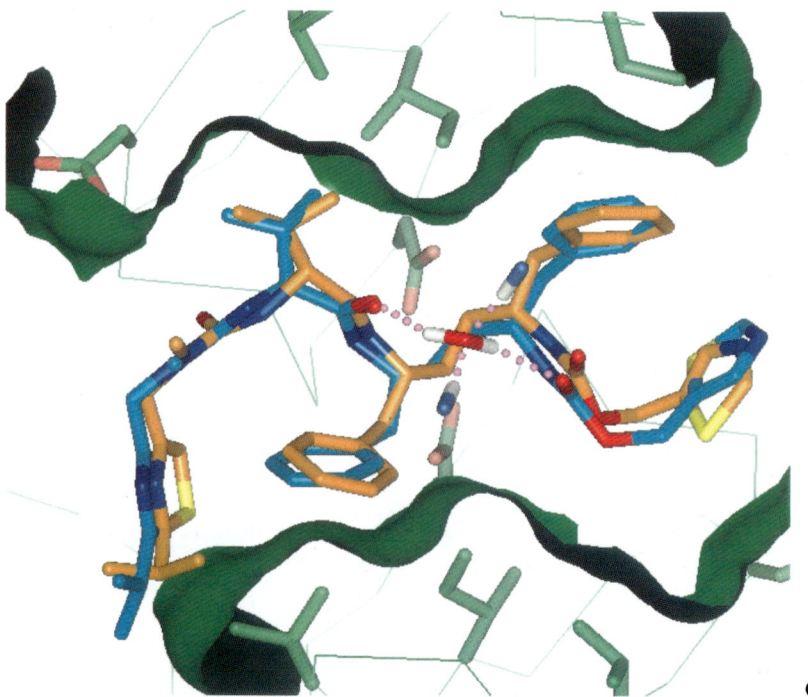

Fig. 4.14 (C) The structure of A-117673 (orange), inverted at P_2, together with ritonavir (blue). Minor conformational changes throughout the structure enable A-117673 to place the Val P_2 side-chain into the P_2 pocket, as does ritonavir, without perturbing the interaction with the enzyme.

These D peptides and proteins are resistant to naturally occurring hydrolytic enzymes, a property that confers potentially enhanced stability and duration of action. This property is doubly exploited in 'mirror-image-phage-display' a genetically encoded combinatorial approach to the identification of peptide ligands (Schumacher *et al.* 1996; Fig. 5.1). In this technique a D protein is used to isolate L peptide ligands—for reasons of symmetry the D enantiomers of these peptides will interact with the target protein of natural handedness. With this technique a cyclic D peptide was synthesized that interacts with the Src homology 3 domain of c-Src, sharing in part the physiological ligand-binding site.

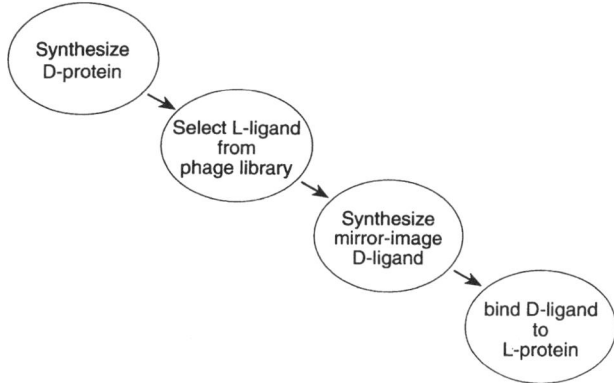

Fig. 5.1 Identification of D peptide ligands through mirror-image phage display. Modified from Schumacher *et al.* (1997).

Although D peptides and proteins are attractively degradation-resistant species, the exclusivity of the natural L world means that such D species cannot interact with the mirror-image surface of the L protein or peptide. Thus, a D peptide might be protease-resistant but be unable to fit a receptor site because the amino acid side chains will be oriented in the opposite direction to that of the natural L-peptide. If, however, the D amino acids run in the opposite sequence to that of the L-peptide the resultant retro-enantiomer will have the same side chain orientation, albeit with a different backbone, and have a similar topochemical surface (Chorev & Goodman 1995).

Several studies have shown that the biological activity of such retro-enantiomers is similar to that of the parent peptide (Guichard *et al.* 1994; Jameson *et al.* 1994; McDonnell *et al.* 1997). A retropeptide analogue of the CD4 protein surface (Fig. 5.2) was as active *in vitro* as the L peptide and had *in vivo* biological activity against experimental allergic encephalomyelitis, a rodent model of multiple sclerosis (Chorev & Goodman 1995). The very similar surface topologies of the cyclic L- and RD-peptides designed to mimic a hair-pin region of the Fc receptor for IgE are shown in Fig. 5.3 (McDonnell *et al.* 1997).

Although chirality of interaction has long been regarded as a primary characteristic of biological recognition processes this might not be an invariant characteristic. Insulin C-peptide is proteolytically cleaved from the insulin prohormone and has long been assumed to be devoid of biological activity (Ido *et al.* 1997; Sleiner & Robenstein 1997). Although C-peptide lacks insulin-like action it does prevent or reduce diabetes-

and hyperglycaemia-induced nerve and vascular dysfunction by a non-chiral process, because both the D amino acid and retro-enantiomers had equal activity. The nature of the underlying process—receptor, ion channel formation or other process—remains to be defined, but the underlying lack of chirality is of considerable interest.

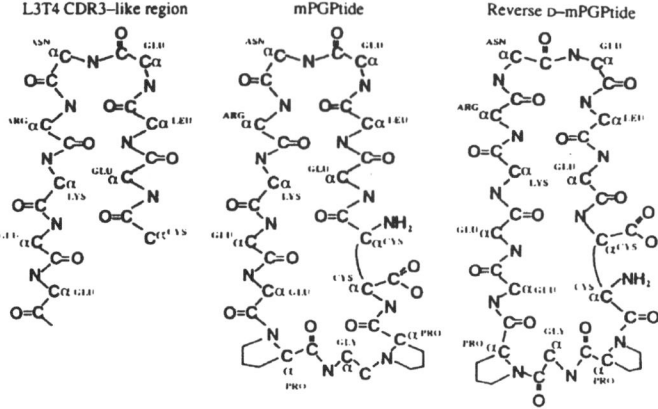

Fig. 5.2 Structures of L amino acid murine PGP (proline–glycine–proline) peptide analogue (mPGPtide) compared with the reverse D amino acid structure of the L3T4CDR3 region of mouse CD4. Reproduced with permission from Jameson *et al.* (1994).

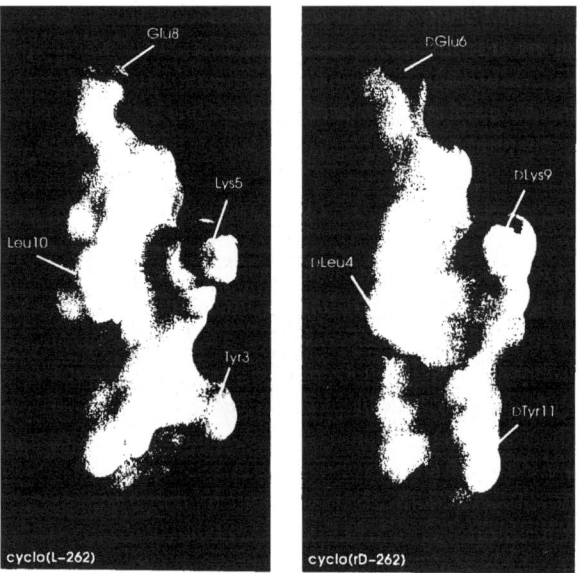

Fig. 5.3 The similarity of topochemical features presented by the retroenantiomeric cyclo(L) and cyclo(RD) peptides. Reproduced with permission from McDonnell *et al.* (1997).

The chirality of interaction is apparent when the two molecules interact, but after dissociation this handedness of recognition is not demonstrated. In appropriately designed systems, however, not only is there chirality of recognition, there is also chiral memory (Furusho *et al.* 1977). The substituted non-planar porphyrin complexes with two molecules of a chiral mandelic acid in a single diastereomeric complex (Fig. 5.4). After dissociation of the complex, however, the porphyrin retains its chiral preference, a process that can be modified by light. The existence of chiral memory, albeit in a simple model of a receptor, might be of substantial importance to drug interactions at pharmacological receptors as a chiral form of 'molecular imprinting'.

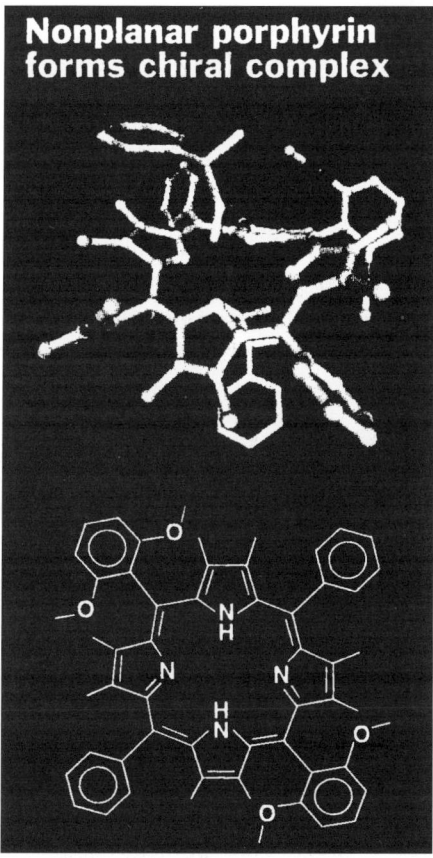

Fig. 5.4 A D2-symmetric octaalkyl-substituted porphyrin as chirality sensor and memory molecule. Reproduced with permission from Chem. Eng. News, 1997.

BIOLOGICAL CONSEQUENCES OF STEREOSELECTIVE DRUG ACTION

Enantiomers may differ both quantitatively and qualitatively in their biological activity. At one extreme one enantiomer might be totally devoid of measurable activity and at the other extreme both enantiomers may have potent, but distinct, biological activity. Thus, the following are possible:

1 Both (all) isomers are equally active and there is no stereoselectivity of action. This situation is not common.
2 The isomers differ quantitatively in their activity. Most drugs that occur as stereoisomers fall into this category. (In the limiting situation one enantiomer is devoid of measurable biological activity.)
3 The isomers differ qualitatively in their activity and have discrete biological activity at the same or different receptors. This discreteness might reflect agonist and antagonist behaviour at the same receptor, receptor subtype selectivity, or activity at distinct receptor classes.
4 The behaviour of the isomers in the racemate is different from that of the single enantiomers alone.

Examples of all of these possibilities are known, and multiple modes of behaviour might be observed for the same molecule (Smith 1983, 1989; Dalgaard 1990; Holmstedt *et al.* 1990; Wainer 1993; Casy 1993a; Aboul-Enein & Wainer 1997; Fig. 5.5). The isomers of promethazine are almost identical in their histamine H_1 antagonism and toxicity (Toldy *et al.* 1959); propranolol is some 40 times more active as the *S* enantiomer at *beta*-adrenoceptors, although the enantiomers are essentially equipotent as local anaesthetics (Barrett & Cullum 1968; Patil *et al.* 1974; Ruffolo 1991a); labetalol, an α,β-adrenoceptor antagonist with two asymmetric centres has β- and α-antagonism (Baum *et al.* 1981; Sybertz *et al.* 1981); picenadol, also with two asymmetric centres, has μ-opiate receptor agonism and antagonism (Zimmerman & Gessellchen 1982); Ca^{2+} channel agonism and antagonism is observed for the enantiomers of the 1,4-dihydropyridine Bay K 8644 (Goldman & Stoltefuss 1991), and the enantiomers of the barbiturate *N*-methyl-5-propylbarbituric acid have convulsant and anticonvulsant properties (Downes *et al.* 1970; Buch *et al.* 1973). Similarly, the *dextro* enantiomer of propoxyphene is a morphine-like analgesic, whereas the *laevo* enantiomer is a therapeutically effective anti-tussive agent (Hyneck *et al.* 1990).

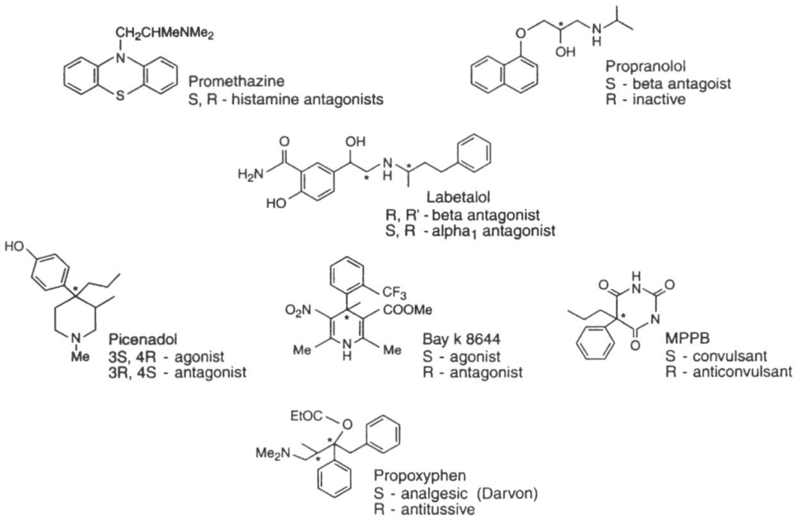

Promethazine
S, R - histamine antagonists

Propranolol
S - beta antagoist
R - inactive

Labetalol
R, R' - beta antagonist
S, R - alpha$_1$ antagonist

Picenadol
3S, 4R - agonist
3R, 4S - antagonist

Bay k 8644
S - agonist
R - antagonist

MPPB
S - convulsant
R - anticonvulsant

Propoxyphen
S - analgesic (Darvon)
R - antitussive

Fig. 5.5 Chiral drugs the enantiomers of which have different pharmacological activity.

Analysis of the chirality of natural, semi-synthetic and synthetic drugs reveals that most of the former are available as single stereoisomers. In the synthetic drug category, however, the availability of single enantiomers is rapidly increasing—in 1982, some 15% were available as single enantiomers but by 1991 this had increased to 40% (Stinson 1993, 1994, 1995). This percentage is likely to increase, an increase fuelled by scientific, economic and regulatory concerns. These chiral drugs amount to a large dollar volume of sales (Table 5.1), a volume likely to increase because of the number of drugs that are candidates for 'chiral switches' (Table 5.2).

Table 5.1 Sales of enantiopure drugs (Stinson 1994).

Category	Sales ($billion)
Cardiovascular	11.3
Antibiotics	10.8
Hormones	4.5
CNS	2.0
Anti-inflammatory	1.5
Anti-cancer	1.0
Other	4.5

Table 5.2 Candidate drugs for chiral switches (Stinson 1993).

Cardiovascular—approximately thirty including:

Acebutolol
Dobutamine
Disopyramide
Nicardipine
Verapamil

CNS—approximately twelve including:

Fluoxetine
Lorazepam
Meclizine

Respiratory—including:

Albuterol
Terbutaline

Anti-inflammatory—approximately sixteen including:

Cicloprofen
Ibuprofen
Ketoprofen

Antihistamines—including:

Terfenadine

THE BASIS OF STEREOSELECTIVITY

In principle, stereoselectivity of drug action is derived from both pharmacodynamic and pharmacokinetic processes and can arise from any and all of the processes involved in drug action, from transport to and from the site of action, storage in depots, or interaction with binding proteins, interaction with receptors, metabolic events, and terminal transport in excretory pathways (Fig. 5.6).

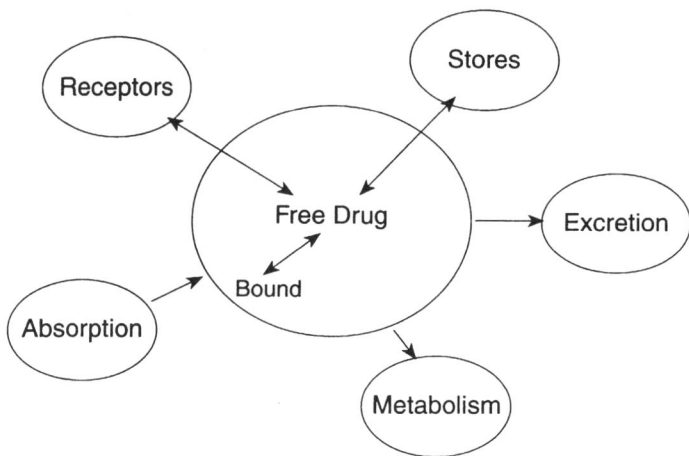

Fig. 5.6 Pathways of drug absorption, action, storage, metabolism and excretion indicating that the stereoselectivity of drug action is dependent upon both pharmacokinetic and pharmacodynamic actions.

Although pharmacodynamic aspects of drug stereoselectivity have been extensively investigated and documented, particularly by *in vitro* receptor-based measurements, to provide information about drug–receptor interactions, the increasing availability of chiral assays for drugs and their metabolites has made it increasingly possible, and necessary, to measure the stereoselectivity of the pharmacokinetic processes—absorption, protein binding, transport and excretion—and their contribution to the observed overall stereoselectivity of drug action (Williams & Lee 1985; Simonyi *et al.* 1986; Testa 1986; Jamali *et al.* 1989; Tucker & Lennard 1990; Levy & Boddy 1991; Wainer 1993). Ariëns, who argued that the non-recognition of stereochemistry was "the basis for sophisticated nonsense in pharmacokinetics and clinical pharmacology" early recognized the importance of these processes, seen predominantly under *in vivo* conditions, to the clinical definition of drug action (Ariëns 1984).

The most important pharmacokinetic parameter is unbound clearance, defined as the product of the free plasma fraction and total drug clearance and typical values for these parameters suggest that stereoselectivity of pharmacokinetic processes will be generally quite low (less than 5) and the contribution to overall stereoselectivity of drug action correspondingly low (Tucker 1991; Casy 1993b). The implications of such pharmacokinetic stereoselectivity with respect to dosage determination, route of administration, drug interactions and clinical importance *can*, however, be very important. Both pharmacodynamic *and* pharmacokinetic processes determine the overall observed clinical pattern of drug stereoselectivity.

ABSORPTION

Most drugs exploit passive diffusion processes where, because of the identical solubility properties of enantiomers, stereoselectivity is not an issue. The crystalline forms and solubilities of racemates might, however, differ from those of the single enantiomers and give rise to differences in dissolution and absorption rates. Where active transport processes are involved, however, stereoselectivity of action is expected and realized. L-DOPA, is more rapidly absorbed, by means of the amino acid transporter, than is the D enantiomer (Wade *et al.* 1973) and chiral β-lactam antibiotics exploit the intestinal dipeptide transporter with D stereoselectivity (Tamai *et al.* 1988).

The stereoselective transport of drugs can be complicated by pharmacodynamic processes. Thus, for chiral local anaesthetics stereoselectivity effects both local anaesthetic and vasoconstrictor activity: differential effects on blood flow result in differences in the rate of systemic absorption (Luduena 1967; Aps & Reynolds 1978).

DISTRIBUTION

Stereoselectivity of plasma binding of drugs is generally quite low (less than 2), reflecting dominantly binding to albumin and α_1-acid glycoprotein (Ariëns 1984; Jamali *et al.* 1989; Tucker & Lennard 1990; Casy 1993b). Propranolol binds with opposing stereoselectivity to these two proteins: the *S* enantiomer binds preferentially to human α_1-acid glycoprotein and the *R* enantiomer to human albumin: the former process dominates in human plasma (Casy 1993b).

Stereoselective tissue binding has been documented for several drugs, although the data are much scarcer than those for protein binding (Ariëns 1984; Tucker & Lennard 1990; Casy 1993b). Propranolol and other β-adrenoceptor antagonists, including atenolol, undergo selective storage in adrenergic nerve terminals with the active *S* enantiomer being selectively bound (Walle *et al.* 1988; Webb *et al.* 1988). Although the *R* enantiomers of some non-steroidal anti-inflammatory drugs are more selectively taken up by fat depots (Williams *et al.* 1986; Sallusho *et al.* 1988), this represents a metabolic process whereby selective formation of the coenzyme A thioester of the *R* enantiomer and its inversion to the *S* thioester is followed by incorporation as unnatural triglycerides (Caldwell *et al.* 1988a; Wechter 1994; Fig. 5.7).

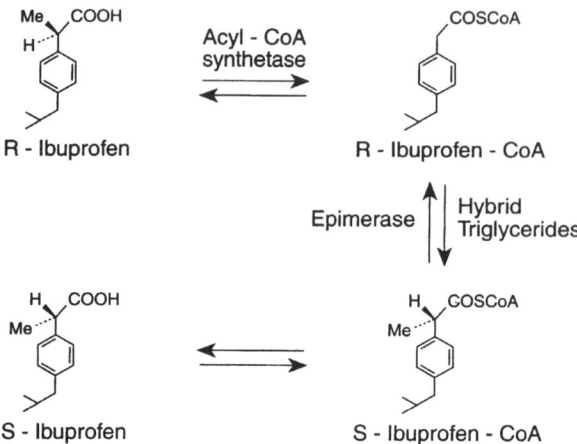

Fig. 5.7 The interconversion of *R*-ibuprofen to the *S* enantiomer by a process of chiral inversion.

METABOLISM

Enantioselective drug metabolism is observed either as differences between the metabolism of chiral drugs (substrate enantioselectivity) or as metabolic generation of chiral metabolites (product stereoselectivity) from achiral drugs (Williams & Lee 1985; Jamali *et al.* 1989; Tucker & Lennard 1990; Levy & Boddy 1991; Casy 1993b).

The principal locus of drug metabolism is the liver, and hepatic blood flow, drug binding and enzyme activity determine the clearance of drugs. For high-clearance drugs given parenterally blood flow is of particular significance whereas for low-clearance drugs given parenterally and for high- and low-clearance drugs given orally drug binding and enzyme activity are more important. Thus, the stereoselectivity of metabolism can be route-dependent *and* be modified by any effects that the drugs themselves have on hepatic blood flow.

The stereoselectivity of clearance of propranolol varies from (+)/(−) = 1.02 to 1.23, the larger ratios being for parenteral administration (Walle *et al.* 1988; Table 5.3). The reduced hepatic blood flow induced by the active *S*-(−) enantiomer after intravenous administration contributes to its lower clearance relative to that of the inactive *R* enantiomer. After oral administration clearance is now largely independent of hepatic blood flow and the lower clearance of the *S* enantiomer is determined by the enantioselectivity of plasma binding and intrinsic clearance.

The measured enantioselectivities of intrinsic clearance can represent the roles of more than one metabolic pathway. In man *S*-(−)-warfarin is 7-hydroxylated, whereas *R*-(+)-warfarin is 6-hydroxylated (Lewis *et al.* 1974). The stereoselectivity of first-pass metabolism of the Ca^{2+} channel antagonist verapamil is a particularly interesting example of intravenous and oral administration resulting in markedly different plasma concentrations of the more active (−) enantiomer (Vogelgesang *et al.* 1984). The metabolism of verapamil is modified in the presence of cimetidine, which reduces the clearance of verapamil, the S-(–) enantiomer preferentially resulting in a more pronounced pharmacodynamic effect of verapamil (Mikus *et al.* 1990).

Table 5.3 Stereoselectivity–propranolol clearance (L min^{-1}) in man (Walle *et al.* 1988).

Clearance	(+)	(–)	Ratio
CLIV	1.21	1.03	1.17
CLUIV	6.0	5.9	1.02
CLPO	2.78	1.96	1.43
CLUPO	13.7	11.1	1.23

CLIV, clearance after intravenous injection.
CLUIV, plasma clearance (based on unbound drug) after intravenous injection.
CLPO, plasma clearance (based on unbound drug) after oral administration.

Achiral drugs can undergo metabolism to form chiral products. Phenytoin in man is selectively metabolized to give S-5-(4-hydroxyphenyl)-5-phenylhydantoin (Butler *et al.* 1976) and debrisoquine is selectively 4-hydroxylated to the S enantiomer (Meese & Eichelbaum 1986). The hypolipidaemic drug fenofibrate is stereoselectively reduced with a marked preponderance of the (–) form (Weil *et al.* 1989; Fig. 5.8).

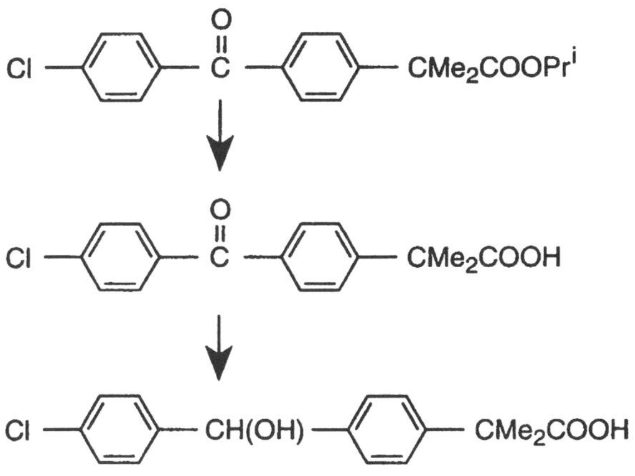

Fig. 5.8 The metabolic pathway of conversion of achiral fenofibrate to chiral metabolites.

An intriguing stereoselective pathway is that mediating the inversion and clearance of the non-steroidal anti-inflammatory 2-arylpropionic acids, including ibuprofen, benoxaprofen, cicliprofen, and thioxaprofen (Caldwell *et al.* 1988a, b; Jamali 1988; Wechter 1994). The differences between the potency of the enantiomers is much greater *in vitro* than *in vivo*. Thus, for ibuprofen the *S* enantiomer is almost 200 times more potent than the *R* enantiomer as an inhibitor of prostaglandin synthase, whereas the *S* enantiomer is less than twice as effective in pain relief (Adams *et al.* 1976). This difference arises from the selective activation of the *R* enantiomer to the coenzyme A thioester, which is then racemized by ibuprofen CoA racemase to release *R*- and *S*-ibuprofen (Fig. 5.7).

EXCRETION

Renal drug clearance is the result of several processes including glomerular filtration, active secretion, and active and passive re-absorption. Modest stereoselectivity of excretion has been found for several drugs including terbutaline, pindolol, metoprolol, chloroquine, and other amines (Jamali *et al.* 1989; Tucker & Lennard 1990). The stereoselectivity ratio did not exceed 2. The factoring of this ratio to renal metabolism, protein binding, or transport remains to be determined.

STEREOSELECTIVITY OF DRUG CLASSES

The general pharmacodynamic and pharmacokinetic principles defining drug stereoselectivity can be usefully illustrated with reference to specific drug classes—including G protein-coupled receptors, ion channels and general anaesthetics.

G PROTEIN-COUPLED RECEPTORS: β-ADRENOCEPTOR ANTAGONISTS

The stereochemistry of drug action at adrenoceptors has been well investigated and comprehensive reviews are available (Patil *et al.* 1974; Ariëns *et al.* 1983; Dalgaard 1990; Ruffolo 1991a; Casy 1993a). Several instances of racemate/enantiomer mechanisms and actions have been documented.

The availability of the sequences of many G protein-coupled receptors has enabled the construction of models of the transmembrane organization of these 7-domain α-helical proteins with the objective of defining drug binding sites and receptor coupling and transduction processes (Strader *et al.* 1994).

The chirality of agonist recognition is defined around the β-OH group of catecholamines, including the selective agonist isoproterenol. With an arrangement of helices in a circular and anti-clockwise manner current models place transmembrane domains III and V as interacting with the amine and catechol functions respectively of the agonist catecholamines. Asn-293 of transmembrane helix VI might be important in defining stereoselectivity, because its mutational replacement in the human $β_2$-receptor by leucine or alanine led to loss of agonist affinity and stereoselective recognition (Weiland *et al.* 1996). This was principally because of loss of affinity of the more active enantiomer and agonists that lack the β-OH group, including dobutamine, showed no loss, nor did *beta*-antagonists (Table 5.4).

Table 5.4 Stereoselectivity of agonists and antagonists at human β_2-adrenoceptors (Strader *et al.* 1994).

Agonist	K_I, 10^{-6} M	
	Wild	Mutant
(−)-Epinephrine	0.17	6.2
(+)-Epinephrine	2.1	3.1
Stereoselectivity	13	5.0
(−)-Norepinephrine	3.9	43
(+)-Norepinephrine	174	800
Stereoselectivity	49	18
(+)-Dobutamine	66	63

Antagonist	Stereoselectivity	
	Wild	Mutant
Propranolol	104	112
Alprenolol	22	19
Metoprolol	21	25

There is excellent correlation between the intrinsic activity of agonists and the loss of affinity in the 293-mutant (Fig. 5.9). This correlation suggests that the correct orientation of the β-OH group is linked both to stereoselective recognition *and* receptor recognition processes.

Most β-adrenoceptor antagonists are available as racemates and they are stereoselective in both pharmacokinetic and pharmacodynamic processes. Propranolol, metoprolol, bufarolol, penbutolol, and other analogues all undergo stereoselective oral clearance in which the less active enantiomer is cleared more rapidly (reviewed by Casy (1993)). In contrast with propranolol and the majority of other β-antagonists, the non-selective agent timolol is marketed as the enantiomeric *S*-(−) isomer. Timolol is used in the treatment of glaucoma where it blocks β_2-adrenoceptors that control the formation of aqueous humour The enantiomeric *R* form, however, although significantly less active as a β-blocker is only slightly less active than the *S* enantiomer in the reduction of intraocular pressure and has fewer cardiac and respiratory side effects (Keats & Stone 1984; Richards & Tattersfield 1985).

Labetalol, with two asymmetric centres (Fig. 5.5), is a mixed α,β-adrenoceptor antagonist with pharmacological activity unevenly distributed across the enantiomers (Table 5.5) (Baum *et al.* 1981; Sybertz *et al.* 1981).

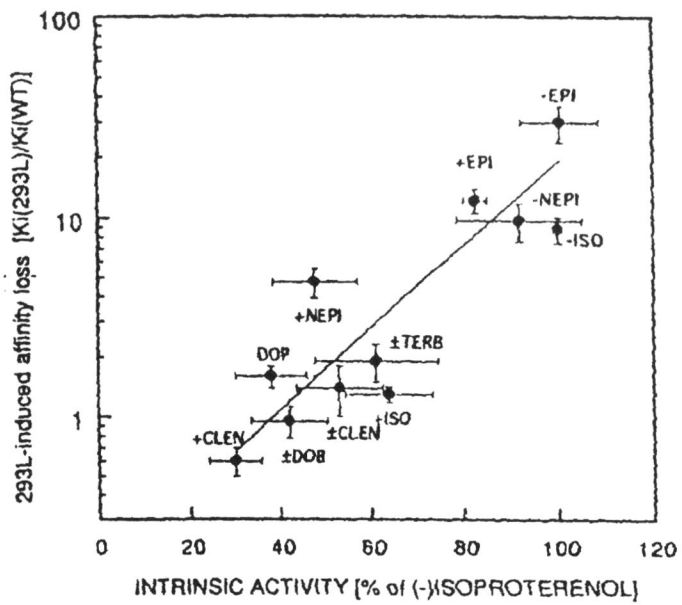

Fig. 5.9 Comparison of the intrinsic activities of a series of catecholamines and the loss of affinity induced in the human β_2-adrenoceptor by the 293-leucine mutation. Reproduced with permission from Wieland *et al.* (1996).

Table 5.5 Relative α- and β-adrenoceptor antagonist potencies of labetalol isomers.

Absolute Configuration	Relative Potency		
	α_1	β_1	β_2
Labetalol 1R,4R;1S,4S;1R,4S;1S,4R	1	1	1
1R,4R	0.15	2.27*	2.18*
1S,4S	0.39	0.03	<0.02
1R,4S	0.23	0.15	0.09
1S,4R	1.74**	0.04	0.02

*Predominantly β_1 and β_2
**Predominantly α_1

The 1R,4R enantiomer is the dominant β-antagonist and the 1S,4R enantiomer is the dominant α-antagonist. Although the 1R,4R enantiomer of labetalol was marketed briefly as dilevalol it was withdrawn from clinical use because of hepatotoxicity (Shah 1993). Apparently, racemic labetalol has an enantiomeric interaction that minimizes hepatotoxicity. In contrast with the differential distribution of α- and β-adrenoceptor antagonism across the four enantiomers of labetalol, both enantiomers of the antihypertensive agent carvedilol, with one asymmetric centre, have α-antagonist behaviour, but potent $β_1$-antagonism is observed for the S-(−) enantiomer only (Nichols et al. 1989; Ruffolo 1991b). Thus, the profile of antihypertensive action seen with racemic carvedilol, β-blockade and peripheral vasodilatation, would not occur in either enantiomer alone.

Sotalol is a non-selective β-adrenoceptor antagonist that has class III antiarrhythmic properties. Sotalol has stereoselective action at β-adrenoceptors, but is a non-stereoselective anti-arrhythmic agent (Gomoll & Bartek 1980; Somani & Watson 1991; Advanc & Singh 1995). Given the essential absence of β-adrenoceptor activity in d-sotalol this agent can be considered almost a pure Class III anti-arrhythmic agent at myocardial IK_R^+ channels (Roden & George 1996). Although the SWORD trial of d-sotalol was negative (Cobbe 1996; Colasky 1995), racemic sotalol is an excellent example of enantiomeric differentiation of discrete pharmacological effects.

ION-CHANNEL RECEPTORS: VOLTAGE-GATED Na^+ AND Ca^{2+} CHANNELS

Voltage-gated ion channels are a major class of pharmacological receptors richly endowed with drug binding sites that mediate local anaesthetic, hypotensive, anti-arrhythmic and other activity (Hille 1992; Godfraind 1994; Roden & George 1994). These ion channels are also a principal locus of offensive and defensive chemical strategies employed by both invertebrates and vertebrates (Ezzel 1995; Gilmore et al. 1995; Narahashi 1996). The interpretation of structure–function relationships, including stereoselectivity, in drugs acting at ion channels presents specific difficulties because drug access and affinity are dependent on the state of the channel as determined by both physiological and pathological conditions (Hille 1977; Hondeghem & Katzung 1984; Kwon & Triggle 1991).

Ion channels exist in states that are defined by membrane and chemical potential (Hille 1977; Hondeghem & Katzung 1984; Hille 1992; Fig. 5.10).

These states represent different conformations of the channel proteins and there might be significant differences between the conformations of the open and closed states. Selective access or interactions of drugs with specific channel states mean that:
1 Different states have different affinities for the same drug.
2 Structure–activity relationships might be quantitatively and qualitatively different for different states.

There are many well documented examples of such state-dependent interactions including local anaesthetic and anti-arrhythmic actions at Na^+ channels (Hille 1977; Hondeghem & Katzung 1984; Kwon & Triggle 1991), and the different cardiovascular profiles of the various structurally distinct Ca^{2+} channel antagonists (Hondeghem & Katzung 1984; Triggle 1989, 1990; Kwon & Triggle 1991). For lidocaine there is an approximately 40-fold affinity difference between the resting and open states of the Na$^+$ channel (Bean *et al.* 1983); for nitrendipine there is an almost 1000-fold affinity difference between the resting and inactivated states of the L-type Ca^{2+} channel (Sanguinetti & Kass 1984); verapamil has increasing affinity for this channel as the frequency of heart rate increases (Eichelbaum & Gross 1996; Ehara & Kaufmann 1978).

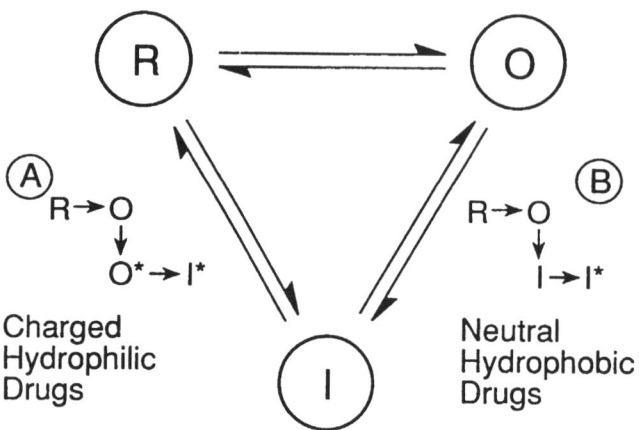

Fig. 5.10 The equilibrium between the open, resting and inactivated states of an ion channel. Drugs can interact selectively with one or other of these channel states and the apparent (measured) affinity of the drug will then depend upon the equilibrium between these states. Drugs might access states selectively by different pathways, as is indicated by hydrophilic and hydrophobic drugs that selectively stabilize the inactivated state of the channel but do so via different pathways.

These state-dependent interactions define the Class I anti-arrhythmic properties of lidocaine, the selective vasodilating capacity of nitrendipine and the Class IV anti-arrhythmic properties of verapamil.

The stereoselectivity of drug interaction at ion channels might be determined by these state-dependent interactions. The stereoselectivity of action of the chiral local anaesthetics RAC-109 and its quaternary derivative RAC-421 (Fig. 5.11) increases with decreasing membrane potential and increasing availability of the inactivated state of the Na$^+$ channel (Table 5.6) (Yeh 1980; Ragsdale *et al.* 1994).

Fig. 5.11 The chiral local anaesthetics RAC-109 and RAC-421.

This stereoselectivity of local anaesthetic interaction is also dependent upon stimulus mode—depolarization or batrachotoxin activation (Lee-Son *et al.* 1992). Advances in molecular biology have enabled determination of the sequences around drug-binding sites at ion channels. The local anaesthetic site at the Na^+ channel is located at segment S6 of domain IV of the α-subunit (Ragsdale *et al.* 1994). At this site specific amino acid residues contribute to the drug-binding site; phenylalanine 1764 is particularly critical because its replacement by alanine reduces local anaesthetic affinity by a factor of approximately 100 and abolishes use- and frequency-dependence (Ragsdale *et al.* 1994). This region seems to be a common molecular determinant for anti-arrhythmic, anti-convulsant and local anaesthetic interactions (Ragsdale *et al.* 1996).

Table 5.6 Stereoselectivity (EC_{50}, mM) of local anaesthetic action at Na^+ channels of squid axon (Yeh 1980).

	Conditioned block at:		
	Resting state	0 mV	+80 mV
(−)-RAC 109	1.40	0.14	0.034
(+)-RAC 109	1.45	0.79	0.49
(−)-RAC 421	1.99	0.10	0.042
(+)-RAC 421	2.35	0.98	0.42

This region also contains the binding site for the L-type Ca^{2+} channel antagonist verapamil, which has stereoselectivity of interaction and prominent use- and frequency-dependence (Schuster *et al.* 1960; Hockerman *et al.* 1995). Verapamil is marketed as the racemate and has clinical uses as an anti-anginal, anti-hypertensive and anti-arrhythmic agent (Triggle 1990, Eichelbaum & Gross 1996). The S-(−) enantiomer is more potent than the R-(+) isomer in both cardiac and vascular preparations, although the stereoselectivity is higher in cardiac preparations (van Amsterdam & Zaagsma 1988; Kwon & Triggle 1991; Eichelbaum & Gross 1996; Table 5.7).

Thus, *S*-verapamil has both vasodilating and cardiac depressant properties, whereas the *R* isomer is principally a vasodilating agent. The stereoselectivity of verapamil arises from both pharmacokinetic and pharmacodynamic actions that together determine its therapeutic profile. Verapamil is less potent after oral administration than after a single intravenous administration. This difference, of obvious importance in the use of verapamil in the control of supra-ventricular tachycardias, is a result of stereoselective first pass metabolism (Eichelbaum *et al.* 1984; Vogelgesang *et al.* 1984; Echizen *et al.* 1985). Plasma clearance of the *S* isomer is approximately twice that of the *R* enantiomer and the bioavailability of the pharmacodynamically more active *S* enantiomer is correspondingly lower.

Table 5.7 Cardiovascular activity of verapamil and gallopamil in Langendorff-perfused rat heart (van Amsterdam & Zaagsma 1988).

	Coronary Flow		Maximum systolic left ventricular pressure (MSLVP)		
	pEC_{50}	Ratio R/S	Ratio Flow/MSLVP	pEC_{50}	Ratio R/S
R-(+)-Verapamil	7.31		43.6		5.67
S-(−)-Verapamil	7.20		3.4		7.17
		2.45		31.6	
R-(+)-Gallopamil	7.12		15.8		5.92
S-(−)-Gallopamil	7.64		0.93		7.67
		3.30		56.2	

Although verapamil has a common stereoselective pharmacodynamic interaction at the Ca^{2+} channels of both cardiac and vascular tissues ($S > R$), the ratio differs significantly between tissues (Table 5.7). This difference is probably a result of several factors. Differences between state-dependent interactions arise from different equilibria between states in the cardiac and vascular preparations—a greater fraction of channels in the inactivated state in vascular tissue. In addition, the pharmacological behaviour of different L-channel subtypes, including stereoselectivity, is probably different (Welling et al. 1993).

Of particular interest is that the experimental cardioprotective effects of the enantiomers of the verapamil analogue gallopamil accord with cardiac and vascular stereoselectivity during the ischaemic and reperfusion phases respectively in the Langendorff preparation (van Amsterdam & Zaagsma 1988; Table 5.8). These data suggest a possible enantiomer-specific therapeutic strategy in cardioprotection: during the reperfusion phase a vascular-selective enantiomer is appropriate and during the ischaemic phase a negatively inotropic enantiomer is appropriate.

Table 5.8 Cardiovascular stereoselectivity of gallopamil enantiomers in normoxic and ischaemic Langendorff perfused heart (Zheng et al. 1992).

	LVP	Ratio	Flow	Ratio	tEDPm	Ratio	EDPm	Ratio	tR90	Ratio
S-(−)-Gallopamil	7.4		7.8		7.8		7.6		8.9	
R-(+)-Gallopamil	5.6		6.7		6.1		6.0		8.0	
		63		12.6		50		40		8

LVP, left ventricular pressure; tEDPm, time to maximum end-diastolic pressure after the onset of ischaemia; EDPm, maximum end-diastolic pressure reached during the ischaemic period; tR90, time from the start of reperfusion to 90% recovery from the maximum post-ischaemic end-diastolic pressure.

The 1,4-dihydropyridines are therapeutic and experimental cardiovascular agents that interact at discrete receptors as activators and antagonists of the L-type Ca^{2+} channel. Substantial stereoselectivity of interaction is observed for both activator and antagonist 1,4-dihydropyridines. For some species stereoselectivity is discrete for activator and antagonist properties (Fig. 5.12)—the 1,4-dihydropyridine Bay K 8644 has S and R stereoselectivity for activator and antagonist properties, respectively (Triggle et al. 1989; Goldman & Stoltefuss 1991; Zheng et al. 1992).

These opposing interactions are believed to be mediated through a single 1,4-dihydropyridine binding site (Peterson & Catterall 1995). Of absorbing interest is the observation that a single 1,4-dihydropyridine enantiomer can 'switch' activity from activator to antagonist according to membrane potential (Kass 1987; Fig. 5.13). This property might be related to the weak state-dependence of action of the 1,4-dihydropyridine activators whereby they can bind to the open and inactivated channel states with little discrimination.

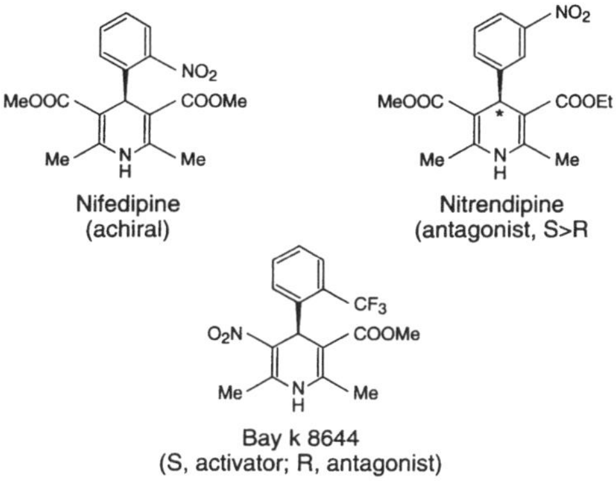

Fig. 5.12 Achiral and chiral 1,4-dihydropyridines active at L-type Ca^{2+} channels.

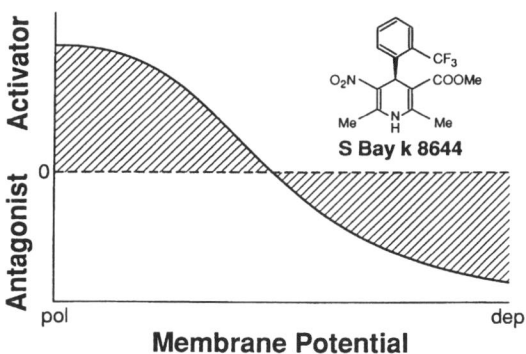

Fig. 5.13 The S enantiomer of Bay K-8644 behaves as an activator at polarized levels of membrane potential and as an antagonist at depolarized membrane potentials. The null point is that level of membrane potential at which no detectable channel activity is measured.

GENERAL ANAESTHETIC RECEPTORS: GENERAL ANAESTHETICS

There has long been an assumption that general anaesthesia is a non-specific process driven by anaesthetic partitioning into lipid membranes to exert a general depressant effect on cellular excitability. The close correlation of general anaesthetic potency with oil–water partition coefficients has been a principal component of this evidence (Miller 1985); this process would not be expected to be significantly stereoselective. The demonstration of the close correlation of anaesthetic potency with the effectiveness of inhibition of the firefly enzyme luciferase has, however, stimulated consideration of protein sites of interaction for general anaesthetics (Frantz & Lieb 1994).

Stereoselectivity of general anaesthesia has now been demonstrated for the volatile agents isoflurane and halothane (Harris *et al.* 1992; Hall *et al.* 1994; Fig. 5.14). Thus S-(+)-isoflurane is ca 50% more active than the R enantiomer in sleep-inducement in mice. Similarly small differences in enantiomeric potency have been observed in genetic mutants of *Caenorhabditis elegans* (Sedensky *et al.* 1994). Such differences, unremarkable in other drug interactions, might, however, be of clinical significance and expand the safety margin of general anaesthesia (Moody *et al.* 1994).

$$F_3C\overset{*}{C}HBrCl \qquad F_3C\overset{*}{C}HClOCF_2H$$
Halothane   Isoflurane

Fig. 5.14 Chiral general anaesthetics.

REGULATORY REQUIREMENTS FOR CHIRAL DRUGS

Currently both racemic and chiral drugs are marketed. Irrespective of stereochemistry drugs enjoy great clinical and commercial success. Diltiazem is marketed as a single enantiomer whereas fluoxetine is a racemic mixture; both are billion-dollar drugs. Although there are, currently, no restrictions on the development and marketing of racemic drugs, it is likely that there will be an increasing trend toward the development of chiral drugs. This will, however, have to be an individual decision— the regulatory guidelines are still evolving (Laganiere 1997).

Current regulatory requirements have been documented (FDA 1992; Gross *et al.* 1993; Heydorn 1995). General requirements outlined by the Food and Drug Administration include the following statements:

> Despite the problems identified with some racemates, the common practice of developing racemates has resulted in few recognized adverse consequences. Although it is now technologically feasible to prepare purified enantiomers, development of racemates may continue to be appropriate. However, currently available information suggests that the following should be considered in product development:
> 1 Appropriate manufacturing and control procedures should be used to assure stereoisomeric composition of a product with respect to identity, strength

quality and purity. Manufacturers should notify compendia of these specifications and tests.

2   Pharmacokinetic evaluations that do not use a chiral assay will be misleading if the disposition of the enantiomers is different. Therefore, techniques to quantify individual stereoisomers in pharmacokinetic samples should be available early. If the pharmacokinetics of the enantiomers are demonstrated to be the same or to exist as a fixed ratio in the target population, an achiral assay or an assay that monitors one of the enantiomers may be used, subsequently.

Pharmacological activities of the individual enantiomers should be characterized for the principal pharmacologic effect and any other important pharmacological effect, with respect to potency, specificity, maximum effect etc.

To monitor *in vivo* interconversion and disposition, the pharmacokinetic profile of each isomer should be characterized in animals and later compared to the clinical pharmacokinetic profile obtained in phase I.

It is ordinarily sufficient to carry out toxicity studies on the racemate. If toxicity other than that predicted from the pharmacologic properties of the drug occurs at relatively low multiples of the exposure planned for clinical trials, the toxicity study where the unexpected toxicity occurred should be repeated with the individual isomers to ascertain whether only one enantiomer was responsible for the toxicity.

Where little difference is observed in activity and disposition of the enantiomers, racemates may be developed.

In some situations, development of a single enantiomer is particularly desirable (e.g., where one enantiomer has a toxic or undesirable pharmacologic effect and the other does not).

In general, it is more important to evaluate both enantiomers clinically and consider developing only one when both enantiomers are pharmacologically active, but differ significantly in potency, specificity or maximum effect, than when one isomer is essentially inert. Where both enantiomers are fortuitously found to carry desirable, but different, properties development of a mixture of the two, not necessarily the racemate, as a fixed combination might be reasonable.

Because of the number of drugs that are candidates for 'chiral switching', the regulatory guidelines for developing a single enantiomer from a racemate are of particular importance. In principle, this development will require a completely new application. In practice the requirements will probably be defined, at least in the interim, on a case-by-case basis; 'bridging studies' that enable the use of existing racemate data are likely to be involved. The FDA document notes that:

> To develop a single stereoisomer from a mixture that has already been studied non-clinically, an abbreviated, appropriate pharmacology/toxicology evaluation could be conducted to allow the existing knowledge of the racemate available to the sponsor to be applied to the pure stereoisomer.

Although the current requirements for the development of chiral drugs are very similar in the United States, Canada, Europe and Japan they are still undergoing evolution; further harmonization will be required, particularly on toxicological issues and the question of bridging studies (Laganiere 1997).

DECISION PROCESSES

The available scientific and clinical data on racemate and single enantiomer drugs does not enable simplistic conclusions. It *cannot* be argued that all racemates are satisfactory or that all single enantiomers are better, because there are simply too many well-established exceptions to either generalization. Some instances where either a single enantiomer or a racemate is preferable have already been noted. Several additional examples are worthy of note (Aboul-Enein & Abou-Bashia 1997; Baillie & Schultz 1997).

S-(+)-ketamine is an anaesthetic and analgesic; side effects of psychosis and hallucination are associated with the R-(+) enantiomer (White *et al.* 1980). The S,S enantiomer of the antituberculosis drug ethambutol has the desired activity but is associated with the production of serious, and possibly irreversible, eye damage (Blessington 1997). The R-(+) enantiomer of indancrinone is the active diuretic and also promotes uric acid retention whereas the S-(−) enantiomer promotes uric acid secretion and the racemate (or other fixed combination) is more desirable than either enantiomer alone (Tobert *et al.* 1981). Terodiline, marketed as the racemate as a drug for incontinence was withdrawn because it caused cardiac arrhythmias in a small number of patients; this effect is associated principally with the R-(+) enantiomer (Hartigan-Go *et al.* 1996).

Accordingly, Testa has formalized the process of racemate/enantiomer choice as a decision tree (Fig. 5.15) in which a set of questions is posed and the answers used as a guide to subsequent steps (Testa & Trager 1990). These answers might well not be binary 'yes/no', but rather encoded by 'fuzzy logic'. It remains true that a drug with a chiral centre is, by definition, a mixture of two entities, but in the absence of further and specific information this statement remains one of fact rather than one of value (Daniels *et al.* 1997).

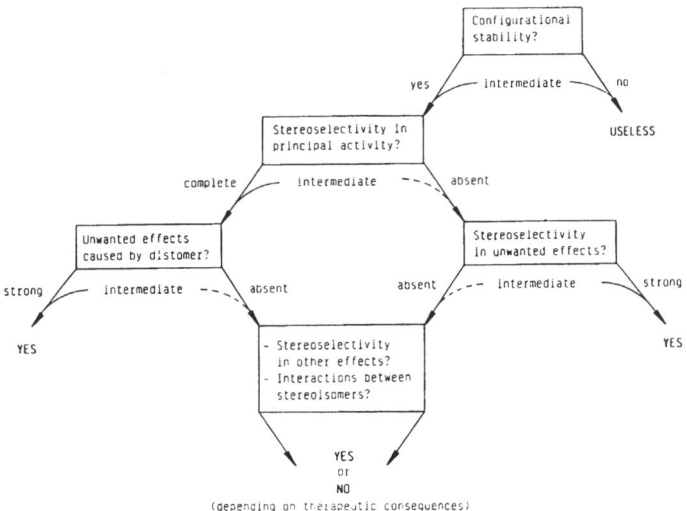

Fig. 5.15 A logical decision scheme that provides a basis for the question of developing a racemate or a single enantiomer. Reproduced with permission from Testa & Trager (1990).

ACKNOWLEDGEMENTS

Parts of this chapter draw upon my previous review 'Stereoselectivity of Drug Action', published by Elsevier Press in Drug Discovery Today (Volume 2: 138–147, 1997).

REFERENCES

Aboul-Enein, H. Y. & Abou-Bashia, L. I. (1997) Chirality and drug hazards. In: *The Impact of Stereochemistry on Drug Development and Use* (Ed. by H. Y. Aboul-Enein & I. W. Wainer), pp. 1–19. Wiley–Interscience, New York.

Aboul-Enein, H. Y. & Wainer, I. W. (Eds.) (1997) *The Impact of Stereochemistry on Drug Development and Use.* Wiley–Interscience, New York.

Adams, S. S., Bresloff, P. & Mason, C. G. (1976) Pharmacological differences between the optical isomers of ibuprofen: evidence for metabolic inversion of the (–)-isomer. *J. Pharm. Pharmacol.* **28**, 256–257.

Advanc, S. V. & Singh, B. N. (1995) Pharmacodynamic, pharmacokinetic and antiarrhythmic properties of d-sotalol, the *dextro*-isomer of sotalol. *Drugs*, **49**, 664–679.

Aps, C. & Reynolds, F. (1978) An intradermal study of the local anaesthetic and vascular effects of the isomers of bupivacane. *Brit. J. Clin. Pharmacol.*, **6**, 63–68.

Ariëns, E. J. (1984) Stereochemistry, a basis for sophisticated nonsense in pharmacokinetics and clinical pharmacology. *Eur. J. Clin. Pharmacol.*, **26**, 663–668.

Ariëns, E. J., Soundijn, W. & Timmermans, P. B. M. W. M. (Eds) (1983) *Stereochemistry and Biological Activity of Drugs.* Blackwell Scientific Publications, Oxford.

Baillie, T. A. & Schultz, K. M. (1997) Stereochemistry in the drug development process: role of chirality as a determinant of drug action, metabolism, and toxicity. In: *The Impact of Stereochemistry on Drug Development and Use* (Ed. by H. Y. Aboul-Enein and I. W. Wainer, pp. 21–44. Wiley–Interscience,.

Barrett, A. M. & Cullum, V. A. (1968) The biological properties of the isomers of propranolol and their effects on cardiac arrhythmias. *Br. J. Pharmacol.*, **34**, 43–55.

Baum, T., Watkins, R. W., Sybertz, E. J., Vermulapalli, S., Pula, K. K., Eynon, E. Nelson, S., van de Vliet, G., Glennon, J. & Morse, R. M. (1981) Anti-hypertensive and haemodynamic actions of SCH 19927, the *RR* isomer and labetalol. *J. Pharmacol. Exp. Ther.*, **218**, 444–452.

Bean, B. P., Cohen, C. J. & Tsien, R. W. (1983) Lidocaine block of cardiac sodium channels. *J. Gen. Physiol.*, **81**, 613–642.

Blessington, B. (1997) Ethambutol and tuberculosis, a neglected and chiral puzzle. In: *The Impact of Stereochemistry on Drug Development* (Ed. by H. Y. Aboul-Enein and I. W. Wainer, pp. 235–261. Wiley–Interscience, New York.

Broccardo, M., Erspamer, V., Falconieri-Erspamer, G., Improta, G., Linari, G., Melchiorri, P., Montecucchi, P. C. (1981) Pharmacological data on dermorphins, a new class of potent opioid peptides from amphibian skin. *Br. J. Pharmacol.*, **73**, 625–631.

Buch, H. P., Schneider-Affeld, F. & Rummel, A. (1973) Stereochemical dependence of pharmacological activity in a series of optically active *N*-methylated barbiturates. *Naunyn-Schmied. Arch. Pharmacol.*, **277**, 191–198.

Butler, T. C., Dudley, K. H., Johnson, D. & Roberts, S. B. (1976) Studies on the metabolism of 5,5-diphenylhydantoin relating principally to the stereoselectivity of the hydroxylation reactions in man and dog. *J. Pharmacol. Exp. Ther.*, **119**, 82–92.

Caldwell, J., Hutt, A. J. & Fasnel-Gigleux, Y. (1988a) The metabolic chiral inversion and dispositional enantioselectivity of the 2-aryl propionic acids and their biological consequences. *Biochem. Pharmacol.*, **37**, 105–114.

Caldwell, J. C., Winter, S. M., Hutt, A. J. (1988b) The pharmacological and toxicological significance of the stereochemistry of drug disposition. *Xenobiotica*, **18**, 59–70.

Casy, A. F. (1993a) *The Steric Factor in Medicinal Chemistry. Dissymetric Probes of Pharmacological Receptors.* Plenum Press, New York.

Casy, A. F. (1993b) Pharmacokinetics. Chapter 3 in: *Factors in Medicinal Chemistry*, pp. 49–71. Plenum Press, New York.

Chorev, M. & Goodman, M. (1995) Recent developments in retro peptides and proteins—an ongoing topochemical exploration. *Trends Biotech.*, **13**, 438–445.

Cobbe, S. M. (1996) Class III antiarrhythmics put to the sword? *Heart*, **75**, 111–113.

Colatsky, T. J. (1995) Antiarrhythmic drugs. Where are we going? *Pharm. Res.*, **2**, 17–23.

Cronin, J. R. & Pizzarello, S. (1997) Enantiomeric excesses in meteoritic amino acids. *Science*, **275**, 951.

Cushny, A. R. (1926) *Biological Relation of Optically Isomeric Substances.* Balliere, Tindall and Cox, London.

Dalgaard, L. (Ed.) (1990) Racemates and Enantiomers in Drug Research and Development. *Acta Pharm. Nord.*, **2**, 129–230.

Daniels J. M., Nestmann, E. R. & Kerr, A. (1997) Development of stereoisomeric (chiral) drugs: a brief review of scientific and regulatory considerations. *Drug Inf. J.*, **31**, 639–636.

deL Milton, R. C., Milton, S. C. F. & Kent, S. B. H. (1992) Total chemical synthesis of a D-enzyme: the enantiomers of HIV-1 protease show demonstration of reciprocal chiral substrate specificity. *Science*, **256**, 1445–1448.

Downes, H., Perry, R. S., Ostlund, R. E., Karler, R. (1970) A study of the excitatory effects of barbiturates. *J. Pharmacol. Exp. Ther.*, **175**, 692–699.

Echizen, H., Vogelgesang, B. & Eichelbaum, M. (1985) Effects of d,l-verapamil on atrioventricular conduction in relation to its stereoselective first pass metabolism. *Clin. Pharmacol. Ther.*, **36**, 71–76.

Ehara, T. & Kaufmann, R. (1978) The voltage- and time-dependent effects of (–)-verapamil on the slow inward current in isolated rat ventricular myocardium. *J. Pharmacol. Exp. Ther.*, **207**, 49–55.

Eichelbaum, M. & Gross, A. S. (1996) Stereochemical aspects of drug action and disposition. *Adv. Drug Res.*, **28**, 2–64.

Eichelbaum, M., Mikus, G. & Vogelgesang, B. (1984) Pharmacokinetics of (+)-, (–)- and (±)-verapamil after intervenous administration. *Brit. J. Clin. Pharmacol.*, **17**, 453–458.

Enquist, M. & Arak, A. (1994) Symmetry, beauty and evolution. *Nature*, **372**, 169–172.

Ezzel, C. (1995) Snail toxins yield fast-paced advances in drug research. *J. NIH Res.*, **7**, 30–32.

FDA (1992) FDA's policy statement for the development of new stereoisomeric drugs. *Chirality*, **4**, 338–340.

Frantz, N. P. & Lieb, W. R. (1994) Molecular and cellular mechanisms of general anaesthesia. *Nature*, **36**, 607–614.

Furusho, Y., Kimura, T., Mizuno, Y. & Ada, T. (1977) Chirality-memory molecule: a D2-symmetric fully substituted porphyrin as a conceptually new chirality sensor. *J. Am. Chem. Soc.*, **119**, 5267–5268.

Gilmore, J., Dell, C., Bowman, D. & Lodge, D. (1995) Neuronal calcium channels. *Ann. Rep. Med. Chem.*, **30**, 51–60.

Godfraind, T. (1994) Calcium antagonists and vasodilation. *Pharmacol. Ther.*, **64**, 37–75.

Goldman, S. & Stoltefuss, J. (1991) 1,4-Dihydropyridines: effects of chirality and conformation on the calcium antagonist and calcium agonist activities. *Angew. Chem. Int. Ed. Engl.*, **30**, 1555–1578.

Gomoll, A. W. & Bartek, M. J. (1980) Comparative β-blocking activities and electrophysiological actions of racemic sotalol and its optical isomers in anaesthetized drugs. *Eur. J. Pharmacol.*, **122**, 123–135.

Gross, M., Cartwright, A., Campbell, B. *et al.* (1993) Regulatory requirements for chiral drugs. *Drug Inf. J.*, **27**, 453–457.

Guichard, G., Benkirane, N., Zeder-Lutz, G., van Regenmortel, M. H., Briand, J. P., Muller, S. (1994) Antigenic mimicry of natural L-peptides with retro-inverse peptidomimetics. *Proc. Natl. Acad. Sci. USA*, **91**, 9765–9769.

Hall, A. C., Lieb, W. R. & Franks, N. P. (1994) Stereoselective and nonstereoselective actions of isoflurane at the $GABA_A$ receptor. *Br. J. Pharmacol.*, **112**, 906–910.

Harris, B., Moody, E. & Skolnick, P. (1992) Isoflurane anaesthesia is stereoselective. *Eur. J. Pharmacol.*, **217**, 215–216.

Hartigan-Go, K., Bateman, D. N., Daly, A. K. & Thomas, S. H. L. (1996) Stereoselective cardiotoxic effects of terodiline. *Clin. Pharmacol. Ther.*, **60**, 89–98.

Heck, S. D., Siok, C. J., Krapch, K. J., Kelbaaugh, P. R., Thadeio, P. F., Welch, M. J., Williams, R. D., Ganong, A. H., Kelly, M. E., Lanzetti, A. J., Gray, W. R., Phillips, D., Parks, T. N., Jackson, H., Ahlijanian, M. K., Saccomano, N. A., Volkmann, R. A. (1994) Functional consequences of post-translational isomerization of Ser46 in a calcium channel toxin. *Science*, **266**, 1005–1068.

Heydorn, W. E. (1995) Developing racemic mixtures vs. single isomers in the US. *Pharm. News*, **2**, 19–21.

Hille, B. (1977) Local anaesthetics: hydrophilic and hydrophobic pathways for the drug–receptor reaction. *J. Gen. Physiol.*, **69**, 497–515.

Hille, B. (1992) *Ion Channels*, 2nd edn. Sinauer Associates, Sunderland, MA.

Hockerman, G. H., Johnson, B. D., Scheuer, T. & Catterall, W. A. (1995) Molecular determinants of high-affinity phenylalkylamine block of L-type calcium channels. *J. Biol. Chem.*, **270**, 22119–22122.

Holmstedt, B. (1990) The use of enantiomers in biological studies: an historical review. In: *Chirality and Biological Activity* (Ed. by B. Holmstedt, H. Frank & B. Testa), pp. 1–14. Alan R. Liss, New York.

Holmstedt, B., Frank, H. & Testa, B. (Eds) (1990) *Chirality and Biological Activity*. Alan R. Liss, New York.

Hondeghem, L. M. & Katzung, B. C. (1984) Anti-arrhythmic agents: the modulated receptor mechanism of sodium and calcium channel-blocking drugs. *Ann. Rev. Pharmacol.*, **24**, 387–423.

Hyneck, M., Deut, J. & Hook, J. B. (1990) In: *Chirality in Drug Design and Synthesis* (Ed. by C. Brown), pp. 1–27. Academic Press, London and New York.

Ido, Y., Vindigni, A., Chang, K., Stramm, L., Chance, R., Heath, W. F., DiMarchi, R. D., Di Cera, E., Williamson, J. R., Schumacher, T. N., Mayr, L. M., Minor, D. L., Milhollen, M. A., Burgess, M. W., Kim, P. S. (1997) Prevention of vascular and neural dysfunction in diabetic rats by C-peptide. *Science*, **277**, 563–566.

Jamali, F. (1988) Pharmacokinetics of enantiomers of chiral non-steroidal anti-inflammatory drugs. *Eur. J. Drug. Metab. Pharmacokinet.*, **13**, 1–9.

Jamali, F., Mehvar, R. & Pasitto, F. M. (1989) Enantioselective aspects of drug action and disposition: therapeutic pitfalls. *J. Pharm. Sci.*, **78**, 695–715.

Jameson, B. A., McDonnell, J. M., Marini, J. C. & Korngold, R. (1994) A rationally designed CD4 analogue inhibits experimental allergic encephalomyelitis. *Nature* **368**, 744–746.

Jung, G. (1992) Proteins from the D-chiral world. *Angew. Chem.*, **31**, 1457–1459.

Kass, R. S. (1987) Voltage-dependent modulation of cardiac calcium channel current by optical isomers of Bay K 8644: implications for channel gating. *Circ. Res.*, **61** (Suppl. 1), 1–5.

Keats, E. U. & Stone, R. (1984) The effect of D-timolol on intraocular pressure in patients with ocular hypertension. *Am. J. Ophthalmol.*, **98**, 73–78.

Kreil, C. (1994) Conversion of L- to D-amino acids: a post translational reaction. *Science*, **266**, 996–997.

Kreil, G. (1997) D-Amino acids in animal peptides. *Ann. Rev. Biochem.*, **66**, 337–345.

Krings, M. (1997) Neanderthal DNA sequences and the origin of modern humans. *Cell*, **90**, 19–30.

Kuwada, M., Tetsuyuki, T., Kumagaye, K. Y., Nakajima, K., Watanabe, T., Kawai, T., Kawakami, Y., Niidome, T., Sawada, K., Nishizawa, Y., Katayama, K. (1994) ω-Agatoxin-TK containing D-serine at position 46, but not synthetic ω-[L-Ser46]agatoxin-TK, exerts blockade of P-type calcium channels in cerebellar Purkinje neurons. *Mol. Pharmacol.*, **46**, 587–593.

Kwon, Y.-W. & Triggle, D. J. (1991) Chiral aspects of drug action at ion channels. A commentary on the stereoselectivity of drug actions at ion channels with particular reference to verapamil actions at the Ca^{2+} channel. *Chirality*, **3**, 393–404.

Laganiere, S. (1997) Current regulatory guidelines of stereoisomeric drugs: North American, European and Japanese points of view. In: *The Impact of Stereochemistry on Drug Development and Use* (Ed. by H. Y. Aboul-Enein & I. W. Wainer) pp. 545–565, Wiley–Interscience, NY.

Lee-Son, S., Wong, G. K., Concus, A., Crill, E. & Strichartz, G. (1992) Stereoselective inhibition of neuronal sodium channels by local anaesthetics. *Anaesthesiology*, **77**, 324–335.

Levy, R. H. & Boddy, A. V. (1991) Stereoselectivity in pharmacokinetics: a general theory. *Pharm. Res.*, **8**, 551–556.

Lewis, R. J., Trager, W. F., Chan, K. K., Breckenridge, A. M., Orme, M., Rowland, M. & Schary, W. (1974) Warfarin: stereochemical aspects of its metabolism and the interaction with phenylbutazone. *J. Clin. Invest.* **53**, 1607–1617.

Luduena, F. P. (1967) Duration of local anaesthesia. Ann. Rev. Pharmacol. **9**, 503–520.

McDonnell, J. M., Fushman, D., Cahill, S. M., Sutton, B. J. & Cowburn, D. (1997) Solution structures of FceRI α-chain mimics: a hairpin peptide and its retroenantiomer. *J. Am. Chem. Soc.*, **119**, 5321–5328.

Meese, C. O. & Eichelbaum, M. (1986) Stereochemical aspects of polymorphic 4-hydroxylation of debrisoquine in man. *Naunyn-Schmid. Arch. Pharmacol.*, **332** (Suppl.), R95.

Mikus, G., Eichelbaum, M., Fischer, F., Gumulka, S., Klotz, U. & Kroemer, H. K. (1990) Interaction of verapamil and cimetidine: stereochemical aspects of drug metabolism, drug disposition and drug action. *J. Pharmacol. Exp. Ther.*, **253**, 1042–1048.

Miller, K. W. (1985) The nature of the site of general anaesthesia. *Int. Rev. Neurobiol.*, **27**, 1–62.

Montecucchi, P. C., de Castiglione, R. & Erspamer, V. (1981) Identification of dermorphin and hyp6-dermorphin in skin extracts of the Brazilian frog *Phyllomedusae rhodei. Int. J. Prot. Res.*, 7, 316–321.

Moody, E. J., Harris, B. D. & Skolnick, P. (1994) The potential for safer anaesthesia using stereoselective anaesthetics. *Trends Pharmacol. Sci.*, **15**, 387–390.

Narahashi, T. (1996) Neuronal ion channels as the target sites for insecticides. *Pharmacol. Toxicol.*, **78**, 1–14.

Nichols, A. J., Sulpizio, A. C., Ashton, D. J., Hieble, J. P. & Ruffolo, R. R. (1989) The interactions of the enantiomers of carvedilol with α_1- and β_1-adrenoceptors. *Chirality*, **1**, 265–270.

Pasteur, L. (1860) On the asymmetry of naturally occurring organic compounds. (Two lectures delivered before the Chemical Society of Paris, 20th January and 3rd February, 1860). In: *Memories on Stereochemistry* (Ed. by M. Richardson), pp. 1–33. American Book Company, London, New York, 1901.

Patil, P. N., Miller, D. D. & Trendelenburg, U. (1974) Molecular geometry and adrenergic drug activity. *Pharmacol. Rev.*, **26**, 323–392.

Peterson, B. Z. & Catterall, W. A. (1995) Calcium binding in the pore of L-type calcium channels modulates high affinity 1,4-dihydropyridine binding. *J. Biol. Chem.* **270**, 18201–18204.

Petsko, G. A. (1992) On the other hand.... *Science*, **256**, 1403–1404.

Ragsdale, D. S., McPhee, J. C., Scheuer, T. & Catterall, W. A. (1994) Molecular determinants of state-dependent block of Na^+ channels by local anaesthetics. *Science*, **265**, 1724–1728.

Ragsdale, D. S., McPhee, J. C., Scheuer, T. & Catterall, W. A. (1996) Common molecular determinants of local anaesthetic, anti-arrhythmic and anti-convulsant block of voltage-gated Na^+-channels. *Proc. Natl. Acad. Sci. USA*, **93**, 9270–9275.

Richards, R. & Tattersfield, A. R. (1985) Bronchial β-adrenoceptor blockade following eye-drops of timolol and its isomer L-714,465 in normal subjects. *Br. J. Clin. Pharmacol.*, **20**, 459–462.

Roden, D. M. & George, A. L. (1994) The cardiac ion channels. *Ann. Rev. Med.*, **47**, 135–148.

Roden, D. M. & George, A. L. (1996) The cardiac ion channels; relevance to management of arrhythmias. *Ann. Rev. Med.*, **47**, 135–148.

Ruffolo, R. R. (1991a) Chirality in α- and β-adrenoceptor agonists and antagonists. *Tetrahedron*, **47**, 9953–9980.

Ruffolo, R. R. (1991b) *Drugs of Today*, **28**, 465.

Sallusho, B. C., Meffin, P. J. & Knights, K. M. (1988) The stereospecific incorporation of fenoprofen into rat hepatocyte and adipocyte triacylglycerols. *Biochem. Pharmacol.*, **37**, 1919–1923.

Sanguinetti, M. C. & Kass, R. S. (1984) Voltage dependent block of calcium channel current in calf cardiac Purkinje fibres by dihydropyridine calcium channel antagonists. *Circ. Res.*, **55**, 336–348.

Schumacher, T. N. M., Mayr, L. M., Minor, D. L., Jr. *et al.* (1996) Identification of D-peptide ligands through mirror image phage display. *Science*, **271**, 1854–1857.

Schuster, A., Lacinova, L., Klugbauer, N., Ito, H., Birnbaumer, L. & Hofmann, F. (1960) The IVS6 segment of the L-type calcium channels is critical for the action of dihydropyridines and phenylalkylamines. *EMBO J.* **15**, 2365–2370.

Sedensky, M. M., Cascorbi, H. F., Meinwald, J., Radford, P. & Morgan, P. G. (1994) Genetic differences affecting the potency of stereoisomers of halothane. *Proc. Nat. Acad. Sci. USA* **91**, 10054–10058.

Shah, R. R. (1993) Clinical pharmacokinetics: current requirements and future perspectives from a regulatory point of view. *Xenobiotica*, **23**, 1159–1193.

Simonyi, M., Fitos, I. & Visy, J. (1986) Chirality of bioactive agents in protein binding storage and transport processes. *Trends Pharmacol. Sci.*, **7**, 112–116.

Sleiner, D. F. & Robenstein, P. H. (1997) Proinsulin C-peptide biological activity. *Science*, **277**, 531–532.

Smith, D. F. (Ed.) (1983) *Handbook of Stereoisomers: Drugs in Psychopharmacology*. CRC Press, Boca Raton, FL.

Smith, D. F. (Ed.) (1989) *Handbook of Stereoisomers: Therapeutic Drugs*. CRC Press, Boca Raton, FL.

Somani, P. & Watson, D. L. (1991) Antiarrhythmic activity of the *dextro-* and *levo-*rotating isomers of 4-(2-isopropyl(amino-1-hydroxyethyl)methanesulphonanilide (MJ 1991). *J. Pharmacol. Exp. Ther.*, **164**, 317–325.

Stinson, S. C. (1993) Chiral Drugs. *Chem. Eng. News*, Sept. 27, pp. 58–65.

Stinson, S. C. (1994) Chiral Drugs. *Chem. Eng. News*, Sept. 19, pp. 38–71.

Stinson, S. C. (1995) Chiral Drugs. *Chem. Eng. News*, Oct. 9, pp. 44–74.

Strader, C. D., Fong, T. M., Tota, M. R. & Underwood, D. (1994) Structure and function of G-protein coupled receptors. *Ann. Rev. Biochem.*, **63**, 101–132.

Sybertz, E. J., Sabin, C. S., Pula, K. K., van de Vliet. G., Glennon, R. J., Gold, E. H. & Baum T. (1981) *Alpha-* and *beta-*adrenoceptor blocking properties of labetalol and its *RR* isomer SCH 19927. *J. Pharmacol. Exp. Ther.*, **218**, 435–443.

Tamai, I., Ling, H.-Y., Timbul, S.-M., Nishikodo, J. & Tsuji, A. (1988) Stereospecific absorption and degradation of cephalexin. *J. Pharm. Pharmacol.*, **40**, 320–324.

Testa, B. (1986) Chiral aspects of drug metabolism. *Trends Pharmacol. Sci.*, **7**, 60–64.

Testa, B. & Trager, W. F. (1990) Racemates *versus* enantiomers in drug development: dogmatism or pragmatism. *Chirality*, **2**, 129–133.

Tobert, J. A., Cirillo, V. J., Hitzenberger, G., James, J. Pryor, J., Cook, T., Buntinx, A. Holmes, J. B. & Lutterbeck, P. M. (1981) Enhancement of uricosuric properties of indacrinone by manipulation of the enantiomer ratio. *Clin. Pharmacol. Ther.*, **29**, 344–350.

Toldy, I., Wargha, L., Tolh, I. & Borsy, J. (1959) Uber untersuchungen von promethazin, I. *Acta Chem. Acad. Sci. Hung.*, **19**, 273–277.

Triggle, D. J. (1989) Structure–activity relationships of 1,4-dihydropyridine calcium-channel activators and antagonists. In: *Molecular and Cellular Mechanisms of Antiarrhythmic Agents* (Ed. by L. M. Hondeghem), pp. 269–291. Futura Publishing, Mt. Kiscoe, NY.

Triggle, D. J. (1990) Calcium antagonists. In: *Cardiovascular Pharmacology* (Ed. by M. Antonnacio), pp. 102–160. Raven Press, New York.

Triggle, D. J. (1997) Stereoselectivity of drug action. *Drug Discovery Today*, **2**, 138–147.

Triggle, D. J., Langs, D. A. & Janis, R. A. (1989) Calcium channel ligands: structure–function relationships of the 1,4-dihydropyridines. *Med. Res. Rev.*, **9**, 123–180.

Tucker, G. T. & Lennard, M. S. (1990) Enantiomer-specific pharmacokinetics. *Pharmacol. Ther.*, **45**, 309–329.

Tucker, G. T. (1991) A clinical pharmacologist's perspective. *Biochem. Soc. Trans.*, **19**, 460–462.

van Amsterdam, F. T. M. & Zaagsma, J. (1988) Stereoisomers of calcium antagonists discriminate between coronary vascular and myocardial sites. *Naunyn-Schmied. Arch. Pharmacol.*, **337**, 213–219.

Vogelgesang, B., Echizen, H., Schmidt, E. & Eichelbaum, M. (1984) Stereoselective first pass metabolism of highly cleared drugs: studies of the bioavailability of Land D-verapamil examined with a stable isotope technique. *Br. J. Clin. Pharmacol.*, **18**, 733–740.

Wade, D. N., Mearrick, P. T. & Morris, J. L. (1973) Active transport of L-dopa in the intestine. *Nature*, **242**, 463–465.

Wainer, I. W. (Ed.) (1993) *Drug Stereochemistry. Analytical Methods and Pharmacology*, 2nd edn. M. Dekker, New York.

Walle, T., Webb, J. G., Bagwell, E. E., Walle, U. K., Daniell, H. B. & Gaffney, T. E. (1988) Stereoselective delivery and actions of β-receptor antagonists. *Biochem. Pharmacol.*, **37**, 115–124.

Webb, J. G., Sheet, J. A., Bagwell, E. E., Walle, T. & Gaffney, T. E. (1988) Stereoselective secretion of atenolol from PC12 cells. *J. Pharmacol. Exp. Ther.* **247**, 958–964.

Wechter, W. J. (1994) Drug chirality: on the mechanism of *R*-aryl propionic acid class NSAIDs. Epimerization in humans and the clinical implications for the use of racemates. *J. Clin. Pharmacol.* **34**, 1036–1042.

Weil, A., Caldwell, J., Guichard, J.-P. & Picot, G. (1989) Species differences in the chirality of the carbonyl reduction of [^{14}C]fenofibrate in laboratory animals and humans. *Chirality*, **1**, 197–201.

Weiland, K., Zuurmond, H. M., Krasel, C., Ijzerman, A. P. & Lohse, M. J. (1996) Involvement of Asn-293 in stereospecific agonist recognitions and in activation of the β_2-adrenergic receptor. *Proc. Natl. Acad. Sci. USA*, **93**, 9276–9281.

Welling, A., Kwan, Y.-W., Bosse, E., Flockerzi, V., Hofmann, F. & Kass, R. S. (1993) Subunit-dependent modulation of recombinant L-type calcium channels. Molecular basis for 1,4-dihydropyridine selectivity. *Circ. Res.*, **73**, 974–980.

White, P. F., Ham, J., Way, W. L. & Trevor, A. J. (1980) Pharmacology of ketamine isomers in surgical patients. *Anaesthesiology*, **52**, 231.

Williams, K. & Lee, E. (1985) Importance of drug enantiomers in clinical pharmacology. *Drugs* **30**, 333–354.

Williams, K., Day, R., Knihinicki, R. & Duffield, A. (1986) The stereoselective uptake of ibuprofen enantiomers into adipose tissue. *Biochem. Pharmacol.*, **35**, 3403–3405.

Yeh, J. Z. (1980) Blockade of sodium channels by stereoisomers of local anaesthetics. In: *Molecular Mechanisms of Anaesthesia* (Ed. by B. R. Fink). Raven Press, New York.

Zheng, W., Stoltefuss, Goldman, S. & Triggle, D. J. (1992) Pharmacologic and radioligand binding studies of 1,4-dihydropyridines in rat cardiac and vascular preparations; stereoselectivity and voltage-dependence of antagonist and activator interactions. *Mol. Pharmacol.*, **41**, 535–541.

Zimmerman, D. M. & Gessellchen, P. D. (1982) Analgesia (peripheral and central), endogenous opioids and their receptors. *Ann. Rep. Med. Chem.*, **17**, 21–30.

Chapter 6
Chirality in Medicinal Chemistry

Popat N. Patil
Division of Pharmacology, College of Pharmacy, The Ohio State University, Parks Hall, 500 West 12th Avenue, Columbus, OH 43210, USA

Duane D. Miller
Department of Medicinal Chemistry, College of Pharmacy, University of Tennessee, Memphis, TN 38163, USA

HISTORICAL ASPECTS

The natural plant products used in ancient times were the forerunners of present day synthetic medicinal drugs. As chemical sciences advanced, substances from plants were isolated. In France, by 1822 (the birth date of Louis Pasteur), P. J. Pelletier and J. B. Caventou, who taught at the Ecole de Pharmacie in Paris, characterized quinine, strychnine, brucine, veratradine and emetine (Holmstedt & Liljestrand 1963; Partington 1964). The asymmetric carbon atoms of these alkaloids were to be discovered later.

A remarkable historical event occurred, however. The famous physiologist, F. Magendie, and the pharmaceutical chemist, Pelletier, collaborated in an investigation of the pharmacological action of newly discovered alkaloids. Structure–activity studies were initiated and Magendie discovered the dorsal sensory neuronal inflow and ventral motor neuronal outflow of the vertebrate spinal chord. His student, Claude Bernard, identified neuromuscular junctions as the site of action of the arrow poison curare. Langely, a British physiologist, used the ganglionic stimulant action of nicotine to identify the location of the cell bodies of the neurones.

Thus by 1901 the basic functional outline of vertebrate neurones was established. The symmetrical molecule acetylcholine was proved to be the transmitter released by the cholinergic nerve endings of the heart. The physiological action of endogenous or exogenous acetylcholine on many organs was similar to that of muscarine that was isolated from mushrooms. It is, therefore, customary to refer to this in pharmacology as the muscarinic action of acetylcholine. Nicotine not only stimulated autonomic ganglia but also stimulated adrenal medulla, skeletal muscle and carotid body chemosensitive structures. Because acetylcholine also stimulates these structures, the responses are referred to as nicotinic action of acetylcholine. Specific blockers proved to be valuable substances in the classification of receptors. Various muscarinic and nicotinic receptor subtypes have been discovered in the body. Physiological and pharmacological actions are mediated by drugs interacting with these cellular receptors.

Von Euler (1946) proved that noradrenaline rather than adrenaline was the transmitter released by the postganglionic sympathetic neurones. At that time, the actions of (±)-noradrenaline were compared with the physiological effects produced by stimulation of the sympathetic nerve. Two years later, Tullar (1948) meticulously resolved pure diastereoisomers, namely (−)-noradrenaline-(+)-bitartrate and (+)-noradrenaline-(+)-bitartrate from the racemate. Exactly 100 years earlier Pasteur (1848) had separated the optical isomers of tartaric acid. The potency of the resolved (−)-noradrenaline was similar to that produced by nerve stimulation (Tainter *et al.* 1948).

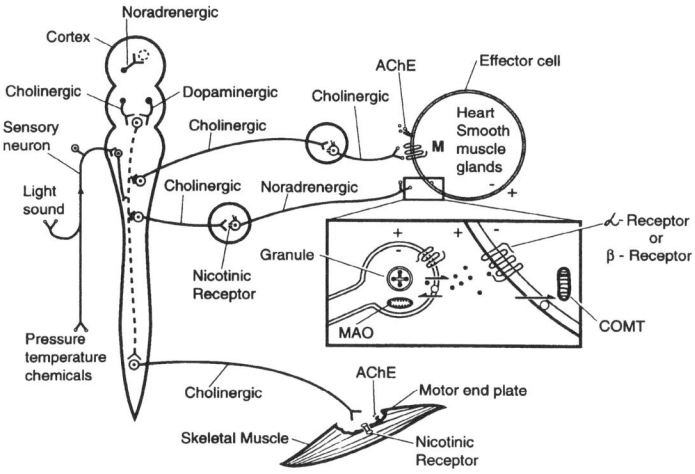

Fig. 6.1 An outline of the sensory inflow and motor outflow neurones of the nervous system. Cholinergic, dopaminergic and noradrenergic neurones release acetylcholine, dopamine and (−)-noradrenaline, respectively, as transmitters. The muscarinic (M) receptor is associated with cholinergic transmission in the heart, smooth muscles and glands. The nicotinic receptors in the ganglia and at the motor end plate are also activated by acetylcholine but these ion-channel receptors are of a different subtype. Acetylcholine esterase (AchE) is found at all cholinergic neuroeffector junctions and plasma.
Noradrenergic nerve terminals release noradrenaline to activate α- or β-adrenoceptors. Excess transmitter is transported in the neurone by uptake$_1$ and extraneurally in the effector cell by uptake$_2$. Monoamine oxidase (MAO) and catechol-*O*-methyltransferase (COMT) control the metabolism of the catecholamine. Prejunctional receptors also control the release of the transmitter. Details of the transmitters and receptors of sensory neurones are slowly being elucidated.

Nearly 50 years previously, Abel and Takamine chemically characterized the hormone of the adrenal glands as adrenaline (Holmstedt & Liljestrand 1963). The natural product was identified as the (−) enantiomer when the resolved isomers were examined for unequal pressor effects in the anaesthetized dog (Abderhalden & Müller 1908; Cushny 1908). Two types of adrenoceptor (α and β) were postulated. The chemical anatomy of the nervous system is presented in Fig. 6.1.

STEREOSELECTIVITY OF THE BIOSYNTHESIS OF TRANSMITTERS AND RELATED MEDICINALS

Details of biosynthesis of acetylcholine and (−)-noradrenaline have been elucidated (Hardman *et al.* 1996). The enzymes tyrosine hydroxylase, Dopa decarboxylase and dopamine-β-oxidase, rapidly synthesize (−)-noradrenaline from L-tyrosine. The transmitter is stored in the granules. The depolarizing stimulus releases the transmitter by exocytosis at the neuroeffector junction. Either α- or β-adrenoceptors can be activated. Excess transmitter is reabsorbed in the neurone by the active uptake$_1$ process. Cytoplasmic monoamine oxidase can partly metabolize the transmitter to keep the optimum equilibrium between granule-bound and free cytoplasmic transmitter. At the postjunctional membrane (−)-noradrenaline is also transported by extra neuronal carrier-mediated influx called uptake$_2$ (Trendelenburg 1980).

Synthetic L-(−)-dopa is used therapeutically. It is converted to dopamine in the dopaminergic neurons of patients with Parkinson's disease, who produce less transmitter in specific parts of the brain. D-(+)-dopa is not decarboxylated to dopamine. It is also toxic. Exogenous L-α-methyldopa can also enter neuronal biosynthetic pathways where it can be converted to (−)-α-methylnoradrenaline (Patil *et al.* 1974; Bartholini *et al.* 1975; Fujiwara *et al.* 1976; Fig. 6.2).

This pseudo transmitter can dilute natural (−)-noradrenaline in the storage granules. When a mixture of the transmitter and *pseudo*-transmitter is released at the vascular neuroeffector junction there is less vasoconstriction, partly because α-methylnoradrenaline is a relatively weak receptor activator. Antihypertensive action is observed as a result. Remarkable chiral selectivity is observed in the decarboxylation and β-hydroxylation of the drug. Exogenous (−)-α-methyldopamine does not seem to be converted into the corresponding hydroxylated amine.

The stereoselectivity of this noradrenergic biosynthetic pathway is further indicated by the fact that only a single enantiomer of 3,4-dihydroxyphenylserine (DOPS) can be decarboxylated. In rats, specificity in the decarboxylation might be lost because *erythro*-DOPS is converted to (+)-noradrenaline. The latter amine cannot be stored as a false transmitter. (Patil *et al.* 1974; Bartholini *et al.* 1975; Fujiwara *et al.* 1976; Fig. 6.2).

The transmitter (−)-noradrenaline serves as a substrate for monoamine oxidase and catechol-*O*-methyltransferase. Circulating (−)-adrenaline is metabolized more quickly than the (+) form. Thus, at pharmacologically equi-effective doses the duration of action of (+)-adrenaline will seem longer than that of the natural (−) enantiomer (Patil *et al.* 1974; Henseling & Trendelenburg 1978).

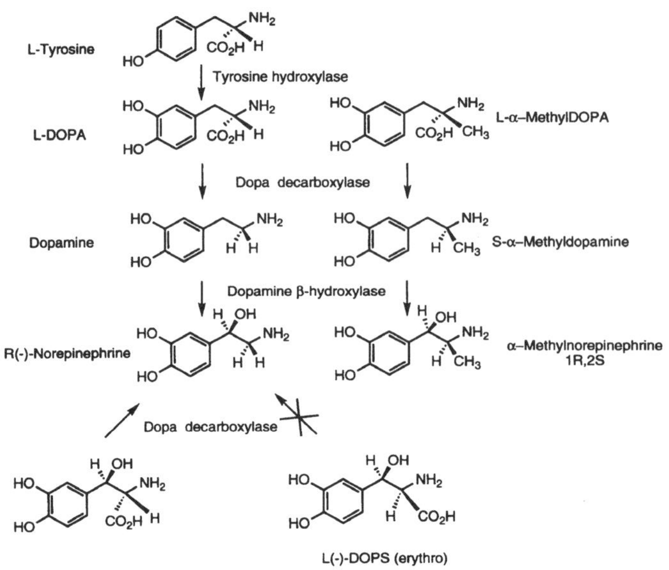

Fig. 6.2 Stereospecific biosynthesis of *R*-(−)-noradrenaline and metabolic biotransformation of related medicinals. Oral L-dopa can be converted to dopamine in the dopaminergic neurones of the brain. Exogenous α-methyldopa forms α-methylnoradrenaline, a pseudo transmitter which activates $α_2$-adrenoceptors in brain neurones. As a result the blood vessels dilate. This is, in part, the basis of antihypertensive action of the drug. The conversion of exogenous 3,4-dihydroxyphenylserine (DOPS) is also stereoselective.

The achiral molecule tyramine, which acts by displacing neuronal noradrenaline, is rapidly metabolized by neuronal monamine oxidase. The enzyme is stereoselective for the deuterated tyramine enantiomer (Belleau *et al.* 1960). These experiments with (−)- and (+)-α-deuterotyramine are identical, in concept, with Ogston's proposal that the enzyme can distinguish between two similar groups of the substrate (Ogston 1948). All α-methylated β-phenylethylamines such as amphetamine are not metabolized by the monoamine oxidase and, as inhibitors, differences between the enantiomers are small. Conformationally restricted (±)-*cis* and (±)-*trans* forms of tranylcypromine show preference for the (+)-*trans* form as the inhibitor of the monoamine oxidase (Horn & Snyder 1972). Therapeutic antidepressant efficacy is in part related to one isomer.

Activation of the receptor is the most important physiological function of the neuronally released transmitter. Excess (−)-noradrenaline can be transported, by carrier-mediated extraneuronal uptake$_2$, across the effector cell membrane (Trendelenburg 1980). Intracellular catechol-*O*-methyltransferase can '*O*' methylate (−)-noradrenaline or injected catecholamines. Compared with neuronal uptake, uptake$_2$ has low selectivity for the transport of enantiomers of noradrenaline (Iversen 1963; Hellman *et al.* 1971; Creveling *et al.* 1972; Garg *et al.* 1973; Grohmann & Trendelenburg 1984; Hensling 1984). Enantiomers of (−)-isoproterenol and adrenaline, however, have two- to fourfold selectivity for the (–) form (Hellman et al. 1971; Grohmann & Trendelenburg 1984; Table 6.1).

Table 6.1 Initial neuronal uptake$_1$ and extraneuronal uptake$_2$ and neuronal retention of enantiomers of noradrenaline in the heart (Iversen 1963; Hellman et al. 1971; Garg et al. 1973; Grohmann & Trendelenburg 1984; Hensling 1984).

Amine	Isolated rat heart Neuronal uptake$_1$		Mouse heart* (*In vivo*)		Uptake$_2$ ID$_{50}$ (μmol L^{-1})
	K_m (μM)	V_{max} (nmol min^{-1} g^{-1})	Rate constant (h^{-1})	$t^{1/2}$ (h^{-1})	
(±)-noradrenaline	0.67	1.36	0.27	2.19	
			0.06	10.10	
(–)-noradrenaline	0.27	1.18	0.07	9.32	224
(+)-noradrenaline	1.39	1.72	0.17	3.92	173
dopamine	0.69	1.45			498

*After intravenous 40 µcg kg^{-1} of the labelled amine.

The selectivity of '*O*' methylation by the enzyme is also small (Creveling et al. 1972). Although '*O*' methylation of the *meta* hydroxy group is prominent, under certain conditions *para* hydroxy '*O*' methylation is observed. Studies of conformationally rigid analogues of noradrenaline indicate that *trans* orientation of the catechol phenyl group to the amino group is preferred (Patil et al. 1974). Steric orientation of the catechol groups of *S*-(–)- and *R*-(+)-salsolinol resulted in small yet interesting preference of the latter isomer for the enzyme (Hötzel & Thomas 1997). Methylation of the hydroxy group of *R*-(+)-salsolinol corresponding to the *meta* hydroxy group of the catecholamines was about eight times greater than that of the *S*-(–)-salsolinol (Table 6.2). The isoquinoline salsolinol has a semi-rigid structure and the catechol groups are unequally '*O*' methylated by the enzyme. The substrate alignment of the hydroxy group and the methyl-NH$_2$ function of the *R*-(+) isomer must be in the *trans* conformation.

Circulating corticosterone is an effective inhibitor of extraneuronal uptake$_2$ of the catecholamines. The process of '*O*' methylation by the intracellular enzyme is blocked. Details of the steric factors which affect the functions of the enzyme need to be elucidated.

Table 6.2 Kinetic parameters for the formation of 'O' methylated products of catechol derivatives by catechol 'O' methyltransferase (Creveling *et al.* 1972; Hötzel & Thomas 1997).

Catechol derivative	K_m (mM)	V_{max} (nmol mg^{-1} min^{-1})	m-'O'CH$_3$/p-OCH$_3$ Ratio
(−)-noradrenaline	0.26	0.66	5.3
(+)-noradrenaline	1.62	0.30	7.4
dopamine	0.2	0.38	1.0[a]
	(0.46)	(0.13)	(2.0)[b]
(−)-*erythro*-α-methylnoradrenaline	0.73	0.77	8.4
(+)-*erythro*-α-methylnoradrenaline	0.55	0.46	13.5
(−)-isoproterenol	0.04	0.12	2.3
(+)-isoproterenol	1.94	0.80	5.3
R-(+)-salsoline	0.14	0.20	7.6[c]
S-(−)-salsoline	0.16	0.20	0.9[c]

[a]*ortho*/*meta* ratio, rat liver, [b]pig brain, [c]7OCH$_3$/6'O'CH$_3$ ratio of isoquinoline alkaloid.

AFFINITY AND INTRINSIC ACTIVITY

E. J. Ariens (1964) introduced the concept of affinity and intrinsic activity of receptor-activator drugs called 'agonists'. Acetylcholine added exogenously to isolated tissue *in vitro* under physiological conditions will diffuse to the cellular receptors. The activated receptor will produce a concentration-dependent response. The maximum response is referred to as the 'intrinsic activity' of the agonist. Intrinsic 'efficacy' of the agonist at the receptor produces so-called intrinsic mechanical activity. A concentration–response curve of the acetylcholine on the tissue can be generated.

Because exogenous acetylcholine is also destroyed by the tissue acetylcholinesterase, a stable agonist such as (+)-muscarine can be used to investigate the concentration–response relationship for the series of molecules. Thus, two medicinals can be compared for their affinity and intrinsic activity in tissue preparations containing specific physiological receptors. For practical purposes the concentration which produces 50% of the maximum response (EC$_{50}$ mol L^{-1}) and the maximum response are equated with affinity and intrinsic activity. In many organs a potent agonist will produce 100% of the mechanical response without saturating all the receptors in the organ (Stephenson 1956).

Only a small fraction of total receptors are activated either by nerve stimulation or by exogenous potent agonist to produce maximum response in some organs. If these spare or excess receptors are partially inactivated by an irreversible blocker, under these receptor saturation conditions, EC_{50} and the maximum response indicate the affinity and the intrinsic activity or efficacy (Furchgott & Bursztyn 1967), because under these conditions the agonist can saturate remaining tissue receptors at the maximum response (Fig. 6.3).

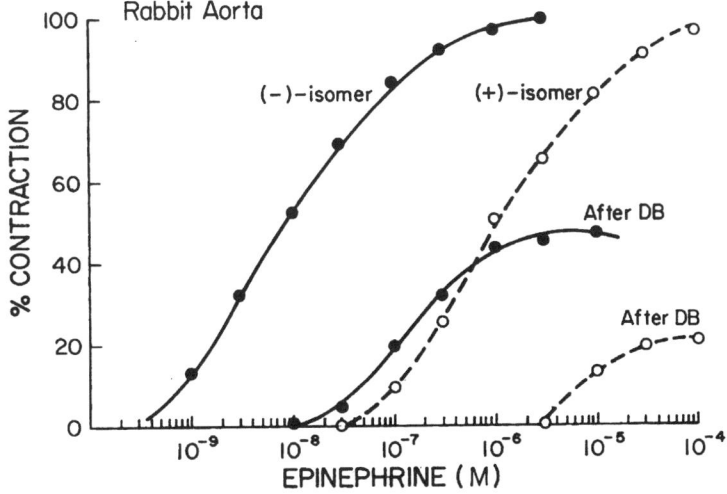

Fig. 6.3 Concentration–response curves of the enantiomers of adrenaline in an organ with receptor reserve. The different affinities are generally indicated by 50% of the maximum, denoted the EC_{50} values of the isomers. Intrinsic efficacy differences as seen by the maximum response can, however, be differentiated only after the fractions of the receptors in the organ are irreversibly inactivated by dibenzamine (DB).

Enantiomers at muscarinic receptors or adrenoceptors can vary in their affinity and intrinsic activity. As far as the receptor-interacting properties of a pair of enantiomers are concerned, various possibilities exist:

(1) One enantiomer has more affinity and intrinsic activity at a given receptor than its enantiomer.
(2) Only one enantiomer has affinity and intrinsic activity at a receptor and the other enantiomer is 'inert'.
(3) The enantiomer without apparent biological activity blocks the pharmacological activity of the other competitively or non-competitively.
(4) (−)-Dobutamine is a potent agonist at α-adrenoceptors and a weak agonist at β-adrenoceptors whereas the (+) enantiomer is a potent agonist at β-adrenoceptors and a weak antagonist at α-adrenoceptors. The racemate induces an interesting summation of action in cardiovascular systems which contain both types of adrenoceptor (Ruffolo *et al.* 1981).
(5) Enantiomers of organophosphate inhibitors of acetylcholine esterase are the most important example of the selective toxicological action of the pesticide. The enantiomer which is mainly toxic to pests is preferred to the racemate, which is equally toxic to pests and mammals (Ariens *et al.* 1988).

The affinity and maximum binding capacity (β_{max}) of an agonist can be investigated in the subcellular receptor-rich fraction isolated from a tissue. The competition between ligands (agonist or blockers) can be compared on the basis of equilibrium dissociation constants (K_i). The method does not provide a parameter for the intrinsic activity of agonists, which can be obtained from functional pharmacological experiments (Patil 1996). Various receptors in the body are sub-classified on the basis of the relative order of potency of agonists and the dissociation constants (K_b) of the competitive reversible blockers (Arunlakshana & Schild 1959).

STERIC ASPECTS OF DRUG ACTION AT THE CHOLINERGIC NEUROEFFECTOR JUNCTION

Acetylcholine was discovered by Otto Loewi in the 1920s and was the first chemical neurotransmitter reported. The actions of acetylcholine were divided into muscarinic and nicotinic because of to the similarity of the actions of acetylcholine to those of muscarine on smooth muscles and glands and nicotine on voluntary muscle and ganglia. Much progress has been made in the identification of new drugs which interact with these receptor systems. Synaptic acetylcholine is quickly metabolized by acetylcholine esterase (AchE). Biosynthesis of the transmitter has been reported. Acetylcholine does not have a chiral centre but both muscarine and nicotine are chiral molecules. Casey & Dewar (1993) have published a very good report on the stereochemical synthesis and biological actions of muscarinic and nicotinic agents.

MUSCARINIC AGONISTS

Muscarine, which has three chiral centres, can be regarded as a cyclic analogue of acetylcholine; of the eight possible isomers it is the (+)-5*S*,4*R*,2*S* isomer that is the most active on muscarinic receptors. The agonist potency is twice that of acetylcholine. Cholinergic agonists with muscarinic activity have been used therapeutically to increase the tone and motility of the intestine, to relieve urinary retention, to treat glaucoma, and a major effort has recently been directed towards finding such agents to treat Alzheimer's disease and other cognitive disorders.

In muscarine a critical stereochemical relationship is that between the 5-methyl- and the 2-trimethylammoniummethyl group. These functional groups should have a *cis* stereochemical relationship about the tetrahydrofuran ring for optimum muscarinic receptor activity. A host of chemical variants of acetylcholine, muscarine and nicotine have been studied. The addition of a methyl group to either the α or β segment of acetylcholine gives rise to optically active molecules. The isomers of methacholine, the β-methyl analogue of acetylcholine, have been prepared by stereospecific synthesis and the *S*-(+) isomer is 240 times more active than the isomer *R*-(−) isomer on guinea-pig ileum. The *S*-(+) isomer of methacholine also has greater intrinsic activity on the muscarinic receptor.

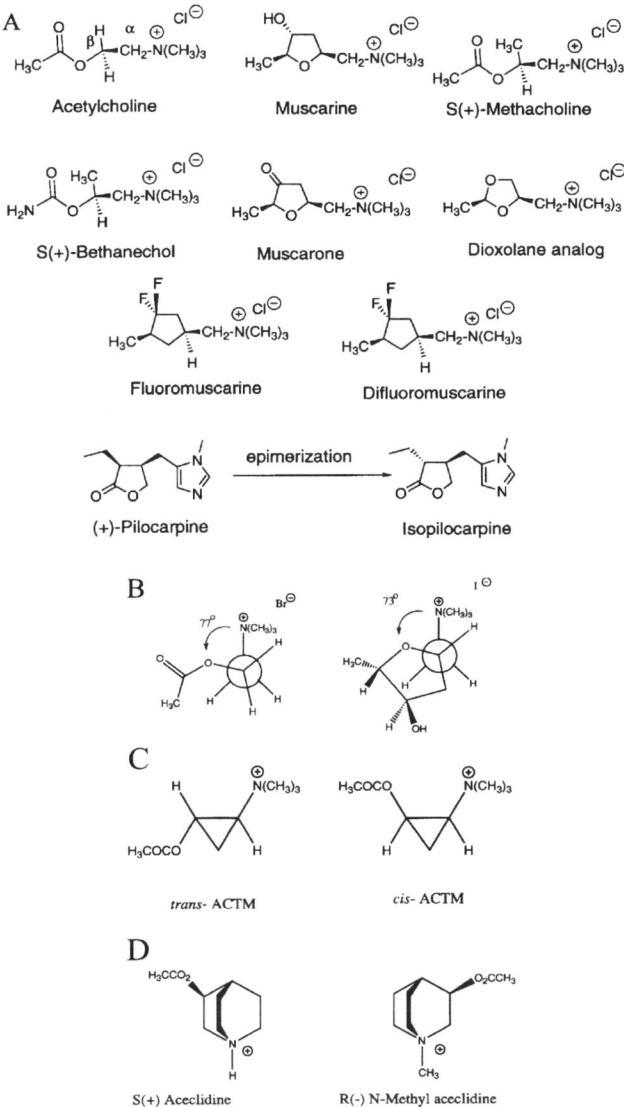

Fig. 6.4 A. The three-dimensional chemical structures of muscarinic cholinergic agonists. B. Solid-state conformations (Neuman projection formulae) of acetylcholine and muscarine indicating the nearly equal dihedral angle of 73–77° of the muscarinic receptor. C. Conformationally restricted analogues of acetylcholine. D. Aceclidine and its *N*-methyl analogue.

The important relationship of the chiral centre of S-(+)-methacholine and the 5-S chiral carbon of muscarine is illustrated in Fig. 6.4. Also shown in the figure is the most active isomer, S-(+)-bethanechol. Again, large differences, up to 915 fold on guinea-pig ileum are observed between the isomers of bethanechol. Thus, it seems that the H atom on the *beta* face of the structures must play a significant role in not impeding the binding to the muscarinic receptor.

Oxidation of the hydroxyl group of muscarine to a keto group gives rise to (+)-muscarone, the activity of which is dramatically different—it is a more potent nicotinic agonist than acetylcholine. Although differences between the effects of the isomers of muscarone on muscarinic receptors are small compared with those of the isomers of muscarine, replacement of the CHOH group of muscarine with an oxygen to give the 1,3-dioxolane furnished very potent muscarinic agents. A host of dioxolane analogues with sulphur, sulphoxide and even carbon are potent muscarinic agonists.

The oxygen of muscarine can also be replaced with a carbon and it retains potent agonist activity. The CHOH group has also been replaced with CHF and CF_2 groups to give, respectively, fluoromuscarine and difluoromuscarine that have high potency and different activity at different muscarinic receptor subtypes (Jean & Davis 1994; Angeli *et al.* 1997). The chiral centre in all the most potent muscarinic agents correlates with the 5-*S* chiral centre for muscarine.

Pilocarpine is an important natural product with muscarinic agonist activity. The drug is used therapeutically for the treatment of glaucoma. It is unusual compared with the other molecules discussed because it does not contain a quaternary nitrogen—the quaternary salt of pilocarpine is pharmacologically inert. Stereochemistry plays a major role in the activity of this agent. Base-induced epimerization of the alkyl chain to give isopilocarpine provides a molecule with no activity at muscarinic receptors.

In conformational studies of acetylcholine several reports (Casey & Dewar 1993) have been directed towards gaining a better understanding of the torsional angle between two important functional groups of acetylcholine, the trimethylammonium and the acetate, as illustrated in Fig. 6.4B, and the biological activity of these functional groups (Armstrong *et al.* 1968). The dihedral angle of 77° is probably a result of a compromise between the steric repulsion of the two groups and the electrostatic attraction between the quaternary salt positive charge and the electronegative oxygen of the ester group. It is interesting that the separation between the ether oxygen and the quaternary salt of muscarine is 73°.

In an attempt to learn more about the preferred conformation of acetylcholine several researchers have turned to the preparation of conformationally restricted or conformationally rigid analogues. These studies were directed towards finding the preferred conformation at the muscarinic and nicotinic sites. The use of conformationally restricted analogues might provide some insight into the relationship between the preferred conformation in the solid state/solution state and the conformational requirements for agonist activity at the receptor.

After the use of several different ring systems the cyclopropane conformationally rigid analogues prepared by Cannon and his research team produced potent agonist activity at muscarinic receptors (Armstrong *et al.* 1968). They found that the activity of (+)-*trans*-2-acetoxycyclopropyltrimethylammonium iodide (ACTM) at muscarinic sites was equal to that of acetylcholine. The use of the decalin and bicyclic analogues to restrict the conformation showed that the use of too much skeleton interferes with interaction with the muscarinic receptor.

The conformationally restrained analogue of acetylcholine, *S*-(−)-aceclidine is of interest because the *N*-methyl quaternary analogue is less active than the protonated tertiary amine. Another interesting relationship is that the *S*-(+) isomer is the most active of the protonated series whereas *R*-(–)-*N*-methylaceclidine is the most active of the *N*-methyl isomers on guinea-pig ileum muscarinic receptors. (Figs. 6.4C and D).

NICOTINIC AGONISTS

Although nicotine patches are used to assist individuals stop smoking, no nicotinic agonist is used for its therapeutic action against a disease. Because the actions of nicotine are understood, however, new therapeutic potential is suggested for selective nicotinic agonists including Alzheimer's disease, Parkinson's disease, analgesia, cognitive and attention disorders and ulcerative colitis (McDonald *et al.* 1995). A variety of natural products, including lobeline, cytisine, Anatoxin-a, and epibatidine, have been described as having nicotinic agonist activity. Lobeline was found in Indian tobacco, it is said to have nicotine-like activity, and has been used as a nicotine substitute (Fig. 6.5).

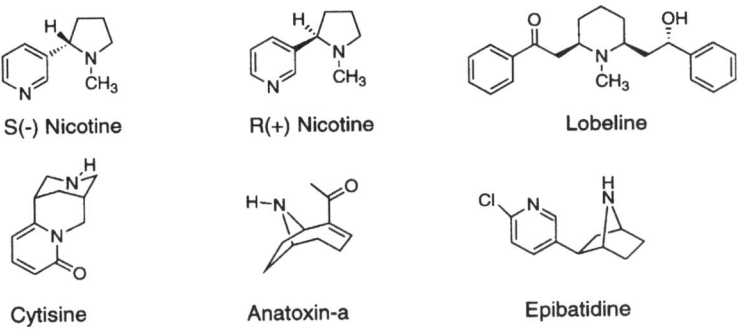

Fig. 6.5 The enantiomers of nicotine and related agonists.

The natural product cystisine has been reported to be up to four times as potent as nicotine at nicotinic receptors. The (+) isomer of anatoxin-a, a natural product found in blue–green algae, is thought to be a cause of death of waterfowl and livestock. Epibatidine was isolated from extracts of the skin of a poisonous Ecuadoran frog, *Epipedobates tricolour*. This material was shown to be a very potent analgesic, up to 200 times more powerful than morphine, and its action could not be blocked by narcotic antagonists. It has been shown to be a highly potent nicotinic receptor agonist, and has a close relationship to nicotine—having both a pyridine and aliphatic nitrogen in a ring system.

Natural epibatidine, obtained from the skin extracts, is laevorotatory as the free base and dextrorotatory as the salt. Epibatidine has been found to be more potent with a higher affinity and more intrinsic activity at almost all nicotinic receptors and it seems there is little difference between effects of the isomers of epibatidine on nicotinic receptors. Several number of synthetic schemes have been published for epibatidine and many analogues will be studied in the future of this nicotinic agonist (Fletcher *et al.* 1994).

ACETYLCHOLINESTERASE INHIBITORS

The polymeric forms of the enzymes involved have been elucidated. At the cholinergic neuroeffector junction acetylcholinesterase limits the concentration of the transmitter. The structure of the catalytic amino acid residues of the centrosymmetric 20 Å gorge are known. A single molecule of the esterase can hydrolyse 6×10^5 molecules of substrate acetylcholine min^{-1}. The product of the hydrolysis, choline, is inactive.

Edinburgh pharmacologists are credited for the introduction of Calabar bean extract, containing physostigmine, to medicine. Stedman elucidated the structure of physostigmine and developed the synthetic analogue neostigmine. The natural alkaloid (+)-physostigmine is a potent inhibitor of the acetylcholine esterase, k_i 10^{-9} M. The non-natural enantiomer also inhibits the enzyme but the k_i value is low. This work led to a number of achiral molecules being used as acetylcholine esterase inhibitors including neostigmine. More recently the tacrine prepared in the 1930s has been used to treat Alzheimer's disease. Clinical trials in China indicate that huperzine A, a constituent of a Chinese herbal medicine, is an effective acetylcholinesterase inhibitor; this has stimulated research on analogues of this agent (Hosea et al. 1995; Lattin 1995; Fig. 6.6).

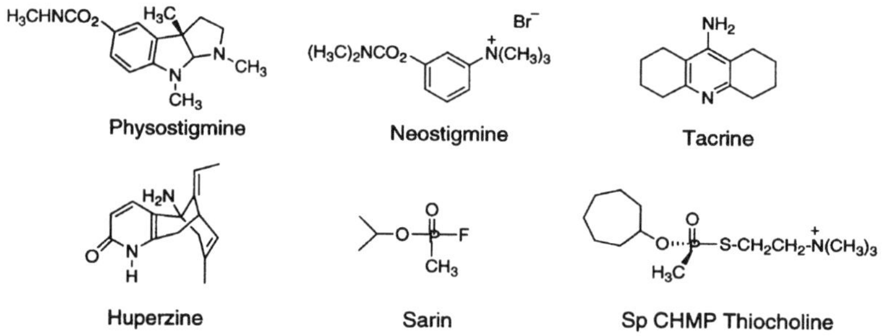

Fig. 6.6 Reversible and irreversible acetylcholinesterase inhibitors.

Acetylcholine interacts with the anionic cavity and esteratic site of the histidine and serine residue of the catalytic gorge. The pattern of interaction of the enantiomers of methacholine with the enzyme is interesting. The S-(+) form is hydrolysed at a rate nearly equal to that of acetylcholine whereas the R-(−) form is a weak inhibitor of the enzyme. In the racemate the latter form will preserve the former enantiomer.

Succinylcholine, a chiral acetylcholine analogue, is rapidly hydrolysed by a variant esterase. Individuals who lack this enzyme show prolonged neuromuscular blocking activity to succinylcholine.

The irreversible inhibitors organophosphates have found application as agricultural pesticides and chemical warfare agents. In the presence of the enzyme inhibitor, excess acetylcholine accumulates at the central and peripheral synapses leading to initial stimulation and paralysis of the skeletal muscle. The latter utility must be banned. Boter and Dijk (1969) found that (−)-sarin reacts at least 4200 times faster than the (+) form with acetylcholinesterase, but both isomers inhibit horse serum butyryl cholinesterase at nearly equal rates. The interacting site geometry of the two esterases must differ.

Taylor and associates (Hosea *et al.* 1996; Wong *et al.* 1997) investigated the stereoselectivity of chiral phosphoryl esters on mouse cholinesterase. The reaction was 200 times faster with the Sp enantiomer of cyclopentyl methylphosphoryl thiocholine (CHMP thiocholine) than with the Rp isomer. When inverted stereoselectivity was observed for a mutant, the enantiomers were used to map steric hindrance in different esterases. An aspartate residue near the gorge entrance resulted in enhanced reactivity with the cationic organophosphate. When this charge was removed in the mutant esterase, the rates of reaction of both Rp and Sp were reduced equally. These studies provide detailed understanding of the molecular mechanism of inhibition of the esterase.

CHOLINERGIC MUSCARINIC BLOCKERS

A. R. Cushny had the first opportunity to examine the pharmacological actions of (−)- and (±)-hyoscyamine. The racemate is called atropine. The antagonist shifts the concentration–response curve of agonist to the right. (Fig. 6.7) *In vivo* physiological effects of neurally released acetylcholine are also competitively blocked at the muscarinic site. The natural (−)-hyoscyamine is found in various species of *Atropa belladonna*. This plant has been used for centuries to treat various gastrointestinal disorders and in eye makeup as a cosmetic agent.

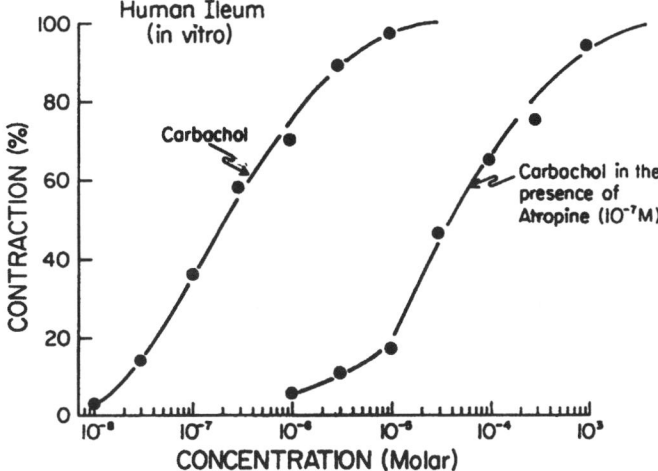

Fig. 6.7 Competitive inhibition of the muscarinic agonist by atropine (0.1 μM). Note the parallel displacement of the curve by the blocker. A dissociation constants, K_b, of 1 nM, was calculated from the data. (P. J. Rice and P. Patil, unpublished data.)

The alkaloid is moderately soluble in the lipid phase (pK_a 9.8). Crystal structure analysis reveals a tilting of the phenyl ring toward the tropane moiety, which can assume boat and chair conformations. *S*-(−)-Hyoscyamine is a competitive reversible antagonist of the muscarinic receptor in the heart, smooth muscle and glands. The equilibrium dissociation constant K_b is approximately 0.5 nM. The receptor-blocking activity of the non-natural isomer R-(+)-hyoscyamine is weak ($K_b \geq 14$ nM). No quantifiable racemization of the enantiomers was observed.

Atropine, and related synthetic analogues cyclopentolate and tropicamide, are used in ophthalmology. The pK_a of the latter antimuscarinic agent is 5.3, thus relatively more non-ionized drug crosses the corneal barrier for the quick onset of mydriatic activity. The iris muscarinic receptor-blocking activities of the enantiomers of tropicamide differ by a factor of approximately 50.

The pharmacokinetics of topical application of racemates (*e.g.* atropine) to the eye are complicated. Transcorneal movement of the atropine might be stereoselective. The melanin pigment binds drugs with little or no stereoselectivity. The esterases in aqueous humour might hydrolyse isomers stereoselectively. These problems in pharmacology must be addressed.

Paton and Rang (1965) thoroughly examined the kinetics of interaction of atropine with smooth muscle. Two binding sites with affinity constants of 1.1 and 450 nM were discovered. Only the former correlates with the functional blockade of the muscarinic receptor. The significance of the high capacity–low affinity site is unknown. Quaternization of atropine increases its affinity for the receptor approximately twofold; the cationic head of the molecule does not seem to be essential for the antimuscarinic activity (Lattin 1995).

One of the most important factors determining muscarinic antagonist activity is the length of the molecule. Molecules longer than eight atoms attached to a quaternary nitrogen or a bulky tertiary amine will often have blocking activity.

Extensive studies of structure–activity relationships have been conducted with muscarinic antagonists (Barlow 1964; Dahlbom 1983; Casey & Dewar 1993). It has been shown for benzylic esters of β-methylcholine that the benzylic chiral centre is very important for muscarinic activity whereas the β-methylcholine segment is not. It has been speculated that agonists and antagonists do not occupy the same binding site but if they do it is the acyl portion that binds more stereoselectively (Fig. 6.8). Interest in the relationship between steric structure and activity continues, with the result that muscarine M_1–M_5 receptor subtypes in different organs can be characterized. Assuming that one subtype of muscarinic receptor predominates in a given system, a selective blocker should reduce the cholinergic hyperactivity in the organ without producing side-effects.

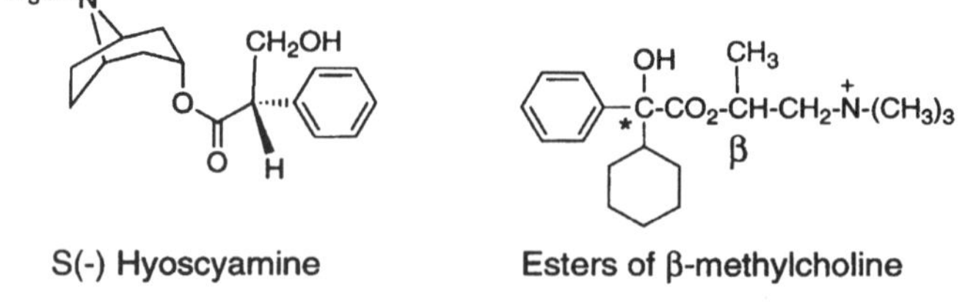

Fig. 6.8 Antimuscarinic drugs. Note the steric similarities of the molecules.

Dexetimidine, a potent muscarinic blocker with a high eudismic ratio was used to locate receptor-related binding in the rat brain caudate nucleus and in bovine tracheal smooth muscle. Distinct stereoselective binding to the muscarinic receptor was demonstrated (Soudijn *et al.* 1973; Beld & Ariens 1974). Excellent correlation was obtained between results from functional pharmacological experiments with (+)-benzetimide and binding to the subcellular receptor ($K_i \sim 1$ nM for both experiments).

The binding constants (negative log K_i values) of (+)-telenzepine for the muscarinic receptors of rat cerebral cortex, heart and lachrymal gland were 9.48, 7.80 and 8.4, respectively, and the eudismic ratios were 510, 75 and 160, respectively. The enantiomer has muscarinic M_1 receptor selectivity, and racemization of the inactive enantiomer to the active form was very slow.

Such stability of enantiomers might be of practical therapeutic value (Eveleigh *et al.* 1989). Lambrecht *et al.* (1997) synthesized *p*-fluorotrihexypendyl enantiomers. The *R* form was more potent than the *S*; negative log molar K_i values for the M_1 receptor were 9.1 and 6.6, respectively. Use of the methiodide salt of each increased K_i to 10.3 and 7.8, respectively. The affinity of each enantiomer for the M_2 receptor was less, indicating high M_1 receptor selectivity for potential use. Similarly, some enantiomers of glycopyrronium analogues were highly potent M_1 receptor blockers (Czeche *et al.* 1997).

NICOTINIC CHOLINERGIC BLOCKERS

Acetylcholine stimulates three major types of nicotinic receptor at: (i) the neuronal sensory carotid sinus, (ii) autonomic ganglia, including adrenal chromaffin cells, and (iii) the motor end plate of the skeletal muscle. Hexamethonium and (+)-tubocurarine are considered to be competitive reversible blockers of the ganglionic and motor end plate, respectively. Selective blockers of nicotinic receptors of the carotid sinus receptors are not known. Neuronal synaptic nicotinic receptors in the brain are partly blocked by tertiary nicotinic ganglionic blockers such as mecamylamine.

All nicotinic receptors seem to be ion channels. At high doses of nicotine, the receptors are desensitized and the membrane is persistently depolarized leading to the blockade of the transmission. α-Bungarotoxin, a cobra venom, interacts irreversibly with the motor end-plate receptor and cholinergic transmission fails. The mechanics of respiration associated with skeletal muscle are inhibited. Reversible (+)-tubocurarine-like molecules are used to relax skeletal muscle during surgery. The stereochemistry of the molecule is quite complex. The non-natural isomer is no more than one tenth as active as (+)-tubocurarine. The skeletal muscle-relaxant activity of (+)-*N*-dimetuylcurine is 2–3 times higher than that of the (–) enantiomer.

Curare alkaloids have provided templates for the synthesis of many useful muscle relaxants (Fig. 6.9). At high plasma concentrations (+)-tubocurarine will compete with the nicotinic receptor in the ganglia leading to hypotension as a result of the blocking action. Simple molecules like tetraethylammonium chloride block ganglionic nicotinic receptors but little effort has been focused on understanding the stereochemistry of ganglionic blockers. Although nicotinic blockers of the carotid sinus receptors are not in high demands in therapy, selective blockers of central nicotinic receptors might be of value in the treatment of nicotine addiction. The stereochemistry of nicotinic stimulants and blockers at different sites is, therefore, worthy of attention.

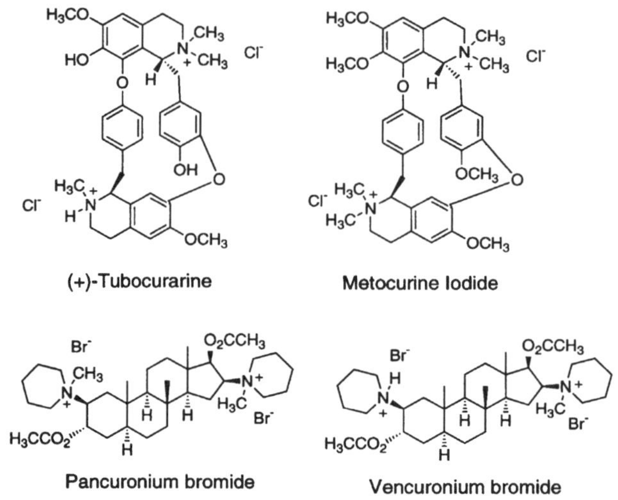

Fig. 6.9 Some blockers of the nicotinic receptors. EC_{50} of (+)-tubocurarine-like compounds for blocking neuromuscular cholinergic ion-channel transmission is ~1 μM.

STERIC ASPECTS OF DRUG ACTION AT THE NORADRENERGIC AND DOPAMINERGIC NEUROEFFECTOR JUNCTION

(−)-Noradrenaline released from the nerve terminal causes profound cardiovascular changes by activating specific receptors. Excess transmitter is re-absorbed in the nerve terminal by an active uptake₁ transport process. Even infused (−)-noradrenaline is rapidly taken up by the nerve terminal. This is a transmitter economy. K_m and V_{max} for the initial uptake of (−)-noradrenaline are summarized in Table 6.1. Many sympathomimetic drugs have substrate stereoselectivity at the transport receptor.

Intracellular noradrenergic storage granules retain amines that are transported intraneuronally. β-Phenethylamines with 1R stereochemistry are better retained than the corresponding enantiomers. Mole-for-mole displacement of the transmitter (−)-noradrenaline can occur with the exogenous sympathomimetic amine with 1R stereochemistry. The alcoholic OH group contributes to binding with ATP in the noradrenergic storage granules. If displacement of the transmitter from receptors by the exogenous amine is rapid, the amine is called an 'indirectly' acting sympathomimetic drug. In the absence of the nerve terminal these drugs do not activate adrenoceptors.

Tyramine, enantiomers of amphetamine, and methamphetamine are indirectly acting amines which induce pharmacological effects in various organs by quickly displacing neuronal noradrenaline. Amphetamines and methamphetamines produce CNS stimulation by releasing noradrenaline and partly by inhibiting the re-uptake of synaptic noradrenaline (LaPidus *et al.* 1963). Tyramine and (+)-amphetamine, can be selectively β-hydroxylated and retained in the noradrenergic storage granule but, because the total amount of the so-called false transmitter is small, these substances are not exploited as false transmitters in medicine.

Natural (−)-ephedrine alkaloid is regarded as a mixed acting sympathomimetic drug which produces pharmacological effects partly by releasing (−)-noradrenaline and partly by activating the postjunctional adrenoceptors directly. (+)-Ephedrine, (+) pseudoephedrine, and (−)-pseudoephedrine are indirectly acting amines. The rate of neuronal entry and the rate of displacement of the transmitter by these drugs determine the order of potency of indirectly acting sympathomimetic drugs. The mechanism of action of four stereoisomers of norephedrine is similar to that of the ephedrines. 'Khat', a substance of abuse, contains *nor*-pseudoephedrine and 'cathinone'. Synthetic *N*-methylcathinone is also abused as a psychostimulant (Sparago *et al.* 1996). In the body, these substances can be hydroxylated at the β-carbonyl group to produce long-lasting stimulant action.

INHIBITION OF NORADRENALINE TRANSPORT

The normal physiological effect of sympathetic nerve stimulation and the pharmacological effects of injected (−)-noradrenaline are known to be potentiated by (−)-cocaine. At higher doses (−)-cocaine also induces slowly developing sympathomimetic effects in the body. The competitive inhibition of uptake$_1$ of noradrenaline by (−)-cocaine has been established. (+)-Noradrenaline which quickly saturates the neuronal transport is not potentiated by (−)-cocaine. The transport receptor also has distinct selectivity in the inhibition of the catecholamine. (+)-Pseudococaine does not potentiate the pharmacological responses to (−)-noradrenaline (Komiskey *et al.* 1977).

Structural similarity between the functional groups of the substrate, and some transport inhibitors was proposed (LaPidus *et al.* 1963; Hendley *et al.* 1972; Maxwell *et al.* 1976; Miller *et al.* 1976; Komiskey *et al.* 1977, 1978; Ferris & Tang 1979); subsequently, the preferred chirality of a new class of anti-depressant molecules was discovered. The molecular geometry of some inhibitors of uptake$_1$ is presented in Fig. 6.10.

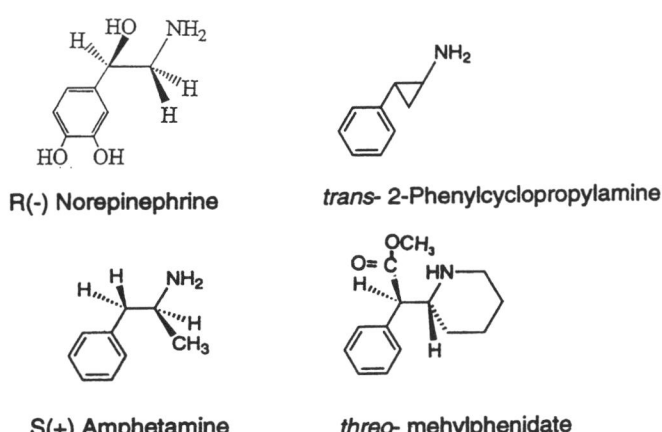

Fig. 6.10 Steric similarity of (−)-noradrenaline, the natural substrate for noradrenergic neuronal uptake$_1$ and the inhibitors of the transport.

Dopaminergic neurons of the brain were also recognized in high affinity uptake$_1$ transport of dopamine. X-ray crystallographic studies and conformational analysis of the amine in solution indicated that the catechol-phenyl group and the amino group are in the *trans* conformation. Molecular competition for the transport at the dopaminergic and noradrenergic synaptosomes revealed some conformational similarities and steric differences. At both types of neurone, *trans* conformation of two important functional groups, catechol-phenyl and -NH$_2$ group is preferred. Data in Table 6.3 summarize the relative neuronal uptake$_1$ inhibitory potency of sympathomimetic class of drugs. The inhibitors of the uptake also potentiate other pharmacological effects of the injected noradrenaline on various organs.

Table 6.3 The inhibition of neuronal uptake$_1$ of noradrenaline (NE) and dopamine by indirectly acting sympathomimetic class of substances (Hendly *et al.* 1972; Miller *et al.* 1976; Maxwell *et al.* 1976; Komiskey *et al.* 1977, 1978; Ferris *et al.* 1979).

Inhibitor	NE IC$_{50}$ (μM)	Relative potency index	Dopamine IC$_{50}$ (μM)	Relative potency index
(+)-amphetamine	0.65	13	2.10	11
(−)-amphetamine	8.70	1	23.00	1
(−)-ephedrine	0.72	~100	6.20	7
(+)-ephedrine	2.80	25	12.0	3
(+)-pseudoephedrine	21.00	3	20.5	2
(−)-pseudoephedrine	70.00	1	41.0	1
(±)-*trans*-2-phenyl-cyclopropylamine	1.20	600	1.7	324
(±)-*cis*-2-phenyl Cyclopropylamine	720.0	1	550.0	1
(+)-*trans*-2-Phenyl Cyclopropylamine	1.00	0.5	4.0	0.25
(−)-*trans*-2-Phenyl cyclopropylamine	0.50	1	1.0	1
trans-2-Amino-1-(3:4 dihydroxyphenyl) cyclobutane	8.60	9	141.0	>2
cis-2-amino-1-(3:4 dihydroxyphenyl) cyclobutane	73.0	1	>300.0	1
(±)-*threo*-methylphenidate	0.85	118	0.12	142
(±)-*erythro*-methylphenidate	100.00	1	17.00	1
(−)-cocaine	3.60	20	2.50	64
(+)-pseudococaine	72.00	1	160.00	1

CHIRALITY AT POSTJUNCTIONAL RECEPTORS

Discovery of asymmetric carbon in adrenaline and isolation of the orally effective bronchodilator alkaloid ephedrine with two asymmetric centres provided great initiative for the synthesis of sympathomimetic amines. Synthetic racemates were resolved (Chen *et al.* 1929; Jarowski & Hartung 1943; Barlow *et al.* 1972; Hawkins & Klease 1973; Waldeck 1993). The direct adrenoceptor activating properties and indirect sympathomimetic effects of the enantiomers were analysed.

Purity of enantiomers

For much of the time since the days of Pasteur, the resolution of enantiomers has remained an art. Absolute purity through chemical procedures was difficult to obtain, with each pair demanding a different art of separation. Chemical assays and highly sensitive quantitative pharmacological bioassays were necessary. Beyond certain experimentally defined limits, purity cannot be guaranteed. If the absolute difference between the activation of a receptor by a pair of enantiomers is a factor of ten, a purity of the resolved enantiomers beyond 90% is technically possible. If, however, the absolute difference is a factor of 1000, in all probability neither chemical assay nor bioassay would be able to give an indication of the 'contamination' related response of the 'inert' enantiomer. In a significant number of instances racemates have half the pharmacological activity of the active enantiomer. Cushny (1908) observed half the pressor effects of synthetic (±)-adrenaline when it was compared with natural (−)-adrenaline. He deduced that pharmacological activity of (+)-adrenaline was weak.

On isolated tissue preparations containing mainly α- or β-adrenoceptors, under proper experimental conditions where the routes of drug disposition were markedly reduced, the isomeric activity ratio for (−)- and (+)-adrenaline was 60. Several arguments indicate that (+)-adrenaline might also activate the adrenoceptor. The weaker isomer obtained from different sources, when different procedures were used to resolve the isomers, is less likely to contain identical residues of the active isomer. In these samples equivalent pharmacological activity was observed (Patil & Ruffolo 1980).

In other words, during resolution of isomers, constant pharmacological activity might provide a practical index of the purity of the enantiomers. The (−) enantiomer of α-methylnoradrenaline is approximately 1000 times more active than the (+) enantiomer at activating α-adrenoceptors. Experiments with different batches of (+)-α-methylnoradrenaline, however, indicated a large variation in pharmacological activity. The (+) enantiomer seems to be 'inert'. Its capacity to activate the adrenoceptor is questionable.

Although the optical rotation of the (−) and (+) isomers is equal and opposite in nature, the melting point and solubility characteristics of the racemate can differ. Even pK_a, the ionization characteristic of the racemate, can differ from that of an enantiomer (Pfeiffer & Jenney 1967).

Wächter *et al.* (1980) detected as little as 0.001% (−)-bupranolol in a standard mixture of (+)-bupranolol and (−)-bupranolol. They took advantage of the melting point of (±)-bupranolol which was 73.5° above that of the 'pure' forms of the enantiomers. One enantiomer was added to the other in different proportions and different melting points were obtained; the data proved the optical purity to be 99.999%. In pharmacological assays the highest eudismic ratio for blocking the β-adrenoceptor was 120. If (+)-bupranolol had 1/1000th the activity of the (−) enantiomer, an 'inert' status of the enantiomer would have been accepted. Because the observed affinity was much smaller, this proves the low affinity (1/120th) of (+)-bupranolol.

The ratio of α-adrenoceptor blocking on rabbit stomach fundus muscle by R- and S-dibozane is only 1:10, *i.e.* the S enantiomer is the more active. The affinity of *meso*-dibozane corresponds to that of S,S-dibozane. The steric demand of the receptor with regard to the dibozane molecule is marginal. It can be deduced that R-dibozane might also interact with the receptor (Fuder *et al.* 1981).

Isomeric activity ratios or eudismic ratios provide a sensitive index for differentiation of receptors. It has been postulated that similarity of receptors should generate similar isomer-activity ratios, and if the receptors are different ratios should be unequal (Patil *et al.* 1974). Such a postulate assumes not only that both agonists interact at the receptor but also that the enantiomers are 'pure'. On the basis of the postulate, and using several pairs of enantiomers, receptors have been differentiated (Buckner & Abel 1974; Birnbaum *et al.* 1975; Ruffolo 1983; Morris & Kaumann 1984; Walter *et al.* 1984).

Receptor active conformation

Knowledge of absolute configuration and of the conformation of the neurotransmitter-related substances is necessary for understanding of the initial events associated with the activation of the adrenoceptor (Pendersen *et al.* 1971; Ison *et al.* 1973; Moereels & Tollenaere 1975; Grol & Rollema 1977; Tollenaere *et al.* 1979). Enantiomers provide a valuable probe which enables understanding of the molecular mechanism in the activation of transmembrane G-protein-linked receptors.

Because correlation of pharmacological activity with optical activity or absolute configuration or the CIP nomenclature of the enantiomers gives limited useful information about the dynamic process, the solution conformations of dopamine, (−)-adrenaline and (−)-noradrenaline were studied. For dopamine '*trans*' conformation of two important functional groups around C_1 and C_2, namely phenylcatechol and NH_2 groups, is favoured. Gouche rotamers with dihedral angles of approximately 60° conformation are less favoured. Although the interconversion energy from one form to another is small, *trans* conformation of the molecule may be favourable for activation of the dopamine receptor.

The ionization characteristics of the phenol and amino groups of the catecholamines also complicate recognition of the receptor-active chemical species. At the physiological pH 95.5% of (−)-noradrenaline exists as a charged cationic NH_2^+ species which in all probability interacts with the anionic group of the aspartate residue of the α-adrenoceptor (Larsen 1969; Ganellin 1971; Miller 1978). Approximately 3.5% of the amine is present as the phenolic zwitterionic OH catechol which may have preference for interaction with the β-adrenoceptor. Thus, a single molecule of catecholamine can activate two adrenoceptors.

Conformationally restricted *cis*- and *trans*-2-(3,4-dihydroxy)cyclobutylamines were prepared to enable understanding of the geometry of activation. The affinity of the *trans* form for the dopamine receptor was 20 times that of the *cis* form. Conformational analysis of other dopamine receptor agonists also suggest the *trans* form of the enantiomers is preferred (Rice *et al.* 1989).

The optimum structure for activation of the adrenoceptor by the catecholamine has been derived on the basis of results from a variety of physicochemical methods. The distances from the centre of the aromatic ring to the protonated amino group and to the β-OH group are 5.1 and 3.6 Å, respectively. Medetomide has considerable flexibility in its interaction with the α_2-adrenoceptor. Use of conformationally restricted analogues revealed the distinct selectivity of one analogue for the receptor (Zhang *et al.* 1996; Fig. 6.11).

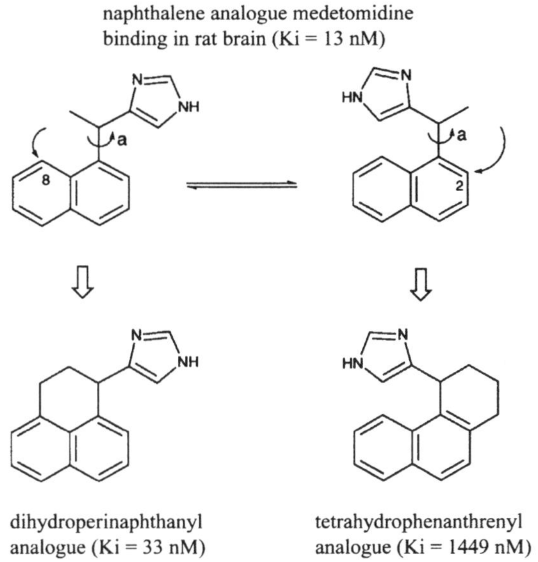

Fig. 6.11 Use of naphthanyl analogues of medetomide enabled differentiation of the α_2-adrenoceptor active conformation of the ligand. Note the K_i values.

The groups of the adrenoceptor with which the transmitter interacts were also postulated. The catechol moiety of (−)-noradrenaline interacts with serine 204 and 207 amino acid residues of the V transmembrane domain. The ASP residue seems to be a common site of interaction for many agonists. This amino acid residue forms a bridge with the Lys-331 residue of the VII *trans* membrane domain of the receptor (Porter *et al.* 1996). Agonists with charged NH_2^+ groups compete for the ASP of the bridge. Specific conformational changes can occur.

Interaction of the benzylic OH group with the receptor domain has been difficult to detect. Nonetheless, site-directed mutagenesis experiments revealed that ser 90 of the second transmembrane domain of α_2-adrenoceptor might be involved in the interaction of (−)-noradrenaline (Li *et al.* 1995). Serine-165 of the transmembrane domain was also postulated as the site of interaction of the benzylic OH group with the receptor. This interaction seems to be stereoselective because the (+) enantiomer has low affinity for interaction with the mutant.

Conformation induction by enantiomers

Because there are large differences in the pharmacological activity of stereoisomers acting on a single type of receptor, it was assumed that selective conformational changes occur when isomers act at the receptor. This process is referred to as conformation induction (Burgen 1981). The rate of change of the conformation seems to be related to the magnitude of the pharmacological effect.

When the β_2-adrenoceptor-rich fraction from the rat lung membrane was combined with (–)-adrenaline a change in the 205–220-nm region of the circular dichroism spectrum was observed (Patil et al. 1996). The (+) enantiomer had little or no effect, indicating the selectivity of the interaction. The spectrum of bovine serum albumin was altered equally by both enantiomers indicating a non-specific change. Changes in the spectral region 205–220 nm indicate that the receptor helices might be perturbed by the enantiomer (Fig. 6.12).

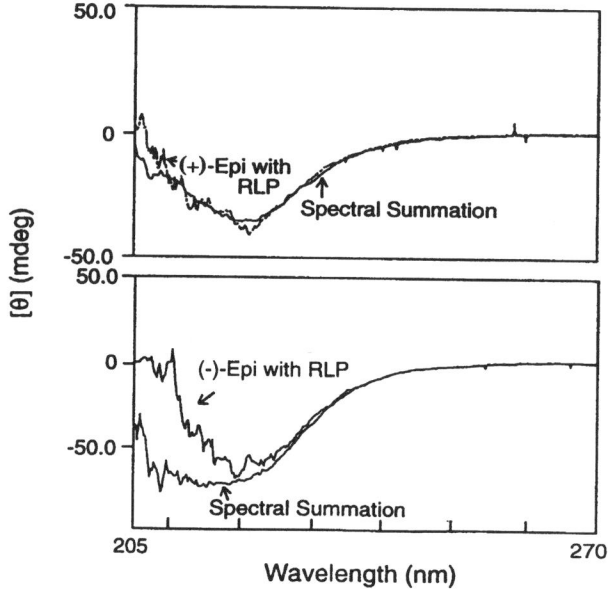

Fig. 6.12 Stereoselective alteration by the enantiomers of adrenaline of the circular dichroism spectra of rat lung membranes containing the β-adrenoceptor.

(–)-Isoproterenol concentration-dependently changed the spectrum of the fluorescence-labelled β_2-adrenoceptor protein. The (+) enantiomer was less effective. These lines of evidence indicate that stereoselective conformational changes occur when the receptor is activated by the asymmetric agonist (Gether et al. 1996).

Circular dichroism spectra of purified recombinant porcine M_2 muscarinic acetylcholine receptor obtained with and without acetylcholine or the antagonist hyoscyamine were indistinguishable. The antagonist R-(–)-quinuclidinyl benzylate, however, altered the spectra slightly in the 208 and 220 nm regions. The blocker might change the conformation of the receptor non-specifically. The binding constants of the drugs on the preparation, however, revealed expected affinity for the receptor (Peterson *et al.* 1995). Related to these results, however, is the whole question of whether purification or harvesting of the receptor in the alien cells can change its specific conformation.

EFFICACY AND ISOMERIC ACTIVITY RATIOS

Steric structure–activity studies with series of adrenoceptor agonists indicate that the rate and sequence in which functional groups of the agonist are recognized by the various sites of the receptor determines the magnitude of the pharmacological response (Patil *et al.* 1991; Patil 1998). The response is far distant from the initial perturbation of the receptor by the agonists. It is believed that the functional groups of the agonist do not interact at random with the active site of the receptor.

In a series of agonists, the sequence of the drug–receptor interaction might or might not follow the CIP sequence rule assignments of groups of the asymmetric molecule. Thermodynamic calculations of the energy changes of (–)- and (+)- adrenaline associated with the activation of the adrenoceptor furnished $\Delta G°$ values of – 8.1 and –6.2 kcal mol^{-1}, respectively (Rice *et al.* 1989). The difference of approximately 2 kcal mol^{-1} is consistent with the formation of a weak hydrogen-bond between the benzylic hydroxyl group of (–)-adrenaline and the receptor site. These small selective and sequential changes are integrated to produce the effect.

Thus, the descriptive term 'pharmacological intrinsic activity' or 'efficacy' on the organ system, expressed as the relative percentage effect of the agonist, might have such a molecular explanation (Patil 1996). For these reasons, isomeric activity ratios of the agonist based either on affinity or efficacy difference should provide valuable data enabling discrimination of receptor subtypes. Similarity between the active sites of the receptor should generate similar activity ratios for a pair of agonists acting on a receptor.

THE EASSON–STEDMAN HYPOTHESIS

At Edinburgh in 1933, Stedman and co-workers conducted an interesting pharmacological investigation on closely related substances (Easson & Stedman 1933). That report remains a classic explanation of the activities of (–)- and (+)- adrenaline. Easson and Stedman proposed that three important functional groups attached to the asymmetric carbon of (–)-adrenaline were involved in interaction with the receptor. These groups are (i) changed NH_2^+, (ii) phenylcatechol, and (iii) the alcoholic OH group. Because (+)-adrenaline is the mirror image of the (–) form, the alcohol OH group might be oriented away from a site of the receptor.

Two functional-group interactions of the molecule with the receptor which they hypothesized would result in the low potency of the (+) form. Epinine, which lacks a hydroxyl group, was expected to have pharmacological activity equal to that of (+)- adrenaline. The rank order of the biological activity of the series of related compounds was (–)-adrenaline > (+)-adrenaline = epinine. The hypothetical interacting sites of the stereoisomers and epinine are illustrated in Fig. 6.13 (Banning *et al.* 1984).

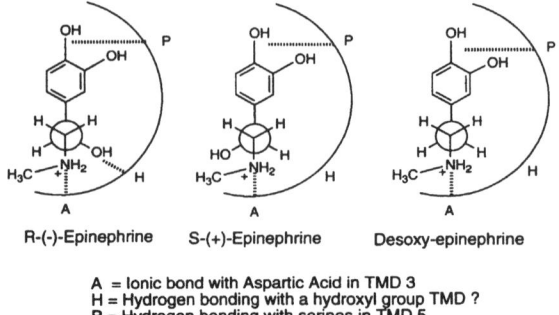

R-(-)-Epinephrine   S-(+)-Epinephrine   Desoxy-epinephrine

A = Ionic bond with Aspartic Acid in TMD 3
H = Hydrogen bonding with a hydroxyl group TMD ?
P = Hydrogen bonding with serines in TMD 5

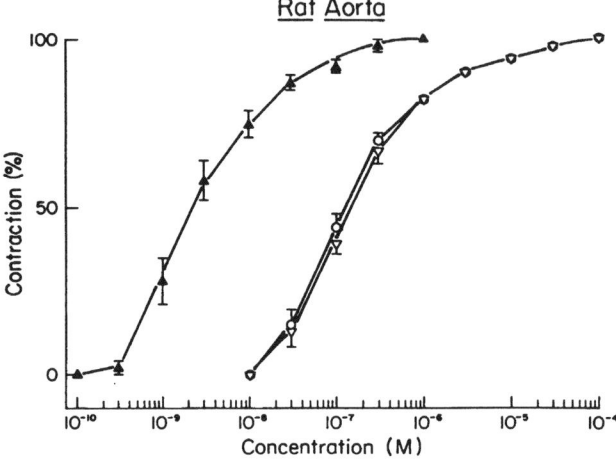

Fig. 6.13 Neuman projection of enantiomers. The Easson–Stedman hypothesis indicated that the greater activity of (–)-adrenaline was because of the sterically correct orientation of the benzylic OH with the receptor. Such an interaction might be missing between the (+) isomer and the desoxy form, resulting in low activity. The dose–response curve for rat aorta (α-adrenergic response) is indicative of the greater higher activity of (–)-adrenaline (▲) and the low but equal potency of (+)-adrenaline (∇) and epinine (O).

This hypothesis was vigorously tested for enantiomers of adrenaline, noradrenaline and isoproterenol. The rank order of potency on the basis of quantitative studies of catecholamines was (–) enantiomer > (+) enantiomer = desoxy analogue; the equal activity of the last two drugs indicated that the (+) isomer acts as if the alcohol hydroxyl group is missing. The hypothesis holds true for activation of α- or β-adrenoceptors (Patil *et al.* 1974).

DEVIATION FROM EASSON–STEDMAN HYPOTHESIS

Sympathomimetic amines with two asymmetric centres provide additional probes for enabling understanding of receptor site topography. α-Methylnoradrenaline contains two asymmetric centres. The agonist activity is (–)-*erythro*-α-methylnoradrenaline >

(+)-desoxy form >> (+)-*erythro*-α-methylnoradrenaline. This ranking does not conform with that predicted by the hypothesis.

The concentration–response curves for (+)-noradrenaline and dopamine deviate from the Easson–Stedman hypothesis that the maximum lipolytic activity of the (+) enantiomer is greater than that of dopamine, which has a bell-shaped dose–response curve. It is possible, therefore, that the desoxy derivative might act on the secondary auto-inhibitory receptor (Lee *et al.* 1974).

Consistent deviation from the Easson–Stedman hypothesis was observed for catecholimidazolines, which are closely related to catecholamines (Miller *et al.* 1983; Rice *et al.* 1987). On a variety of tissues containing α-adrenoceptors the rank order of potency was desoxycatecholimidazoline ≥ (−)-catecholimidazoline > (+)-catecholimidazoline (Fig. 6.14). A similar rank order of potency was observed for catecholamidines, open-chain analogues of catecholimidazolines. Imidazolines or amidines are very poor activators of β-adrenoceptors so the hypothesis cannot be tested for these compounds.

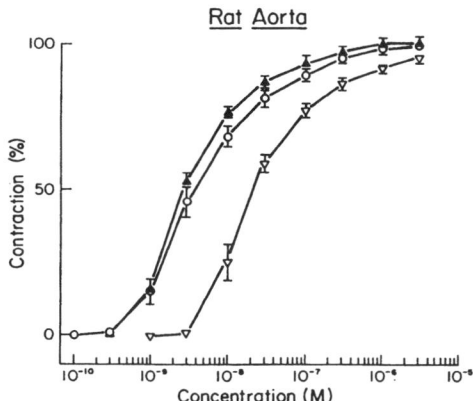

Fig 6.14 The imidazoline class of adrenergic agonists. Isomers of catecholimidazolines and their desoxy analogues do not obey the Easson–Stedman hypothesis. The rank order of activity is desoxy form (○) = (−) enantiomer (▲) > (+) enantiomer (▽). The charged nitrogen of the imidazoline ring might, however, share a common site with the NH_2^+ of adrenaline when interacting with the receptor.

Steric structure–activity investigations with related molecules has provided valuable information about similarities and differences between types of activation of the α-adrenoceptor. Although the charged amino group of (−)-adrenaline and (−)-

catecholimidazoline might act at a common site, the 1R stereochemistry of the alcohol OH group in both types of molecule might result in different orientation during receptor activation, leading to the hindrance of the maximum conformational change of the receptor. Such an analysis explains why desoxycatecholimidazoline is more active than (–)-catecholimidazoline.

CHIRALITY OF ADRENOCEPTOR BLOCKERS

Competitive reversible α-adrenoceptor blockers

Ergot alkaloids are known to block the pressor effects of injected adrenaline in laboratory animals. Ergolines are chemically complex derivatives of the indole alkaloids which interact with both 5-hydroxytryptamine and the α-adrenoceptor (Liang *et al.* 1997). The chemical structures of some blockers are provided in Fig. 6.15.

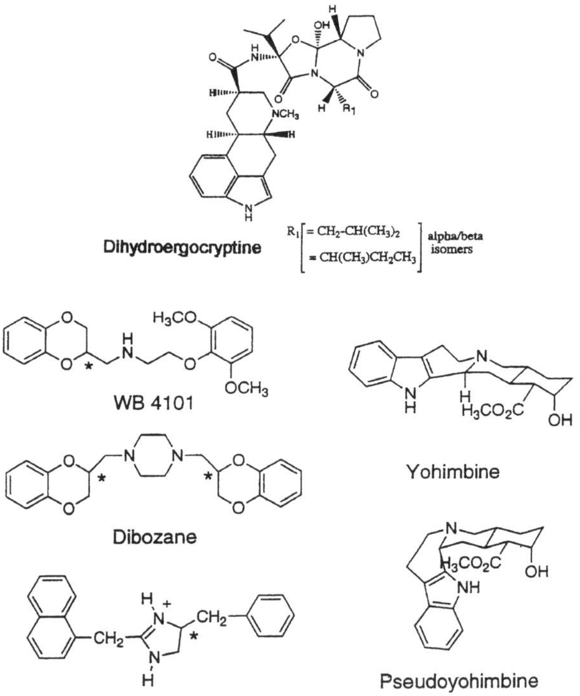

Fig. 6.15 Competitive reversible blockers of α-adrenoceptors. Dissociation constants (nM) as negative log K_b values for α-adrenoceptor blockade were: dihydroergoorptine 8.0; WB4101 *S*, 9.2, *R*, 7.5, *RS* 9.0; dibozane *SS*, 7.6, *RR*, 6.0, *meso*, 7.5; benzylnaphazoline, *S*, 6.7, *R*, 6.6; yohimbine 7.6; pseudoyohimbine ≤ 6.0.

The dissociation constant, K_b, values obtained from the functional pharmacological methods compare well with those obtained from ligand competition studies with tissue homogenates in which radiolabelled ligand competes with the blocker for the receptor. Thus it is possible to classify pre- or postjunctional receptors on the basis of selective antagonists.

There is marked structural diversity in the blockers which block α-adrenoceptor. E. Founeau and D. Bovet examined analogues of Gravitol, which was known for its uterine stimulant activity. Pharmacological screening indicated that prosympal was an adrenaline antagonist (Smeader 1985). Steric structural similarity between (−)-adrenaline and the enantiomers of benzodioxanes has been pointed out. Potent selective competitive α-adrenoceptor blockers at the vascular receptors block the noradrenergic stimuli. Vasoconstrictor effect is blocked, blood pressure is reduced, and antihypertensive effect is observed.

Enantiomeric blocking specificity of various receptors in a single organ can also be readily detected. One type of receptor usually has higher blocking ratios than the other (Table 6.4); *in vitro* the rate of offset of the blocking action of both enantiomers of WB4101 was similar. This indicates that physical diffusion of each competitive reversible enantiomer from the biophase was not related to the stereoselective block at the receptor (Fig. 6.16).

Table 6.4 α-Adrenergic, 5-HTergic and histaminergic receptor blocking actions of enantiomers of WB4101 (from Fuder *et al.* 1981).

Agonist	WB4101 isomer	−log K_B aorta	Eudismic ratio
phenylephrine	S	9.16 ± 0.05 ($n = 5$)	
	R	7.54 ± 0.1 ($n = 5$)	42
5-hydroxytryptamine	S	6.52 ± 0.14 ($n = 4$)	
	R	5.92 ± 0.07 ($n = 4$)	4
histamine	S	< 5.0 ($n = 3$)	
	R	< 4.5 ($n = 3$)	3

The molecular symmetry of dibozane, with two asymmetric carbons, provides one of the best examples of investigations of drug–receptor interactions (Nelson *et al.* 1977). There are no such *meso* compounds with agonist properties. Studies with enantiomers of dibozane indicate the conformational flexibility of the drug–receptor interaction. Compared with antagonists, conformational requirements and restrictions of agonists for induction of effective stimuli are high.

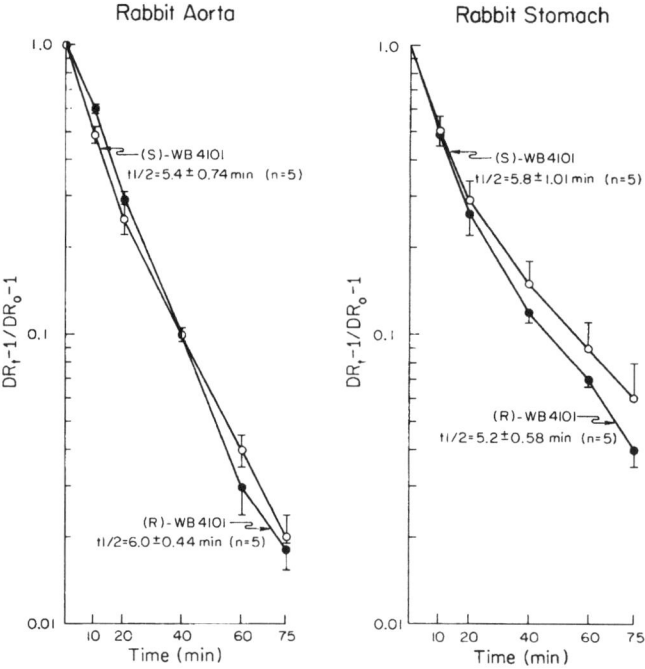

Fig. 6.16 Rate of 'off-set' of α-adrenoceptor blockade in two types of smooth muscle. $t^{1/2}$ varied between 5 and 6 min for each enantiomer, indicating that diffusion out of the biphase is a non-stereoselective process. DR = dose ratio = (EC_{50} with blocker)/(EC_{50} control). From H. Fuder and P. Patil, unpublished observation.

Prejunctional α-adrenoceptors regulate the release of the transmitter at the neuroeffector junction. Exogenous agonists can selectively inhibit the nerve-stimulated release of the transmitter. Antagonists, by blocking the inhibitory receptor, increase the overflow of the transmitter. The affinities of the blocker at prejunctional or postjunctional receptors can differ. Differential blockade of pre and postjunctional receptors by $2R$ and $2S$ enantiomers of WB 4101 a benzodioxane derivative, has been reported (Mottram 1981). Yohimbine and its diastereoisomers have been used as a tool for classification of α-adrenoceptor subtypes (Tanaka & Starke 1980).

Tolazoline and phentolamine among the earliest imidazoline-type competitive antagonists known. Introduction of asymmetry at C_3 of the imidazoline ring reduced agonist potency (Hsu *et al.* 1980). Large ethyl- or phenyl-group substitution at C_3 resulted in competitive inhibition at the α-adrenoceptor. Differences between the affinities of the enantiomers were small. Quaternization of some α-adrenoceptor blocking agents is well tolerated, but the effect of such a change on the eudismic ratio is unknown.

Amosulalol is an N-substituted β-phenethylamine class of α-adrenoceptor blocker. The eudismic ratio is approximately 10. A potent α-adrenoceptor blocker, YM-12617, an analogue related to amosulalol has been synthesized. The *R* enantiomer is one of the most potent α_1-adrenoceptor blockers reported (K_i 0.1 nM). Eudismic ratios in various tissues, including urinary bladder, enabled differentiation of adrenoceptors (Honda & Nakagawa 1986; Honda *et al.* 1986; Yazawa *et al.* 1992).

Irreversible α-adrenoceptor blockers

Cyclization of β-haloalkylamines to the carbonium ions results in irreversible interaction with the receptor (Belleau 1958; Belleau & Triggle 1962; Belleau & Cooper 1963). The specific group of the receptor affected by the blocker is not known. In a tissue containing seven transmembrane domains, receptor families such as the 5-hydroxytryptamine, muscarinic and histaminergic receptors are also blocked irreversibly. Rates of Cyclization of series of β-haloalkylamines affects the onset of the block. Similar blocking behaviour of the enantiomers was explained on the basis that a common molecular species is produced by both substances (Fig. 6.17).

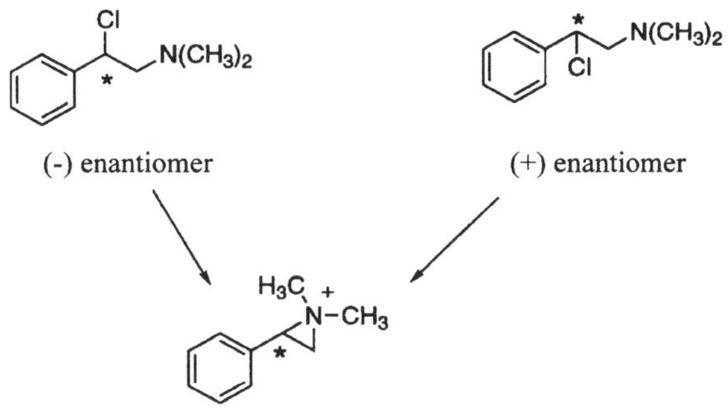

Fig. 6.17 The postulated common cyclized molecular species from the optical isomers of *N,N*-dimethyl-β-chlrophenethylamine, which are equipotent irreversible blockers.

Portoghese *et al.* (1971) synthesized enantiomers of phenoxybenzamine, assigned the absolute configurations, and reported interesting pharmacological data on the substances. The relative order of potency of α-adrenergic blockade was *R*-(+)-phenoxybenzamine > desmethylphenoxybenzamine > *S*-(−)-phenoxybenzamine (Fig. 6.18).

Purified rat liver plasma membrane containing α-adrenoceptors binds [^{3}H]phenoxybenzamine. The pharmacologically active *R*-(+) enantiomer inhibited binding with an EC$_{50}$ value of 0.01 μM whereas *S*-(−)-phenoxybenzamine was 1/100th as protective. Under similar experimental conditions the EC$_{50}$ value of both (−)-adrenaline and (−)-noradrenaline was 100 μM. A concentration of the (+) enantiomers only 2–3 times higher was required to protect against the irreversible binding of the labelled phenoxybenzamine. Data indicate that irreversible blockers interact with a stereoselective site unrelated to the transmitter interacting domain of the α-adrenoceptor (Guellaen *et al.* 1979). Brasili *et al.* (1981) reported tetramine analogues which interact covalently with α-adrenoceptors.

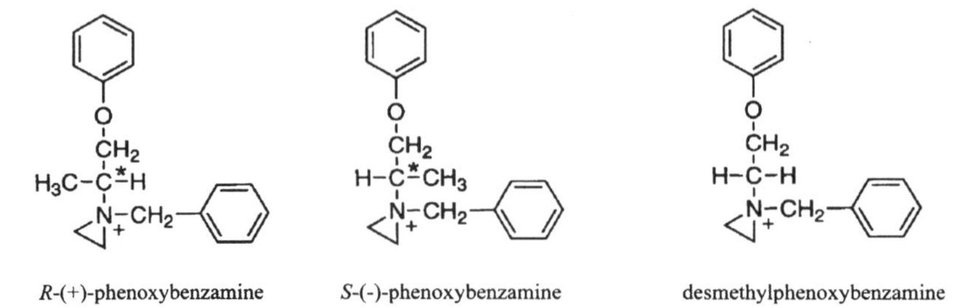

R-(+)-phenoxybenzamine   S-(−)-phenoxybenzamine   desmethylphenoxybenzamine

Fig. 6.18 Differential irreversible blocking action by the enantiomers of phenoxybenzamine is partly related to the retention of the asymmetry. Relative potencies for 50% blockade by (+)-phenoxybenzamine, the desmethyl analogue, and (−)-phenoxybenzamine was 15:7:1, respectively.

Enantiomers of β-adrenoceptor blockers

In 1958, dichloroisoproterenol (in which the catechol groups of isoproterenol had been replaced by dichloro-substitution) was reported as the first β-adrenoceptor blocker; it which blocked the vascular relaxation, uterine inhibition and other pharmacological responses mediated by the agonists (Powell & Slater 1958). The stereochemistry around C_1 was untouched. The marked steric and chemical resemblance between agonists and the antagonists is indicative of syntopic antagonism, i.e. certain groups of agonists and antagonist share a common site of interaction at the receptor (Fig. 6.19).

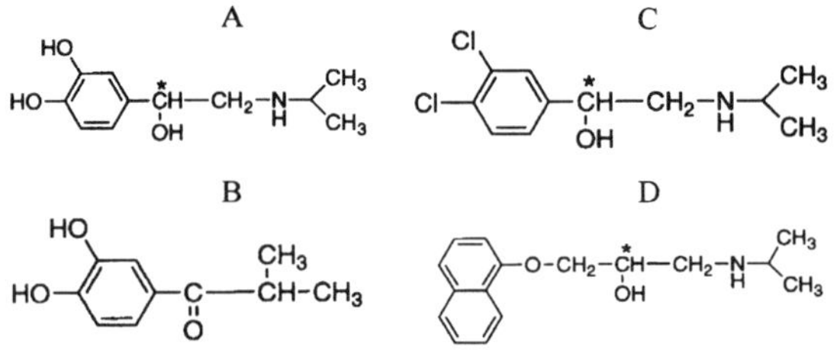

Fig. 6.19 The development of the syntopic antagonist. Note the similar chemical structures of agonists and antagonists. A. *R*-(−)-isoproterenol, potent agonist (EC_{50} 0.5 nM); B. 3,4-dihydroxy-α-methylpropiophenone, weak agonist (EC_{50} 5.5 mM); C. 3,4-dichloroisoproterenol, antagonist (k_B 500 nM); D. propranolol, potent antagonist (k_B 1 nM).

When the racemate was resolved the eudismic ratio was found to be 100. Lipophilicity, pK_a values, and steric factors were invoked to explain the affinity of the antagonists for the receptor. 1*R* Stereochemistry of the different agonists does not necessarily follow the 1*R* stereochemistry of the antagonists. Antagonists might follow a different sequence of orientation so that conformational change of the receptor, if any, might be random or minimum. The dissociation constants of several competitive blockers are summarized in Table 6.5.

Table 6.5 Dissociation constants, presented as $-\log K_b$, of the enantiomers of some β-adrenoceptor antagonists (Patil 1968; Morris & Kaumann 1984; Walter *et al.* 1984; Nathanson 1988).

Name	$-\log K_b$	Isomer ratio
(−)-isopropylmethoxamine	6.53	>1100
(+)-isopropylmethoxamine	<3.50	(T)
(±)-deoxyisopropylmethoxamine	4.85	
(−)-sotalol	6.80	44
(+)-sotalol	5.15	(T)
deoxysotalol	3.88	
(−)-timolol[a]	8.20 (H)	44
	8.26 (CP)	(H)
(+)-timolol	8.55 (H)	1.0
	8.34 (CP)	(CP)
(−)-propranolol[b]	8.71 (H)	110
	9.16 (T)	(H)
(+)-propranolol	6.51 (H)	>100
	7.20 (T)	(T)
(−)-bupranolol[b]	9.2 (H)	63
	9.5 (T)	(H)
(+)-bupranolol	7.4 (H)	30
	8.05 (T)	(T)
(−)-pindolol	9.5 (H)	200
	9.6 (T)	(H)
(+)-pindolol	7.2 (H)	63
	7.8 (T)	(T)

T, guinea-pig trachea; H, heart; CP, ciliary process.
[a] For inhibition of adenylate cyclase.
[b] From ligand-binding experiments.

Honda *et al.* (1986) reported interesting findings with enantiomers of amosulalol and its desoxy form—the α-adrenoreceptor subtypes (α_1 and α_2) favoured the (+) enantiomer and the desoxy form, whereas the β-subtypes (β_1 and β_2) favoured the (−) isomer. These studies indicated the different requirements of two major receptors for interaction with closely related molecules.

Because the β-adrenoceptor-related activity of the blockers were stereoselective, an attempt was made to dissociate receptor-related activity from those related to other receptors (Patil et al. 1974). The local anaesthetic activity of the enantiomers was equal. For timolol, although the ocular hypotensive effect of the S-(−) enantiomer was slightly greater than that of the R-(+) form, the ocular hypotensive mechanism might not be totally dependent on β-adrenoceptor blockade.

Cardiac arrhythmias can be elicited by a variety of drugs, including L-adrenaline. The latter agonist induces cardiac arrhythmia which is quickly terminated by the potent enantiomer with the β-adrenoceptor blocking properties. The local anaesthetic property and weak β-adrenoceptor activity of the distomer provided an important probe enabling understanding of the other receptor-mediated events which contribute to different types of cardiac arrhythmia.

The question arises as to whether β-adrenoceptor blockers interact with α-adrenoceptors Many β-blockers at concentrations approximately 100 times their dissociation constants will competitively block α-adrenoceptors. The enantiomers have equal affinity for the latter type of receptor. Carvedilol has free-radical scavenging properties which are much needed to protect the diseased heart (Ma et al. 1996). The value of the single isomer compared with the racemate needs to be elucidated.

β-Adrenoceptors are widely distributed in the body. Blockade of pineal β-receptors could interfere with sleep cycle, and the stereoselectivity of chronopharmacology has been investigated (Lernmer & Bathe 1982). The enantiomers of a *beta* receptor antagonist do not alter the seizure threshold, indicating that β-adrenoceptors are not involved in the excitability of the central nervous system (Yeoh & Wolf 1970). The R- and S-propranolol glycol metabolites of propranolol have been investigated for β-adrenoceptor-blocking action. The competitive blocking activity of the S form was 1/1000 that of R,S-propranolol (Ogg et al. 1987). The effect of the R form was irreversible. These unexpected findings might have interesting functional implications during chronic exposure *in vivo*.

Studies of labetalol, with two asymmetric centres, have revealed the complexities of recognition of the molecule by the receptor (Riva et al. 1991). Receptor-mediated binding and functional pharmacological methods with enantiomers have provided a useful probe for investigation of differential receptor subtypes, leading to the production of new enantiomeric substances with better therapeutic properties.

CONCLUDING REMARKS

Natural products containing (−)-hyoscyamine, (−)-reserpine and (−)-ephedrine were introduced to medicine a long time ago. Since the discovery of primary transmitters in the nervous system, asymmetric molecules have provided the most valuable probe for enabling understanding of the variety of events associated with the neuroeffector junction.

Stereoisomers of the alkaloid muscarine have furnished important clues about probable conformations required for activation of nicotinic and muscarinic receptors by the symmetrical transmitter acetylcholine. In conformationally restricted cyclopropane analogues the two important functional groups of the charged molecule were *trans* orientated. On the basis of work on (+)-physostigmine new reversible synthetic analogues were prepared resulting in improved cholinergic function in the diseased state. The general availability of the racemate of a organophosphate pesticide is questioned because the vertebrate safely index for one enantiomer might be better than that for the other. Muscarinic receptor subtypes have been investigated with a pair of enantiomers. A selective blocker should reduce the specific hyperactivity of the cholinergic function with minimum side effects.

The elucidation of the steric structure of (–)-reserpine and related alkaloids was not only prize-winning chemistry—the antihypertensive drug was a valuable tool for studying the functions and metabolism of the transmitter associated with the noradrenergic neurone. Independently the stereospecific biosynthetic pathway for (–)-noradrenaline was established. Molecular selectivity in the transport of drugs across the noradrenergic neurone has been investigated. This led to the introduction of L-dopa, L-α-methyldopa and tranylcypromine-like drugs in therapeutics. The central nervous stimulant action of (–)-cocaine, (+)-amphetamine and threomethylphenidate has been explained.

Many α-adrenoceptor blockers were known before the discovery of the first syntopic *beta* adrenoceptor antagonist, dichloroisoproterenol, was discovered in the late fifties, a discovery which led to the introduction of propranolol, the most potent β-blocker. The pharmacological activity resides in a single isomer of many blockers marketed today.

As early as 1933, Stedman and his associates had great insight in the understanding of the activation of receptors in relation to the orientation of the functional groups of the enantiomers. The postulate of the rank order of potency (–)-adrenaline > (+)-adrenaline = the desoxy analogue epinine has been verified for activation of α- or β-adrenoceptors. Stereoisomers of catecholimidazolines provided additional clues about the molecular activation of the α-adrenoceptor. Steric aspects of the action of agonists at the receptor needs additional collaborative study among investigators so that efficacy at the therapeutic level can be understood.

ACKNOWLEDGEMENT

Our investigation included in this chapter was in part supported by a USPHS grant from NIH, GM 29358.

REFERENCES

Abderhalden, E. & Müller, F. (1908) Uber das Verhalten, des Blutdruckes nach Intravenôser Einführung von l-d-und dl-Suprarenin. *Hoppe-Seyler's Z Physiol. Chemie* **58**, 185–188.

Angeli, P., Cantalamessa, F., Cavagna, R., Conti, P., De Amici, M., De Micheli, C. Gamba, A. & Manucci, G. (1997) Synthesis and pharmacological characterization of enantiomerically pure muscarinic agonists: difluoromuscarines. *J. Med. Chem.* **40**, 1099–1103.

Ariens, E. J. (Ed.) (1964) *Molecular Pharmacology*. Academic Press, New York.

Ariens, E. J., Van Rensen, J. J. S. & Welling, W. (Eds) (1988) *Stereochemistry of Pesticides: Biological and Chemical Problems*. Elsevier Science, Amsterdam.

Armstrong, P. D., Cannon, J. G. & Long, J. P. (1968) Conformationally rigid analogs of acetylcholine. *Nature*, **220**, 65–66.

Arunlakshana, O. & Schild, H. O. (1959) Some quantitative uses of drug antagonists. *Brit. J. Pharmacol.*, **14**, 48–58.

Banning, J. W., Rice, P. J., Miller, D. D. *et al.* (1984) Differences in the adrenoceptor activation by stereoisomeric catecholimidazolines and catecholamines. In: Neuronal and Extraneuronal Events in Autonomic Pharmacology (Ed. by Flaeming *et al.*), pp. 167–180. Raven Press, New York.

Barlow, R. B. (1964) *Introduction to Chemical Pharmacology*, 2nd edn, pp. 121–139. John Wiley and Son, New York.

Barlow, R. B., Franks, F. M. & Pearson, J. D. M. (1972) The relation between biological activity and the degree of resolution of optical isomers. *J. Pharm. Pharmacol.*, **24**, 753–761.

Bartholini, G., Constantinidis, J., Puig, M., Tissot, R. & Pletscher, A. (1975) The stereoisomers of 3,4-dihydroxyphenyl serine as precursors of noradrenaline. *J. Pharmacol. Exp. Ther.*, **193**, 523–532.

Beld, A. J. & Ariens, E. J. (1974) Stereospecific binding as a tool in attempts to localize and isolate muscarinic receptors. *Eur. J. Pharmacol.*, **25**, 203–209.

Belleau, B. (1958) The mechanism of action at receptor surfaces. *Can. J. Biochem. Physiol.*, 36, 731–753.

Belleau, B. & Cooper, P. (1963) Conformation of N-(β-chloroethyl)-2-phenoxyethylamines in relation to adrenergic blocking activity. *J. Med. Chem.*, **6**, 579–583.

Belleau, B. & Triggle, D. J. (1962) Blockade of adrenergic α-receptors by a carbonium ion. *J. Med. Pharm. Chem.*, **5**, 636–639.

Belleau, B., Fang, M., Burba, J. & Moran, J. (1960) The absolute optical specificity of the monoamine oxidase. *J. Am. Chem. Soc.*, **82**, 5752–5754.

Bergin, R. (1971) *The Molecular Structure of some Sympathomimetic Amines and Related Substances*, Department of Medical Physics, Karolinska Institute, Stockholm, Sweden.

Birnbaum, J. E., Abel, P. W., Amidon, G. L. & Buckner, C. K. (1975) Changes in mechanical events and adenosine 3',5'-monophosphate levels induced by enantiomers of isoproterenol in isolated rat atria and uteri. *J. Pharmacol. Exp. Ther.*, **194**, 396–409.

Boter, H. L. & Dijk, Van C. (1969) Stereospecificity of hydrolytic enzymes on reaction with asymmetric organophosphorus compounds III, The inhibition of acetylcholinesterase and butyrylcholinesterase by enantiomeric forms of sarin. *Biochem. Pharmacol.*, **18**, 2403–2407.

Brasili, L. Grianella, M., Melchorre, C., Belleau, B. & Benefey, B. (1981) Structural requirements for adrenergic α-receptor covalent occupancy by chiral tetramine disulphides. *Eur. J. Med. Chem.*, **16**, 115–118.

Buckner, C. K. & Abel, P. (1974) Studies on the effects of enantiomers of soterenol, trimetoquinol and salbutamol on *beta* adrenergic receptors of isolated guinea-pig atria and trachea. *J. Pharmacol. Exp. Ther.*, **189**, 616–625.

Burgen, A. S. V. (1981) Conformational changes and drug action. *Fed. Proc.*, **40**, 2723–2728.

Casey, A. A. & Dewar, H. D. (1993) *Cholinergic Agonists in The Steric Factor in Medicinal Chemistry* (Ed. by A. A. Casy), pp. 231–285. Plenum Press, New York.

Chen, K. K., Wu, C. K., & Henrikson, E. (1929) Relationship between the pharmacological action and the chemical constitution and configuration of the optical isomers of ephedrine and related compounds. *J. Pharmacol. Exp. Ther.*, **36**, 363–401.

Creveling, C. R., Morris, N., Schimizu, H., Ong, H. & Daly, J. (1972) Catechol O-methyltransferase. iv. Factors affecting *m*-and *p*-methylation of substituted catechols. *Mol. Pharmacol.*, **8**, 398–409.

Cushny, A. R. (1908) The action of optical isomers III, Adrenaline. *J. Physiol. (London)*, **37**, 130–138.

Czeche, S., Elgen, M., Noe, C., Waelbroek, M., Mutschler, E. & Lambrecht, G. (1997) Antimuscarinic properties of the stereoisomers of glycopyrronium bromide. *Life Sci.*, **60**, Abstracts, p. 1167.

Dahlbom, R. (1983) Stereoselectivity of cholinergic and anticholinergic agents. In: *Stereochemistry and Biological Activity of Drugs* (Ed. by E. J. Ariens, W. Soudijn, & P. B. M. W. M. Timmermans), pp. 127–142. Blackwell Scientific Publications, Oxford, UK.

Easson, L. H., & Stedman, E. (1933) CLXX. Studies on the relationship between chemical constitution and physiological action V. Molecular dissymmetry and physiological activity. *Biochem. J.*, **27**, 1257–1266.

Euler, V. U. S. (1946) A specific sympathomimetic ergone in adrenergic nerve fibres (sympathin) and its relation to adrenaline and nor-adrenaline. *Acta Physiol. Scand.*, **12**, 73–97.

Eveleigh, P., Hulme, E. C., and Birdsall, N. J. M. (1989) The existence of stable enantiomers of telenzepine and their stereoselective interaction with muscarinic receptor subtypes. *Mol. Pharmacol.*, **35**, 477–483.

Ferris, R. M. & Tang, F. L. M. (1979) Comparison of the effects of the isomers of amphetamine, methylphenidate and deoxypipradol on the uptake of l-[^{3}H]noradrenaline and [^{3}H]dopamine by synaptic vesicles from rat whole brain, striatum and hypothalamus. *J. Pharmacol. Exp. Ther.*, **210**, 422–428.

Fletcher, S. R., Baker, R., Chambers, M. S., Herbert, R. H., Hobbs, S. C., Thomas, S. R., Verrier, H. M., Watt, A. P. & Ball, R. G., (1994) Total synthesis and determination of the absolute configuration of epibatidine. *J. Org. Chem.*, **59**, 1771–1778.

Fuder, H., Nelson, W. L., Miller, D. D. & Patil, P. N. (1981) *Alpha* adrenoceptors of rabbit aorta and stomach fundus. *J. Pharmacol. Exp. Ther.*, **217**, 1–9.

Fujiwara, H., Inagaki, C., Ikeda, Y., & Tanaka, C. (1976) Decarboxylation of stereoisomers of 3:4 dihydroxyphenylserine (Dops) *in vivo*. *Folia, Pharmacol., Japan*, **72**, 891–898.

Furchgott, R. F., Bursztyn, P. (1967) Comparison of dissociation constants and of relative efficacies of selected agonists acting on parasympathetic receptors. *Ann. NY Acad. Sci.*, **144**, 882–898.

Ganellin, C. R. (1971) Relative concentrations of zwitterionic and uncharged species in catecholamines and the effect of *N*-substituents. *J. Med. Chem.*, **20**, 579–581.

Garg, B. D., Krell, R. D., Sokoloski, T., & Patil, P. N. (1973) Steric aspects of adrenergic drugs XXII, Retention of (+) and (−) ^{14}C-noradrenaline by mouse heart. *J. Pharm. Sci.*, **62**, 1126–1129.

Gether, U., Lin, S. & Kobilka, B. K. (1996) Fluorescent labelling of purified B_2 adrenergic receptor. *J. Biol. Chem.*, **270**, 28268–28275.

Grohmann, M. & Trendelenburg, U. (1984) The substrate specificity of uptake$_2$ in the rat heart. *Naunyn Schmied. Arch. Pharmacol.*, **328**, 164–173.

Grol, C. J. & Rollema, H. C. (1977) Conformational analysis of dopamine by the INDO molecular orbital method. *J. Pharm. Pharmacol.*, **29**, 153–156.

Guellaen, G., Aggerbeck, M. & Hanoune, J. (1979) Characterization and solubilization of the α-adrenoreceptor of rat liver plasma membranes labelled with [^{3}H]phenoxybenzamine. *J. Biol. Chem.*, **254**, 10761–10768.

Hardman, J. G., Limbird, L. E., Molinoff, P. B., Ruddon, R. W. & Gilman, A. G. (Eds) (1996) *Goodman and Gilman's The Pharmacological Basis of Therapeutics*, 9th edn, pp. 105–248. McGraw–Hill, New York.

Hawkins, C. J. & Klease, G. T. (1973) Resolution of ($\pm$)-adrenaline. *Aust. J. Chem.*, **26**, 2553–2554.

Hellman, G. G., Hertting, G. & Peskar, B. (1971) Uptake kinetics and metabolism of 7-^{3}H-dopamine in the isolated perfused rat heart. *Br. J. Pharmacol.*, **41**, 256–269.

Hendley, E. D., Snyder, S. H., Fauley, J. J. & Lapidus, J. B. (1972) Stereoselectivity of catecholamine uptake by brain synaptosomes: Studies with ephedrine, methylphenidate and phenyl-2-piperidyl carbinol. *J. Pharmacol. Exp. Ther.* **183**, 103–116.

Hensling, M. (1984) Stereoselectivity of extraneuronal uptake of catecholamines in rabbit aorta. *Naunyn Schmied. Arch. Pharmacol.*, **328**, 219–220.

Henseling, M. & Trendelenburg, U. (1978) Stereoselectivity of the accumulation and metabolism of noradrenaline in rabbit aortic strips. *Naunyn Schmied. Arch. Pharmacol.*, **302**, 195–206.

Holmstedt, B. & Liljestrand, G. (1963) *Readings in Pharmacology*. MacMillan, New York.

Honda, K. & Nakagawa, C. (1986) *Alpha*-1 adrenoceptor antagonist effects of the optical isomers of YM-12617 in rabbit lower urinary tract and prostate. *J. Pharmacol. Exp. Ther.*, **239**, 512–516.

Honda, K., Takenaka, T., Miyata-Osawa, A. & Terai, M. (1986) Adrenoceptor-blocking properties of the stereoisomers of amosulalol (YM-09538) and the corresponding desoxy derivative. (YM-11133). *J. Pharmacol. Exp. Ther.*, **236**, 776–783.

Horn, A. S. & Snyder, S. H. (1972) Steric requirements for catecholamine uptake by rat brain synaptosomes studies with rigid analogue of amphetamine. *J. Pharmacol. Exp. Ther.*, **180**, 523–530.

Hosea, N. A., Berman, H. A. & Taylor, P. (1995) Specificity and orientation of trigonal carboxy esters and tetrahedral alkylphosphoryl esters in cholinesterases, *Biochemistry*, **34**, 11528–11536.

Hosea, N. A., Radic, Z., Tsigelny, I., Berman, H. A., Quinn, D. M. & Taylor, P. (1996) Aspartate 74 as a primary determinant in acetylcholinesterase governing specificity to cationic organophosphates. *Biochemistry*, **35**, 10995–11004.

Hötzel, B. K. & Thomas, H. (1997) '*O*' methylation of (+)-(*R*)- and (−)-(*S*)-6,7-dihydroxy-1-methyl-1,2,3,4-tetrahydroisoquinoline (Salsolinol) in the presence of pig brain catechol-*O*-methyltransferase. *Chirality*, **9**, 367–372.

Hsu, F. L., Hamada, A., Booher, M., Fuder, H., Patil, P. N. & Miller, D. D. (1980) Optically active derivatives of imidazolines. α-Adrenergic blocking properties. *J. Med. Chem.*, **23**, 1232–1235.

Ison, R. R., Partington, P. & Roberts, G. C. K. (1973) The conformation of catecholamines and related compounds in solution. *Mol. Pharmacol.* **9**, 756–765.

Iversen, L. L. (1963) The uptake of noradrenaline by the isolated perfused rat heart. *Br. J. Pharmacol.*, **21**, 523–537.

Jarowski, C. & Hartung, W. H. (1943) Amino alcohols XII, optical isomers in the ephedrine series of compounds. *J. Org. Chem.*, **8**, 564–571.

Jean, J. C. & Davis, R. E. (1994) Recent advances in the design and characterization of muscarinic agonists and antagonists. *Ann. Rep. Med. Chem.*, **29**, 23–32.

Komiskey, H. L., Miller, D. D., LaPidus, J. B. & Patil, P. N. (1977) The isomers of cocaine and tropacocaine: Effect of ^{3}H catecholamine uptake by brain synaptosomes. *Life Sci.*, **21**, 1117–1122.

Komiskey, H. L., Hsu. F. L., Bossart, F. J., Fowble, J. W., Miller, D. D. & Patil, P. N. (1978) Inhibition of synaptosomal uptake of noradrenaline and dopamine by conformationally restricted sympathomimetic drugs. *Eur. J. Pharmacol.*, **52**, 37–45.

Lambrecht, G., Pfaff, O., Aasen, A. J., Sjö, P., Mutschler, E. & Waelbroek, M. (1997) Stereoselective interaction of *p*-fluorotrihexyphendyl and its methiodide with four muscarinic receptor subtypes. *Life Sci.*, **60**, Abstracts, p. 1164.

LaPidus, J. B., Tye, A., Patil, P. N. & Modi, B. A. (1963) Conformational aspects of drug action I. The effect of D-(−)-pseudoephedrine on the action of certain pressor amines. *J. Med. Chem.*, **6**, 76–77.

Larsen, A. A. (1969) Catecholamine chemical species at the adrenergic receptors. *Nature*, **224**, 25–27.

Lattin, D. L. (1995) Cholinergic agonists, acetylcholinesterase inhibitors, and cholinergic antagonists. In: *Principles of Medicinal Chemistry* (Ed. by W. O. Foye, T. L. Lemke & D. A. Williams), pp. 321–344. Williams and Wilkins, Baltimore, MD.

Lee, O. S., Bescak, G., Miller, D. D. & Feller, D. R. (1974) Influence of substituted phenethylamines on lipolysis *in vitro*. III Stereoselectivity. *J. Pharmacol. Exp. Ther.*, **190**, 249–259.

Lernmer, B. & Bathe, K. (1982) Stereospecific and circadian-phase-dependent kinetic behaviour of *dl*-, *l*-,and *d*-propranolol in plasma, heart and brain of light- dark-synchronized rats. *J. Cardiovasc. Pharmacol.*, **4**, 635–644.

Li, Y.-O., Hieble, P., Bergsma, D. J., Swift, A. M., Ganguly, S. & Ruffolo, R. R. (1995) The β-hydroxyl group of catecholamines may interact with ser^{90} of the second transmembrane helix of the α_{2A}-adrenoceptor. *Pharmacol. Commun.*, **6**, 125–131.

Liang, J., Miller, D. D. & Patil, P. N. (1998) Alpha-adrenergic and serotonergic actions of oxymetazoline in the rat aorta. *Pharmacol. Rev. Commun.*, **10**, 51–58.

Ma, X.-L., Yue, T.-L., Lopez, B., Barone, F. C., Christopher, T. A., Ruffolo, R. R & Feuerstein, G. Z. (1996) Carvedilol, a new β-adrenoceptor blocker and free radical scavenger, attenuates myocardial ischemia–reperfusion injury in hyper cholesterolaemic rabbits. *J. Pharmacol. Exp. Ther.*, **277**, 128–136.

Maxwell, R. A., Ferris, R. M. & Burscu, J. E. (1976) Structural requirements for inhibition of noradrenaline uptake by phenethylamine derivatives, desipramine, cocaine and other compounds. In: *The Mechanism of Neuronal and Extraneuronal Transport of Catecholamines* (Ed. by D. M. Paton), pp. 95–153. Raven Press, New York.

McDonald, I. A., Cosford, N. & Vemier, J. (1995) Nicotinic acetylcholine receptors: molecular biology, chemistry and pharmacology, *Ann. Rep. Med. Chem.*, **30**, 41–50.

Miller, D. D. (1978) Steric aspects of dopaminergic drugs. *Fed. Proc.*, **37**, 2392–2395.

Miller, D. D., Hsu, F.-L., Salman, K. N. & Patil, P. N. (1976) Stereochemical studies of adrenergic drugs: Diastereomeric 2-amino-1-phenylcyclobutanols. *J. Med. Chem.*, **19**, 180–184.

Miller, D. D., Hamada, A., Craig, C., Christophe, G., Gallucci, J., Rice, R., Banning, J. & Patil, P. (1983) Optically active catecholimidazolines; a study of steric interactions at α-adrenoceptors. *J. Med. Chem.*, **26**, 957–963.

Moereels, H. & Tollenaere, J. P. (1975) A comparison between the solid-state conformation of phenmetrazine and the α-sympathomimetic pharmacophore pattern. *J. Pharm. Pharmacol.*, **27**, 294–295.

Morris, T. H. & Kaumann, A. J. (1984) Different steric characteristics of β_1- and β_2-adrenoceptors. *Naunyn. Schmied. Arch. Pharmacol.*, **327**, 176–179.

Mottram, D. R. (1981) Differential blockade of pre- and postsynaptic α-adrenoceptors by the 2-*R* and 2-*S* enantiomers of WB4101. *J. Pharm. Pharmacol.*, **31**, 767–771.

Nathanson, J. A. (1988) Stereospecificity of *beta* adrenergic antagonists: *R*-enantiomers show increased selectivity for *beta*$_2$ receptors in ciliary process. *J. Pharmacol. Exp. Ther.*, **245**, 94–101.

Nelson, W. L., Wennerstrom, J. E., Dyer, D. C & Engel M. (1977) Absolute configuration of glycerol derivatives. 4. Synthesis and pharmacological activity of chiral 2-alkylaminomethylbenzodioxans, competitive α-adrenergic antagonists. *J. Med. Chem.*, **20**, 880–885.

Ogg, G. D., Neilson, D. G., Stevenson, I. H. & Lyles, G. A. (1987) Comparative studies with enantiomers of the glycol metabolite of propranolol and their effects on the cardiac β-adrenoceptor. *J. Pharm. Pharmacol.*, **39**, 378–383.

Ogston, A. G. (1948) Interpretation of experiments on metabolic processes using isotopic tracer elements. *Nature*, **162**, 963.

Partington, J. R. A (1964) *History of Chemistry*, (Vol. 4). MacMillan, London; St. Martin's Press, New York.

Pasteur, L. (1848) Sur les relations qui peuvent exister entre la forme cristalline, la composition chimique et le sens de la polarization rotoire. *Ann. Chem. Phys.*, **24**, 442–460.

Patil, P. N. (1968) steric aspects of adrenergic drugs. VIII. Optical isomers of *beta*-adrenergic receptor antagonists. *J. Pharmacol. Exp. Ther.*, **160**, 308–314.

Patil, P. N. (1996) Pharmacologic quantitation. *Indian J. Exp. Biol.*, **34**, 615–633.

Patil, P. N. & Ruffolo, R. R. (1980) Evaluation of adrenergic *alpha* and *beta*-receptor activators and adrenergic *alpha* and *beta*-receptor blocking agents. Chapter 3 In: *Handbook of Experimental Pharmacology* (Ed. by L. Szekers), Vol. 54/I, pp. 90–134. Springer, New York.

Patil, P. N., Miller, D. D. & Trendelenburg, U. (1974) Molecular geometry and adrenergic drug activity. *Pharmacol. Rev.*, **26**, 323–392.

Patil, P. N., Feller, D. R. & Miller, D. D. (1991) Molecular aspects of receptor activation by imidazolines: an overview. *J. Neural. Transm.*, **34**, 187–194.

Patil, P. N., Fraundorfer, P. & Dutta, P. K. (1996) Stereoselective modification of circular dichroism spectra of rat lung β-adrenoceptor protein preparation by enantiomers of adrenaline. *Chirality*, **8**, 463–465.

Patil, P. N. (1998) Efficacy in relation to the physical chemical properties and molecular chirality. In: *Principles of Agonism (Abstract), First IUPHAR Conference on Receptor Mechanisms*, 23–25 July, 1998, Merano, Italy, p. 17.

Paton, W. D. M. & Rang, H. P. (1965) The uptake of atropine and related drugs by the intestinal smooth muscle of guinea-pig in relation to acetylcholine receptors. *Proc. Roy. Soc. B.*, **163**, 1–44.

Pendersen, L., Hoskins, R. E. & Cable, H. (1971) The preferred conformation of noradrenaline. *J. Pharm. Pharmacol.*, **23**, 216–218.

Peterson, G. L., Toumadje, A., Johnson, W. C. & Schimerlik, M. L. (1995) Purification of recombinant porcine M_2 muscarinic acetylcholine receptor from Chinese hamster ovary cells. *J. Biol. Chem.*, **270**, 17808–17814.

Pfeiffer, C. C. & Jenney, E. H. (1967) Natural asymmetry isomerism and pharmacological action. *Indian J. Sci. Ind. Res.*, **26**, 29–94.

Porter, J. E., Hwa, J. & Perez, D. M. (1996) Activation of the α_{1B}-adrenergic receptor is initiated by disruption of an interhelical salt bridge constraint. *J. Biol. Chem.*, **271**, 26318–28323.

Portoghese, P. S., Riley, T. N. & Miller, J. W. (1971) Stereochemical studies on medicinal agents, 10. The role of chirality in α-adrenergic receptor blockage by (+) and (−)-phenoxybenzamine hydrochloride. *J. Med. Chem.*, **14**, 561–564.

Powell, C. E. & Slater, I. H. (1958) Blocking of inhibitory adrenergic receptors by a dichloro analogue of isoproterenol. *J. Pharmacol. Exp. Ther.*, **122**, 480–488.

Rice, P. J., Hamada, A., Miller, D. D. & Patil, P. N. (1987) Asymmetric catecholimidazolines and catecholamines. Affinity and efficacy relationships at the alpha adrenoceptor in rat aorta. *J. Pharmacol. Exp. Ther.*, **242**, 121–130.

Rice, P. J., Miller, D. D., Sokoloski, T. D. & Patil, P. N. (1989) Pharmacologic implications of α-adrenoceptor interactive parameters of adrenaline enantiomers in the rat vas deferens. *Chirality* **1**, 14–19.

Riva, E., Mennimi, T., & Latini, R. (1991) The α- and β-adrenoceptor blocking activities of labetalol and its *RR–SR* (50:50) stereoisomers. *Br. J. Pharmacol.*, **104**, 823–828.

Ruffolo, R. R., Jr. (1983) Stereoselectivity in adrenergic agonists and adrenergic blocking agents. In: *Stereochemistry and Biological Activity of Drugs* (Ed. by E. J. Ariens, W. Soudijn, & P. Timmermans), pp. 103–125. Blackwell Scientific, London.

Ruffolo, R. R., Spradlin, T. A., Pollock, G. D., Waddell, J. E. & Murphy, P. J. (1981) *Alpha* and *beta* adrenergic effects of the stereoisomers of dobutamine. *J. Pharmacol. Exp. Ther.*, **219**, 447–452.

Smeader, W. (1985) *Drug Discovery: The Evolution of Modern Medicine*, p. 110. John Wiley & Sons. New York.

Soudijn, W., Van Wijngaarden, I., & Ariens, E. J. (1973) Dexetimide, a useful tool in acetylcholine-receptor localization. *Eur. J. Pharmacol.*, **24**, 43–48.

Sparago, M., Wlos, J., Yuan, J., Hatzidimitriou, G., Tolliver, J., Dal Cason, T., Katz, J. & Ricaurte, G. (1996) Neurotoxic and pharmacologic studies on enantiomers of the N-methylated analogue of cathinone (methcathinone); a new drug of abuse. *J. Pharmacol. Exp. Ther.*, **279**, 1043–1052.

Stephenson, R. P. (1956) A modification of receptor theory. *Br. J. Pharmacol.*, **11**, 379–393.

Tainter, M. L., Tullar B. F. & Luduena, F. P. (1948) *levo*-Arterenol. *Science*, **107**, 39–40.

Tanaka, T. & Starke, K. (1980) Antagonist/agonist preferring α-adrenoceptors or α_1 and α_2-adrenoceptors? *Eur. J. Pharmacol.*, **63**, 191–194.

Tollenaere, J. P., Moereels, H. & Raymaekers, L. A. (1979) *Atlas of Three-Dimensional Structure of Drugs.* Elsevier/North Holland Biomedical Press, New York.

Trendelenburg, U. (1980) A kinetic analysis of extraneuronal uptake and metabolism of catecholamines. *Rev. Physiol. Biochem Pharmacol.*, **87**, 33–115.

Tullar, B. F. (1948) The resolution of *dl*-arterenol. *J. Am. Chem. Soc.*, **70**, 2067–2068.

Wächter, W., Munch, U., Lenoine, H. & Kauman, A. J. (1980) Evidence that (+)-bupranolol interacts directly with myocardial β-adrenoceptors. *Naunyn. Schmied. Arch. Pharmacol.*, **313**, 1–8.

Waldeck, B. (1993) Biological significance of the enantiomeric purity of drugs. *Chirality* **5**, 350–355.

Walter, M., Lemoine, H. & Kaumann, A. J. (1984) Stimulant and blocking effects of optical isomers of pindolol on the sinoatrial and trachea of guinea-pig. Role of β-adrenoceptor subtypes in the dissociation between blockade and stimulation. *Naunyn. Schmied. Arch. Pharmacol.*, **327**, 159–175.

Wong, L., Radic, Z., Hosea, N., Berman, H. & Taylor, P. (1997) Reaction of enantiomeric organophosphoryl conjugates of acetylcholinesterase by bis-quaternary oxime, N. 16. *Pharmacologist*, p. 104, Abstract #451.

Yazawa, H., Takanashi, M., Sudoh, K., Inagaki, O. & Honda, K. (1992) Characterization of [^{3}H] YM 617, R-(–)-5-[2-[[2[ethoxyring(n)-^{3}H](O-ethoxyphenoxy)ethyl]amino]propyl]-2-methoxybenzenesulphonamide HCl, a potent and selective *alpha$_1$* adrenoceptor radioligand. *J. Pharmacol. Exp. Ther.*, **263**, 201–206.

Yeoh, P. N. & Wolf, H. H. (1970) Pharmacological evaluation of seizures induced by electrical stimulation of the hippocampus. *J. Pharm. Sci.*, **59**, 950–954.

Zhang, X., Yao, X.-T., Dalton, J. T., Shams, G., Lei, L., Patil, P., Feller, D. R., Hsu, F. L., George, C. & Miller, D. D. (1996) Medetomide analogues as α_2-adrenergic ligands 2 design, synthesis and biological activity of conformationally restricted naphthalene derivatives of medetomidine. *J. Med. Chem.*, **39**, 3001–3013.

Chapter 7
Separation of Chiral Compounds—From Crystallisation to Chromatography

W. John Lough
Institute of Pharmacy and Chemistry, University of Sunderland, Sunderland SR1 3SD, United Kingdom

CRYSTALLISATION

In appraising the current status of the development of the science of chiral separations in the context of the historical development of the subject area it would be only too easy to drop into *cliché* and draw the analogy of modern day chiral separation scientists standing on the shoulders of giants with Pasteur as the pre-eminent giant. There is, however, no escaping that in hand-picking the individual crystals of paratartaric acid, Pasteur (1848) made the first significant advance in the field. Further, there is no escaping the coincidence that for all Pasteur's subsequent work on chirality it was in separations that he had the most success (Pasteur 1853, 1858).

In just the same way today, for all the advances in the study of chirality it is in separations, above all, where the extent of progress is such that the field can be said to be approaching maturity. Having alluded to the giant's shoulders analogy, however, it would be a distraction to draw it any further. There have been several other major contributors to the field of chiral separations, especially during the boom period for chiral high-performance liquid chromatography in the nineteen-eighties, but it would be only idle conjecture to speculate whether they were true giants or just in the right place at the right time, at a suitable confluence of the needs of the scientific world and the availability of the appropriate emerging technologies.

Pasteur's pioneering work on chiral separations was in the field of crystallisation, and this method of chiral separation was the principal method for resolving enantiomers well into the nineteen-seventies and arguably beyond. In fact Pasteur's method of the fractional crystallisation of a mixture of two distinct diastereomeric salts, as for example between paratartaric acid and an optically active plant-alkaloid base (1853) was among the simplest of the crystallisation methods and became the most widely adopted practice. Nonetheless the level of sophistication acquired by practitioners in the field developed considerably.

These developments were comprehensively documented by Jacques *et al.* (1981). Although in their account of the use of dissociable compounds and complexes they rightly emphasised the predominance of the use of crystalline diastereomeric salts they also pointed out that Lewis acid–base complexes, inclusion compounds and quasi-racemates (*i.e.* crystalline addition compounds; *e.g.* of two organic acids) can be used. All of these are easily used in that they are generally prepared by mixing the constituents in an appropriate solvent, isolation of the individual enantiomers from the resolved complexes is usually very simple, and the resolving agent is almost always recovered in a form that enables its reuse.

Although there are many examples of isolation of individual enantiomers by fractional crystallisation of covalent diastereomeric compounds, this is rarely the most favoured option because of the difficulties in regenerating the enantiomers from the separated diastereomers. This is certainly an issue for amides and can occasionally be a problem even for esters, given the possibility of racemisation during saponification. In their seminal text Jacques *et al.* (1981) also discussed resolution by direct crystallisation (which is less common), subdividing these methods into separation based on simultaneous crystallisation of the two enantiomers (a subset of which is the manual sorting of the conglomerate, or *triage*, as famously first employed by Pasteur 1848), resolution by entrainment, resolution by entrainment in a supercooled melt, and crystallisation in optically active solvents. These methods are worthy of study as interesting phenomena in their own right but, being generally being more difficult to effect, are much less frequently used.

The science and art of resolving enantiomers by crystallisation has continued to progress. Czugler *et al.* (1997), for example, used chiral host molecules involving a bulky terpenoid unit and aromatic ethyne spacer groups to form molecular inclusion complexes to effect complete enantiomer resolution (enantiomeric excess > 99%) of (*R*,*S*)-1-methoxy-2-propanol in one co-crystallisation step. Tamura *et al.* (1997) achieved remarkable enantiomeric enrichment by direct crystallisation of salts formed by [2-[4-(3-ethoxy-2-hydroxypropoxy)phenylcarbamoyl]-ethyl]dimethylsulphonium ions and achiral sulphonate anions. This occurred only for sulphonate ions which gave rise to racemic crystals, however, and not for those that resulted in mixed crystals (solid solutions) composed of the respective optical antipodes.

These relatively recent examples of the effective use of crystallisation techniques illustrate that they remain an attractive, cheap, effective option for the large-scale isolation of individual enantiomers, especially at production levels, despite impressive developments in alternative technologies. For the smaller-scale first chiral resolution of a racemic mixture of a new chemical entity there is, however, little doubt that crystallisation methods have been eclipsed by the scale-up of enantioselective analytical methods employing chromatography and related techniques, which began to take root in the nineteen-seventies and were first commercialised in the nineteen-eighties.

With hindsight it is interesting to note that Jacques *et al.* (1981) were misguided in doubting Pirkle's assertion (1977) that "most 'first time' resolutions of enantiomers will soon be effected almost solely by liquid chromatographic techniques". 'Soon' might not have been quite as soon as Pirkle expected, possibly because the innate conservatism of many organic chemists at the time led them to be reluctant to abandon tried and trusted methodology and to commit themselves to chromatography. As a result of commercial pressures, however, times have changed and the vast majority of the organic chemists of today appreciate the need to employ chromatography when appropriate. This, along with the increasing importance of asymmetric synthesis, is so much so that crystallisation might soon become very much a minority approach even for the production scale resolution of enantiomers.

RUSSIA, ISRAEL, AND NMR

As already indicated, chiral resolution by chromatographic techniques did not begin to emerge as a serious alternative to crystallographic methods until the nineteen-seventies. There were a few individual examples of chiral chromatography before this time, however, and in these, not surprisingly, 'natural' chiral adsorbents were employed (Pryde 1989). For example, as early as 1922 partial resolution of an optically active aniline dye was claimed by Ingersoll and Adams. Later, column chromatography with lactose as the adsorbent was successfully employed by Henderson and Rule (1939) for the resolution of a racemic camphor derivative and by Prelog and Wieland (1944) for the resolution of Troger's base. More unusual was the chromatography of a racemic cobalt complex on optically active quartz (Karagunis & Coumoulos 1938).

The chiral chromatography that was performed in the nineteen-seventies, however, had a much more obvious direct influence on the commercial successes that were to follow in the nineteen-eighties. Although these commercial developments occurred largely in the USA and Japan, some of the earlier work that inspired them can be traced to influential pioneers in Russia and Israel.

Interest in the ligand-exchange mode of HPLC, in which the reversible equilibrium employed is the complexation and decomplexation of analyte molecules with sites on metal–ligand complexes, arose because of its effectiveness for amino acid analysis. It was through early work in this field performed in Russia (Rogozhin & Davankov 1970, Davankov & Rogozhin 1971, Davankov et al. 1973), using L-proline immobilised on macroreticular isoporous polystyrene beads charged with Cu^{2+} and an aqueous ammonia mobile phase, that the direct resolution of enantiomers first became readily accessible to the practising chromatographer.

Davankov continued to be a leading light in this field and was still producing innovative work in the late nineteen-eighties, when he demonstrated the extension of chiral ligand-exchange to non-aqueous systems (Davankov et al. 1987), and beyond. Davankov's original work was ahead of its time and it was not until approximately a decade or more later that chiral stationary phases (CSP) for ligand- exchange HPLC, as a result of the coming together of several factors, became commercially available.

By the nineteen-seventies Gil-Av in Israel had already been responsible for the direct resolution of enantiomers by capillary gas chromatography (GC) on a 100-m column (Gil-Av et al. 1966); arguably his subsequent work on chiral charge-transfer HPLC (Mikes et al. 1976) was more influential. Using the α-(2,4,5,7-tetranitro-9-fluorenylidineaminooxy)propionic acid (TAPA) phase similar to that first introduced by Klemm and Reed (1960), Gil-Av and co-workers were able to resolve the enantiomers of helicenes, e.g. hexahelicene (Fig. 7.1) Subsequent to this Lochmuller (1978) also used chiral charge-transfer HPLC but by far the dominant exponent of this mode of HPLC was Pirkle.

Pirkle was primarily an organic chemist whose initial breakthrough in chiral HPLC was inspired by his results from proton nuclear magnetic resonance (^{1}H NMR) spectroscopy (Pirkle & Sikkenga 1977, Pirkle & House 1979). The chiral solvating reagent, 2,2,2-trifluoro-1-(9-anthryl)ethanol (Fig. 7.2), which had been successfully used in NMR, was subsequently employed as the chiral selector in Pirkle's first chiral stationary phase. Using this CSP it was possible to resolve a wide range of racemates, including sulphoxides, lactones, derivatives of amines, amino acids, and alcohols, which contained π-acidic aromatic groups. Because dinitrobenzoylphenylglycine enantiomers had been very well resolved, the individual enantiomers of *N*-(3,5-dinitrobenzoyl)phenylglycine were used to prepare new CSP which were highly effective for the HPLC resolution of racemates which contained π-basic aromatic groups (Pirkle *et al.* 1980). This effective use of reciprocity and a rational mechanistic approach based on the three-point interaction rule (Fig. 7.3) (Dalgliesh 1952; see also Greer & Wainer, this volume) were Pirkle's trademarks which gave birth to further generations of CSP and many more individual, successful CSP.

Fig.7.1 Enantiomers of hexahelicene; chirality arises because the helix uses either a right-handed or a left-handed twist to overcome molecular overcrowding.

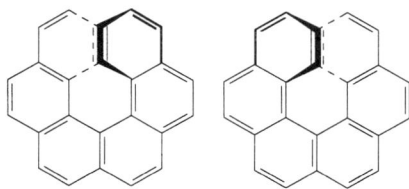

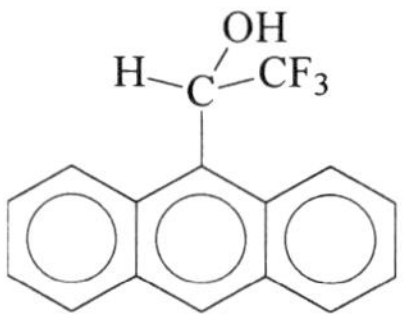

Fig. 7.2 2,2,2-Trifluoro-1-(9-anthryl)ethanol, a chiral solvating agent commonly used in ^{1}H NMR.

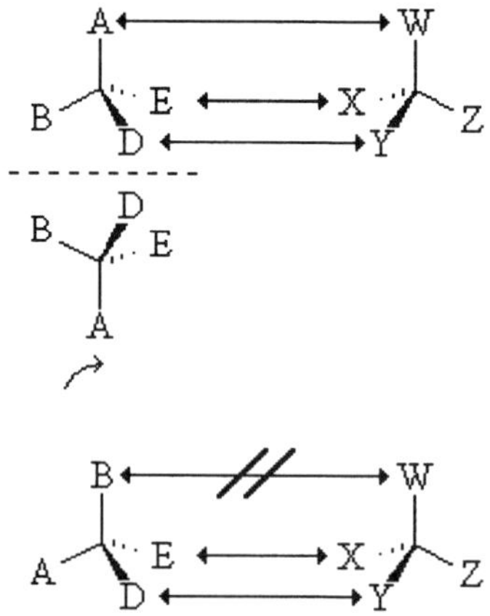

Fig. 7.3 Three-point interaction rule—at least three 'points' of interaction are needed before the chiral selector, CWXYZ, can distinguish between the two mirror-image forms of CABDE.

Although it was not necessarily an influence, it is interesting to note that the use of cyclodextrins for chiral resolution in NMR in the nineteen-seventies (MacNicol & Rycroft 1977) pre-dated their use in the nineteen-eighties as commercial CSP in HPLC. Other key works in the nineteen-seventies which were the harbingers of greater things with related chiral selectors in the nineteen-eighties included the use of cellulose triacetate by Hesse and Hagel (1973), the use of bovine serum albumin by Stewart and Doherty (1973) and the development of host–guest chromatography using chiral crown ethers by Cram and co-workers (1975).

DERIVATISATION

In considering the three-point interaction rule it is worth noting that almost any type of interaction might be involved ranging from, for example, ion–ion attraction, hydrogen-bonding, dipole–dipole interaction, and charge-transfer interaction to a 'negative' interaction such as steric repulsion. It is apparent that a strong interaction such as a covalent bond will be an effective participator in a three-point interaction between individual enantiomers of a pair and a chiral selector. By forming this bond the other parts of the selectand and selector will be brought closer thus facilitating easier 'interaction' between them, and hence easier chiral discrimination.

Table 7.1. Advantages and disadvantages of diastereoisomer formation.

Advantages

A cheap achiral LC phase such as silica or octadecylsilyl (ODS) silica can be used.
A strongly UV-absorbing or fluorescent group can be added to reduce the limit of detection.
With reasonable chromatographic efficiency and appropriate choice of derivatising group it is nearly always possible to achieve a diastereomeric separation.

Disadvantages

An additional step is involved in the analysis, which can increase the overall analysis time.
Decomposition or racemisation might occur during the reaction.
If the reaction does not go to completion the ratio of diastereomers might not be representative of the original ratio of enantiomers, because of kinetic resolution with one enantiomer reacting more quickly.
The ratio of diastereomers might not be representative of the original ratio of enantiomers if the diastereomers have a significantly different detector response.
The ratio of diastereomers might not be representative of the original ratio of enantiomers if the chiral derivatising agent is not enantiomerically pure; this is particularly an issue when determining a trace enantiomeric impurity in a single-enantiomer drug substance.
In preparative applications, recovery of pure enantiomers from the separated diastereomers can be difficult.

The ease with which chiral resolution could be obtained, and the fact that it could be achieved on cheap achiral chromatography columns, were clearly the main advantages of this approach. There were, on the other hand, clearly several very real disadvantages (Table 7.1). It is possible that one of the most significant factors in the move away from chiral derivatisation might have been a perception by many analysts that derivatisation was something non-routine in which only organic chemists or analytical chemists who specialised in derivatisation indulged. Pirkle, one such practitioner of derivatisation, famously remarked at a symposium in Guildford, UK, in 1987 in response to a question from his audience that "Derivatisation is very easy. Why, my five-year old son has just completed his first derivatisation!" Perhaps Pirkle's five-year old son was a bright kid but there is a sneaking suspicion that the use of chemical derivatisation as a possible solution to any analytical problem is an option that is occasionally jettisoned too readily.

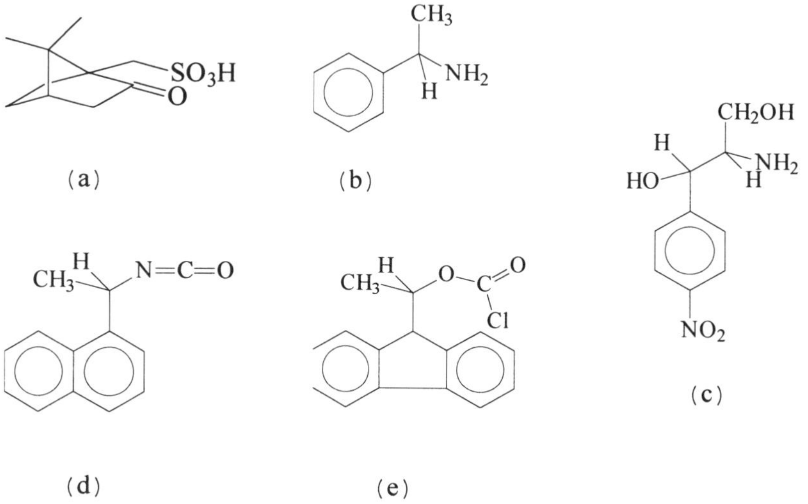

Fig. 7.4 Some commonly-used chiral derivatising reagents. (a) D-10-Camphor-sulphonic acid, a cheap, readily available reagent normally used as its acid chloride. (b) 1-Phenylethylamine, reacts with acid chlorides or can itself be activated to react with acids to give amides. (c) 1-(4-Nitrophenyl)-2-amino-1,3-propanediol, the possibilities of intramolecular hydrogen-bonding can give rise to a greater difference between the physicochemical properties of the diastereomeric derivatives.
(d) 2-Naphthylethyl isocyanate, gives rise to strongly UV-absorbing and fluorescent derivatives without any by-products, reacting with amino groups to yield ureas at ambient temperature and hydroxyl groups to yield carbamates at elevated temperatures in aprotic solvents. (e) 1-(9-Fluorenyl)ethyl chloroformate, reacts with amines under mild alkaline aqueous conditions to give strongly fluorescent carbamates.

COMMERCIALISED CHIRAL HPLC

Despite the many chiral separations that could be achieved by derivatisation there was no stopping the remorseless move towards the routine use of readily-available, commercial CSP that occurred during the nineteen-eighties. As already indicated, much of the groundwork that enabled the development of such CSP had already been done and, because of the increasing awareness of the importance of chirality in drugs that had been building ever since the thalidomide tragedy of the late nineteen-fifties and early nineteen-sixties (de Camp 1989), there was very much work to be done in the pharmaceutical industry to investigate stereochemical aspects of the action of both chiral drug development candidates and racemic drugs already on the market. With such a background and clear market potential the rapid successful development and commercialisation of CSP that occurred was inevitable.

By the end of the nineteen-eighties there were well over one hundred CSP on the market. Despite this, rationalisation of the choice of the most appropriate CSP for a particular chiral separation problem was not too difficult a matter, especially because it was well chronicled that these commercial CSP could be clearly subdivided into specific classifications (Wainer 1987, Krstulovic 1989, Lough 1989). Each of these classifications was characterised by the structure of the chiral selector employed, the mechanism by which chiral recognition was achieved, and the structural types of the racemic compounds which could be resolved by CSP belonging to that classification. These types of CSP are shown in Fig. 7.5, arranged (a) to (g) broadly in order of their introduction to the UK market and including N-benzoxy-glycyl-proline (ZGP) (h), a commonly used chiral ion-pairing agent.

Chiral ligand-exchange LC (Fig. 7.5a) employs a chiral ligand, almost invariably an amino acid, capable of complexing with metal ions in the mobile phase. This can be either immobilised on the LC stationary phase or used as a mobile-phase additive. By virtue of the formation of transient diastereomeric complexes during passage of a racemic analyte down the column, very good separation is often achieved for the enantiomers of amino acids, α-hydroxy acids, and, less commonly, aminoalcohols.

This was impressive at the time chiral ligand-exchange LC was first introduced but clearly it was only useful for a limited range of compound classes. There were also some practical drawbacks. Because the complexation and de-complexation steps could be slow on the chromatographic time-scale, chromatographic efficiency could be compromised. Usually, however, this could be remedied by performing the LC at elevated temperature. Successful separations could generally be achieved only in the pH range 7.5–9.5. Below the lower limit protons compete with the metal ions for the electron-donor groups of the ligand. Above the upper limit metal hydroxides begin to precipitate from the mobile phase.

The structure shown in Fig. 7.5b is that of immobilised N-(3,5-dinitrobenzoyl)-phenylglycine which was introduced as the Pirkle Type-1A CSP. In the same way as for ligand-exchange, a small molecule was used as a natural chiral selector and again it was an amino acid. In this instance, however, the amino acid was modified by chemical derivatisation and 'straight phase' organic mobile phases were used. As discussed earlier, the chiral recognition mechanism involves π–π interactions. This stationary phase was hailed as the first 'broad spectrum' CSP, because it was used to separate the enantiomers of many compounds. It was not, in fact, a genuine 'broad spectrum' CSP, because it could not be used for a wide range of compound classes. The racemates for which it was used generally contained a naphthalene or larger polyaromatic ring system, and were non-polar and neutral. Although this is an apparent weakness which was often "unavoidable" (Doyle 1989) it could be put to good advantage.

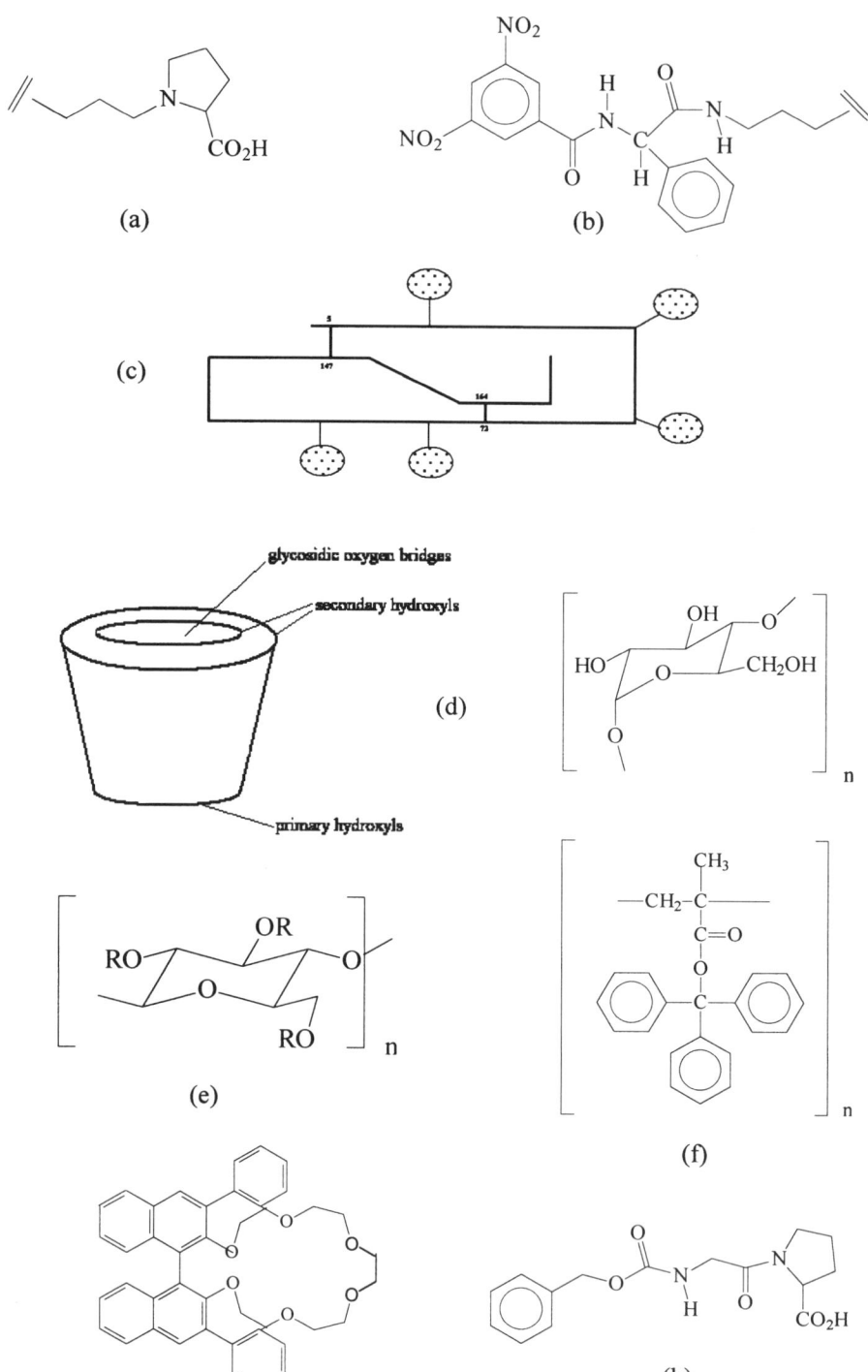

Fig. 7.5 Types or classifications of chiral stationary phase (the means of immobilising the selector to the stationary phase is not shown because often this is not disclosed by the manufacturer and is a subject of conjecture). (a) Ligand exchange, using the amino acid proline as the chiral selector. (b) N-(3,5-Dinitrobenzoyl)phenylglycine, used in the Pirkle Type-1A CSP. (c) Schematic representation of α_1-acid glycoprotein showing the peptide backbone, two disulphide bridges and five carbohydrate units. (d) General cyclodextrin structure (*e.g.* for α-cyclodextrin, $n = 7$); the glucose units form a bucket-like structure suitable for molecular inclusion complexation. (e) Cellulose triacetate and benzoate. (f) Synthetic polymer, poly(triphenylmethyl methacrylate) forms a helical structure similar to that of cellulose; chirality arises from the twist of the helix with one enantiomer being formed because of the use of a chiral initiator in the polymerisation. (g) Crown ether; with this particular example the chirality arises from molecular overcrowding in the binaphthol moiety. (h) Chiral ion-pairing agent, N-benzoxy-glycyl-proline (ZGP), which is used as a mobile phase additive.

Because several of the disadvantages of chiral derivatisation do not apply to achiral derivatisation, when a polar functional group such as amino or hydroxyl is reacted with, say, the acid chloride of 2-naphthalene acetic acid there is almost a guarantee of obtaining very good enantioselectivity when performing LC of the product on the Pirkle Type-1A CSP; the only downsides are slightly longer analysis time and a very slight possibility of racemisation. This approach is especially attractive in chiral drug bioanalysis, given that π-basic groups are readily detectable by UV or fluorescence.

This was applied to good effect by Bourque & Krull (1991) who performed the achiral derivatisation on a pre-column as part of an automated system. It was later pointed out by Lough *et al.* (1996) that the simple, easily-developed methods that could be established by exploiting achiral derivatisation were more compatible than the column-switching methods that had hitherto been popular (Lough & Noctor 1994); the situation at the time was that the decision on which enantiomer to proceed with as a drug candidate was being made in pharmaceutical discovery rather than development.

In principle the high enantioselectivity achievable by the achiral derivatisation/Pirkle approach also makes it attractive for preparative chiral LC. That this opportunity has not been adopted might have something to do with the previously mentioned aversion of many chromatographers to derivatisation reactions (the reasoning being that if it is so difficult that derivatisation is needed we had better bring in our 'chemists'). To be fair, however, many achiral derivatisations involve amide formation, and it is not to easy to regenerate the starting amine or acid.

The next natural chiral selectors to be used in commercial CSP were proteins (Fig. 7.5c). This was a highly logical step, and not just because these much larger molecules had many more potential points for interaction. Such biomacromolecules are responsible for the chiral discrimination of drugs in the body, and it is this that brings about the different pharmacological and toxicological properties of enantiomers that requires the separation of the enantiomers in the first place. It followed that it might be worthwhile to utilise them in chromatographic systems. Schill and Wainer (Schill *et al.* 1986) were prominent in showing that they could be used, and that the commercial α_1-acid glycoprotein CSP, Enantiopac, had genuine 'breadth of spectrum' in that it enabled good enantioresolution for a wide range of both cationic and anionic drug classes.

CSP based on albumin were also successful, but mainly for anionic drugs. This type of success, reported in the literature for marketed racemic drugs, was also being found in drug development in industry. For the first time it was possible to study the full consequences of administering a drug as a racemate. This had quite a profound effect, and almost as much as regulatory pressure was responsible for the current situation that almost all new chiral drugs are administered as a single enantiomer, with the decision on which enantiomer to develop being taken before the development phase.

Protein CSP were immediately widely used but proteins did have inherent weaknesses as selectors in chromatographic systems. Mass transfer into and out of receptor sites is slow and leads to relatively poor chromatographic efficiency; the compounds are also less stable, primarily by virtue of the unfolding of their tertiary structures, than small-molecule chiral selectors. Nonetheless their breadth of spectrum is difficult to match—so much so that they are still widely used today.

The commercial CSP that immediately followed protein CSP did not have the same breadth of spectrum but were useful additions to the 'armoury' of the analytical chemist conducting chiral separations. Developments followed the trend of using cheap, readily available chiral selectors until the most obvious of these had been exhausted.

Cyclodextrins, produced by the action of the enzyme cyclodextrin glycosyl-transferase on starch, are cyclic oligomers of α-(1,4)-linked glucose (Fig. 7.5d). Cyclodextrin CSP were, when first introduced, used solely with polar organic–aqueous mobile phases, which was convenient for the analysis of samples in aqueous solution. Under these conditions chiral recognition depended on steric fit into a hydrophobic pocket in the stable, bucket-shaped cyclodextrin. As such, reasonable breadth of spectrum was obtained (but not approaching that of the α_1-acid glycoprotein CSP) and chromatographic efficiency was good. Armstrong (*e.g.* Armstrong *et al.* 1985) not only performed pioneering work in the use for cyclodextrin CSP for LC, he was also the creator of the commercial Cyclobond phases (from Advanced Separation Technologies, Whippany, New Jersey, USA) and did much to popularise their use.

The carbohydrate cellulose was much less expensive and far more readily available than the cyclodextrins. When used derivatised with ester groups such as acetate and benzoate in a commercial CSP, however (Fig. 7.5e), it differed from cyclodextrins in that (i) organic 'straight phase' mobile phases were used, and (ii) while chiral recognition depended on steric fit, it was a looser fit in a hydrophobic cleft running along the helical polymeric structure. These cellulose ester CSP did little that could not already be achieved on the Pirkle column, although that changed when different functionality was introduced.

The cellulose CSP were the product of the Japanese company Daicel; at approximately the same time Daicel also introduced commercially, for the first time, CSP based on synthetic chiral selectors. These were synthetic polymer CSP (Fig. 7.5f) which had very similar properties to the cellulose CSP. More interesting were crown ether CSP (Fig. 7.5g). Crownpak CR had a very limited breadth of spectrum giving chiral separations only for primary amines. For such analytes, however, the chiral resolution obtained was usually quite large and a very simple mobile phase, dilute aqueous perchloric acid with or without methanol (<5%), was used.

Although these chiral selector types were the basis of commercially available CSP, some—ligand exchange chiral selectors, proteins, crown ethers and, in particular, cyclodextrins—were also used as chiral mobile phase additives. Although the cost of purchasing a range of expensive chiral columns could be avoided by using the chiral selector as a mobile phase additive for LC on a cheaper achiral column, this advantage was offset by the costs involved in the consumption of the chiral selector. Accordingly, this approach was used mainly with cheap chiral selectors or those which could not readily be immobilised. Often the selector would also be non- or poorly UV-absorbing.

Chiral ion-pairing agents fulfilled all these criteria and constituted the other major class of chiral selector. *N*-Benzoxy-glycyl-proline (ZGP), one of the most commonly used chiral ion-pairing agents is shown in Fig. 7.5h. D-10-Camphor sulphonic acid (Pettersson & Schill 1982) was another such commonly-used chiral ion-pairing agent. With the notable exception of the use of a zwitterionic ion-pairing agent by Knox (Knox & Jurand 1982), these chiral ion-pairing agents were used under so-called 'straight-phase' conditions with relatively non-polar organic mobile phases containing either no water or very small amounts. Given the three-point interaction rule, the vast majority of successful examples of successful chiral ion-pairing was with amino-alcohols such as β-blocker drugs (Pettersson & Josefsson 1986) or intermediates in the synthesis of anti-hypertensive drug candidates (de Biasi *et al.* 1989).

CHIRAL HPLC BEYOND THE EIGHTIES

When chiral HPLC had become firmly established by the end of the nineteen-eighties, there followed a period of what can only be described as proliferation. For all the success of the classes of CSP introduced commercially in the nineteen-eighties, there was no one class of CSP that had all the properties of an ideal CSP, *i.e.* complete breadth of spectrum, efficient chromatography, and physical robustness towards a wide range of mobile phases. Research directed towards this elusive, perhaps impossible, ideal continued into the next decade. The strategy employed was usually to build extra complexity into the small-molecule chiral selectors that had already been used, or to use medium-sized molecules which were large enough and complex enough to increase the chances of obtaining sufficient points of interaction without introducing the mass transfer and unfolding problems associated with proteins.

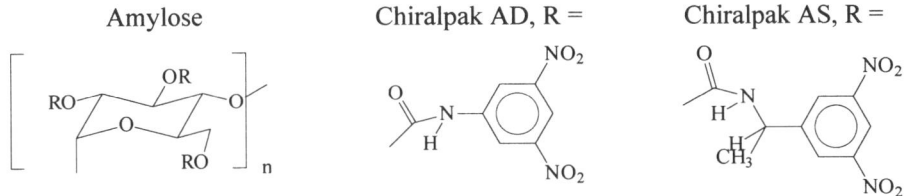

Fig. 7.6 Carbamate derivatives of cellulose and amylose; the three examples shown are amongst the most popular of the Daicel polysaccharide range of CSP.

As already alluded to, there was a marked improvement in cellulose CSP when different functionalities were used. These polysaccharide CSP, carbamate derivatives of both cellulose and amylose, followed earlier acetate and benzoate phases fairly rapidly in an almost seamless transition. The incorporation of a variety of aryl groups through the carbamate link to the monosaccharide rings (Fig. 7.6) was responsible for the dramatic improvement in chiral recognition properties, so much so that these CSP are perhaps the most effective currently in use and are very popular, despite being more expensive than most other types of CSP. It is now a common approach to chiral separation method development to conduct a screening process using a small number of mobile-phase compositions and columns each containing one of a selection of popular, effective CSP. No self-respecting chiral chromatographer would attempt such screening without including at least one derivatised polysaccharide CSP.

Cyclodextrins were also derivatised in much the same way as the carbohydrates but not quite with the same success. The stereochemistry and the fit of enantiomeric analytes into the hydrophobic 'bucket' of the cyclodextrin must be exactly right if chiral recognition is to be obtained. While the addition of functionality to the cyclodextrin might enhance enantioselectivity by giving a complementary 'point-to-point' interaction, it might equally well distort what, before functionalisation, would have been a perfectly good stereochemical fit for chiral recognition. The functionalised cyclodextrin shown in Fig. 7.7 is a good example.

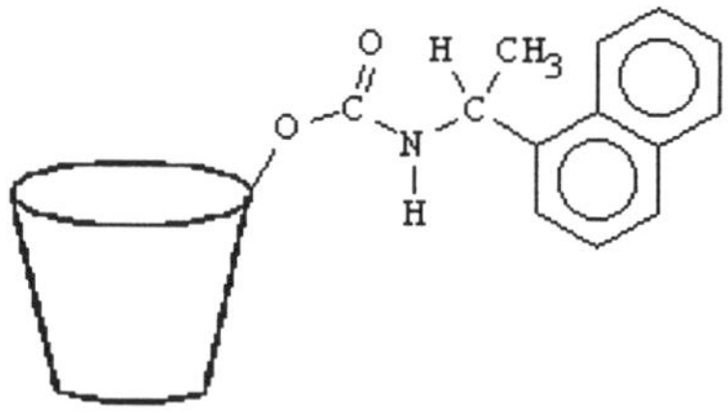

Fig. 7.7 Cyclodextrin derivatised to attach a π-basic naphthalene ring.

This is addition of complexity to one class of CSP to the point where there are two classes of CSP in one selector. A racemate with a π-acidic group an appropriate distance away from, say, a naphthalene ring which fitted into the β-cyclodextrin cavity would be expected to be very well resolved. For most other analytes, however, it would be very difficult to predict whether any enantioselectivity would be better or worse.

A much more significant development in the use of cyclodextrin CSP was the discovery of a 'magic mobile phase' or polar-organic mode (Zukowski *et al.* 1992). The absence of water and the use of acetonitrile with low levels of acetic acid and triethylamine present as the mobile phase suppresses the molecular inclusion of analytes into the cyclodextrin cavity so that hydrogen-bonding interactions become more dominant in the retention mechanism. Under these mobile phase conditions a wide range of chiral separations was possible, and more often than not the compounds which could be resolved into their enantiomers were different from those that were separated under 'reversed-phase' conditions. This mode therefore extended the range and utility of cyclodextrin CSP.

It has already been stated that the early work of Pirkle gave birth to further generations of CSP. Such CSP were given the general descriptive name of synthetic multiple-interaction CSP by Doyle (1989) but they are perhaps better known as Pirkle-concept CSP. This is perhaps not surprising, because it was Pirkle and his co-workers who were responsible for most of the better known of these CSP, *e.g.* Whelk-O 1, α-Burke 2, β-Gem 1, Pirkle 1-J. The Whelk-O 1 phase is shown in Fig. 7.8. It contains two chiral centres but the main addition in complexity is that it contains both a π-acceptor and π-donor. Originally designed to resolve the enantiomers of naproxen (Welch 1993) it can be used under 'reversed phase' conditions but is mainly operated with 'straight phase' mobile phases under which conditions it has good breadth of spectrum, almost on a par with the best of the carbamate-derivatised polysaccharide phases.

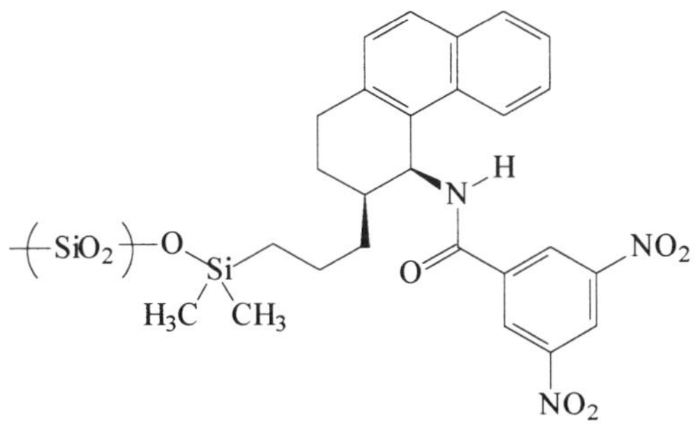

Fig. 7.8 (*S,S*) Whelk-O 1 CSP.

The name Armstrong is inextricably linked with cyclodextrins but during the nineteen-nineties he made a name for himself again with macrocyclic antibiotics (Armstrong *et al.* 1994). These glycopeptides are medium-sized with relative molecular masses in the range 1400–2100 and are suitably complex. They have many stereogenic centres, semi-rigid macrocyclic rings and diverse functionality to provide many opportunities for hydrogen-bonding, dipole stacking, electrostatic interactions, steric repulsion, hydrophobic interaction and, to a lesser extent, π–π interactions. This can be seen from the structure of vancomycin (Fig. 7.9).

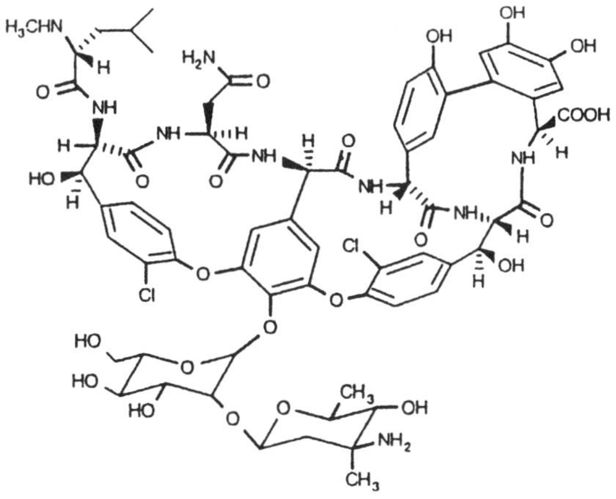

Fig. 7.9 Proposed structure of the macrocyclic antibiotic, vancomycin; it is apparent there are many potential 'points' of interaction with an analyte.

Vancomycin and teicoplanin CSP were commercialised and were followed by a ristocetin CSP. Like the cyclodextrin CSP these macrocyclic antibiotic CSP can be used in straight phase, polar organic, or reversed-phase mode. They are also highly effective. Screening with cyclodextrin and macrocyclic antibiotic CSP in polar organic and reversed phase modes would be expected to be almost as effective as straight-phase screening incorporating the best derivatised polysaccharide CSP. Armstrong continued to work on macrocyclic antibiotics and in 1998 introduced an avoparcin CSP (Ekborg-Ott et al. 1998)

There was also proliferation of protein CSP. Different types of serum albumin phase were produced, e.g. human, rat, rabbit etc., and the original bovine. This is, however, perhaps a trivial example, especially because these phases were more intended as tools for the study of drug–protein binding than as new CSP that might have dramatically different enantiorecognition properties.

The most useful of the newer protein CSP seemed to be that based on ovomucoid, a glycoprotein found in chicken egg-white. In a comparative study of a commercial CSP based on immobilised ovomucoid protein and a second-generation α_1-acid glycoprotein CSP, Kirkland et al. (1991) found that "the ovomucoid column showed generally higher resolution, greater flexibility in operating parameters, and better long-term stability than the acid glycoprotein column." Especially with regard to the issue of generally higher resolution there are points that could be made to suggest that the case in favour of the ovomucoid CSP is not so clear-cut as the impression given. To have discovered another protein CSP which worked for different sets of compounds than those for which α_1-acid glycoprotein CSP was useful, but only had just as good breadth of spectrum and other properties would, however, have been enough of an achievement in itself.

On the commercial front the proliferation of CSP shows no sign of abating, despite the large number of effective materials already on the market. Almost every manufacturer, e.g. Regis, Chirex, Capital HPLC, has its own version of a Pirkle-type CSP. Widening things away from π-acid and/or π-base phases to general synthetic multiple-interaction CSP, the tartramide phases introduced by Lindner have now been commercialised as Kromasil CHI-I and Kromasil CHI-II. Carboxymethyl cyclo-dextrins have been immobilised on polymethacrylate beads. Chiral ion-exchange LC can be performed on chiral ruthenium complexes in combination with basic sodium magnesium silicate particles. A pepsin CSP has been introduced as being complementary to the ovomucoid protein CSP and "ideal for the separation of basic compounds because of its lower isoelectric point" (Commercial brochure, Phenomenex, Torrance, CA, USA). The story continues!

BEYOND HPLC

This proliferation consisted not only of proliferation of new chiral HPLC systems but also a continuation of a trend, that had began in the nineteen-eighties, of employing the chiral selectors that had proved effective in chiral HPLC in other techniques such as nuclear magnetic resonance (NMR) spectroscopy, gas chromatography (GC), supercritical-fluid chromatography (SFC), thin-layer chromatography (TLC), and last, but certainly not least, capillary electrophoresis (CE).

It is the use of popular HPLC chiral selectors as chiral selectors in NMR that is the most ironic, given that, as already discussed, the use of cyclodextrins in NMR pre-dated their popular use in HPLC and Pirkle's ideas on chiral HPLC evolved from his

NMR studies. The use of cyclodextrins for quantitative determination of enantiomers re-emerged (Casy & Mercer 1988) shortly after the time of their introduction as chiral selectors in commercial CSP and has been sustained since that time (*e.g.* Dawson & Black 1995). Not surprisingly, Casy *et al.* (1991) also subsequently looked for correlation between enantioselectivity in HPLC and that in NMR, and also used NMR experiments to attempt to model the interactions between analytes and β-cyclodextrin. This use of HPLC chiral selectors for NMR need not be limited to cyclodextrins or Pirkle-type selectors. As is illustrated in Fig. 7.10, the chiral ion-pairing agent *N*-benzoxy-glycyl-proline (ZGP) can be used to bring about an enantiotopic shift in the proton signals of propranolol.

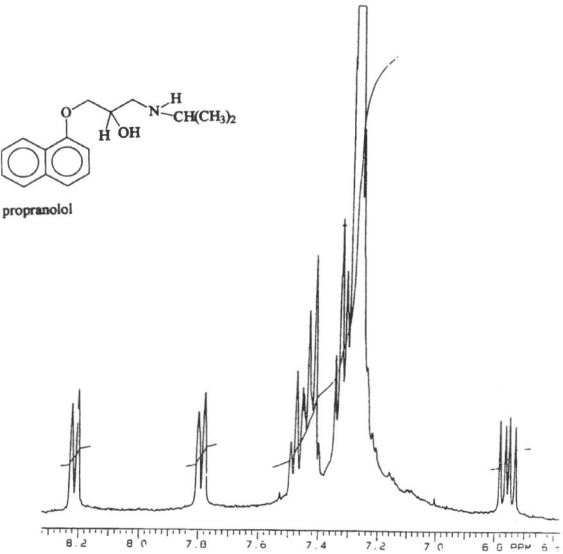

Fig. 7.10. In the aromatic region of the ^{1}H NMR spectrum of propranolol a doublet at 6.7–6.8 ppm is split into two doublets on introduction of *N*-benzoxy-glycyl-proline (ZGP). This was first performed in an attempt to establish whether the aromatic rings of selector and selectand played any part in the chiral recognition mechanism but it was subsequently shown that, after decoupling and optimisation of NMR conditions, the phenomenon could be used to quantify the enantiomers of propranolol.

Chiral GC was well established and in common use before chiral LC (Gil-Av 1975) and Chirasil-Val, the first commercial CSP for GC, was introduced to the market at approximately the same time as some of the earlier commercial CSP for LC. Little is now heard of chiral GC, however, compared with the great interest in the application of chiral LC to pharmaceuticals. Chiral GC is nonetheless still widely used (see, for example, König, this volume) and has experienced cross-over benefits from developments in chiral LC, specifically from the interest in cyclodextrins as chiral selectors. Cyclodextrin CSP are now very popular in chiral GC (König 1992) and with the benefit of the high efficiency and resolving power of capillary GC it is possible to resolve large numbers of components of complex mixtures into their enantiomer pairs in one run. This is usually used in applications such as the study of pheromones, flavour compounds, fragrances, and volatile reaction intermediates.

Supercritical fluid chromatography (SFC) was a technique which attracted much interest briefly in the early eighties and then again in the late eighties. Whatever else might be said about its advantages and disadvantages it was apparent that it could be used for chiral separations, because the most common mobile phase used, supercritical carbon dioxide, was similar in polarity to the 'straight phase' solvents used in chiral LC with synthetic multiple-interaction and derivatised polysaccharide CSP. In principle, dipolar interactions involved in solvation by carbon dioxide could lead to better enantioselectivity than in LC. Van Overbeke *et al.* (1997) found, however, that, at least for their work on benzodiazepines, barbiturates, and non-steroidal anti-inflammatory drugs that the selectivity was disappointingly similar. Even if that were always true, however, it might still be worth considering for slightly improved efficiency, rapid re-equilibration for gradient runs, and the fact that pressure can be used to tune solubility.

For most other applications it is as if SFC has disappeared from sight but it still has its strong advocates within the pharmaceutical industry, not just for analytical chiral separations but also for preparative-scale chiral separations, because carbon dioxide is exceptionally easy to remove from the separated pure enantiomers.

Of all the separative techniques, TLC is the closest in character to HPLC—differing from it in geometry only. Accordingly it was reasonably straightforward to demonstrate that all the classes of chiral selector used in HPLC could also be effective in TLC whether as chiral TLC plates, as in the chiral ligand-exchange product marketed by Merck, or by addition of chiral selectors such as cyclodextrins to the mobile phase. Armstrong (1980) had, in fact, used cyclodextrins as TLC mobile-phase additives before his work which led to the commercial cyclodextrin CSP for LC, and he turned to it again later (Armstrong *et al.* 1988) when working with the next major group of chiral selectors featuring in his research, macrocyclic antibiotics (Armstrong & Zhou 1994a).

It was only a matter of time before, inevitably, cellulose derivatives were employed as selectors in TLC, although Suedee & Heard (1997) found this was not a straightforward exercise. They were successful in resolving the enantiomers of propranolol and bupranolol but during the this work had to deal with problems such as cracking of the phase, poor adhesion to the plate, the need for large quantities of chiral selector, and interference with spot detection from triphenylcarbamate chromophores.

There might seem to be an element of 'handle-turning' in this 'do it by LC then do it by TLC' sequence but it remains, nonetheless, a valid exercise, because there are times when the use of TLC might be preferred, for example, when requiring a quick and easy means of monitoring the progress of a reaction, when investigating mass balance problems in the analysis of a drug substance, and when looking for new degradants in a drug-stability study.

CE was the technique best suited to exploitation of the knowledge of chiral selectors built up in HPLC. It is well known for its high efficiency and, consequently, high resolving power. This applies and is very useful in chiral CE, notwithstanding some loss in efficiency arising from slow mass transfer, because formation and break-up of the transient diastereomeric complexes in the running buffer is generally slower than, *e.g.*, normal adsorption and desorption processes occurring in chromatographic techniques. Apart from the use of proteins as chiral selectors, however, this is not often a significant issue. Perhaps more important than the high resolving power of CE is its speed and the ease with which method development can be performed and, in part related to that, its low cost. With one capillary, the chiral selector can be changed from run to run and very small amounts of, often very expensive, chiral selectors are consumed.

Given the background knowledge of the classes of chiral selector and their mechanisms of chiral discrimination built up in chiral LC, it was a simple matter to apply them to CE. There is virtually no limit to the type of chiral selector that can be used. Good resolution can be obtained by use of low microgram quantities of chiral selector and it is even possible to use UV-absorbing selectors with UV detection, by using the technique of partial-filling (Amini *et al.* 1999). A band of UV-absorbing selector-containing buffer is held up in the capillary or moved toward the anode while the analyte moves through this band toward the cathode.

Not surprisingly, then, the use of CE for chiral resolution has been highly successful and has already been extensively reviewed (Fanali 1996, Soo *et al.* 1999a, b). It is frequently used in the early development phase of the pharmaceutical industry (Soo *et al.* 1999a) but this has not yet fed through to any great extent to the appearance of chiral CE methods in regulatory submissions. Neither has the uptake of chiral CE fed through to production sites to any great extent. This is not because of any particular difficulties involved in chiral CE, however, but more because the CE equipment would not be used for much else other than chiral CE.

MEETING UNFULFILLED NEEDS?

From the description of the developments that have occurred in the field of chiral resolution it should be transparent that it is likely that most of the important advances in the science of chiral resolution have already been made. In assessing whether or not a field such as this has fully matured it is, however, necessary not just to look in isolation at the equilibrium systems that have been used to separate enantiomers but also to consider how they have been used and, more importantly, the extent to which they meet the needs of the applications to which they are applied.

In addressing the issue of whether or not there is scope to develop significantly more effective chiral selectors than those already available it certainly does seem there is little room for improvement. It has long been apparent, as suggested by the strategies adopted during the post-nineteen-eighties period of proliferation (see *CHIRAL HPLC BEYOND THE EIGHTIES*, above) that:

(1) A simple small molecule chiral selector with only a few different types of 'point' available for interacting with selectands will have limited 'breadth of spectrum' in that it is likely to only be able to work well for a few classes of compound.
(2) Although very large multi-'point' globular molecules such as proteins afford breadth of spectrum they are generally less stable than other chiral selectors, and chromatographic systems employing them suffer inherently from poor efficiency arising from slow mass transfer of the analyte diffusing to and from the site of interaction.
(3) The best compromise is, therefore, a medium-size selector with multiple potential points of interaction.

The highly successful derivatised cellulose chiral selectors can be regarded as fitting into the last category, even though they are polymeric in nature. This is because the repeating unit is functionalised sugar molecules forming a cavity in a similar manner to functionalised cyclodextrins, except that the cavity is different in nature and more flexible.

It is difficult to imagine a yet-to-be developed chiral selector that exceeds the exceptional performance of derivatised polysaccharides in straight-phase HPLC systems or vancomycin in reversed-phase HPLC systems. Some exotic yet-to-be-discovered cyclic glycopeptide might fit the bill but the discovery of such a beast is hardly essential, because there is already a wide variety of effective chiral selectors and the screening process to find the right selector for the racemate being studied is no longer difficult. Illustrative of this is the very high success rate achieved for basic enantiomeric pairs by Liu & Nussbaum (1999) when screening by chiral CE with modified cyclodextrins.

The question of whether the chiral selectors currently available to scientists can be used in such a way as to be able to deal successfully with all chiral resolution applications that might be encountered is not such a straightforward matter. This can be illustrated by considering the types of application that are typically found in pharmaceutical R&D, the field in which chiral resolution methodology is most used.

In the determination of traces of an enantiomeric impurity in a chiral drug it is not sufficient simply to achieve chiral resolution. The resolution and limit of detection must be adequate for accurate and precise measurement of a small peak (or signal in, *e.g.*, NMR) of the enantiomeric impurity in the presence of a much larger peak for the main component of the sample. This is often a problem in HPLC when the impurity peak is on the tail of the main peak and, therefore, not easy to integrate. Fortunately there are generally now sufficient options for it to be possible either to produce a separation good enough to resolve the impurity beyond the tail of the main peak or to use the opposite enantiomer of the chiral selector so that the order of retention is reversed thus enabling the now first-eluting impurity peak to be more easily integrated.

This is one area in which chiral liquid chromatography is using 'electro-drive' rather than pressure drive, in other words capillary electrochromatography (CEC) rather than high-performance liquid chromatography would have a distinct advantage, because the chromatographic peaks are usually much sharper. This also applies to CE and, further, peak-tailing is less likely to be encountered in CE.

Another perspective is that as long as the clarion cry from the pharmaceutical industry remains that all is doom and gloom unless analysts, and everyone else for that matter, produce 'more for less', there will always be room for improvement in the field of chiral resolution. Bearing this in mind it is worth considering the work of Wu *et al.* (1990) who determined the enantiomeric ratio in ephedrine and pseudoephedrine samples by achiral chromatography with UV and polarimetric detectors in series, rather than by performing a chiral separation. In 1990 this work might have been considered something of an interesting oddity but now it holds out the prospect of its being possible to determine an enantiomeric impurity in a drug substance in the same chromatographic run as that being used to determine all the other structurally-related impurities.

Cram *et al.* (1998) showed that this method of comparing the ratio of the UV detector response with that of the polarimetric detector for the sample, with the ratio found for an enantiomerically pure standard could be used for comfortable detection of the presence of a 1% enantiomeric impurity in (*S*)-naproxen. Further work is now in progress to explore the limits of this approach given that since 1990 there have been significant improvements in polarimetric detectors and in the precision of LC methods.

Also, along similar lines, Angelaud *et al.* (2000) used the ratio of the responses of optical rotation and refractive index detectors so that this detector ratio approach could be extended to the determination of enantiomeric excesses in non-UV-absorbing compounds such as carbohydrates and aliphatic carbinols. The simultaneous determination of enantiomeric and all other structurally-related impurities would be as useful in drug formulation analysis as in drug substance analysis. A different approach would be required, however, because of the lower concentrations of the drug and the fact that many excipients used in formulations are chiral and, therefore, also rotate plane-polarised light.

Again, as has already been mentioned, the decision on which enantiomer of a chiral new chemical entity should be investigated as a drug candidate is now taken during the discovery rather than late development phase of pharmaceutical R&D. This means that, rather than chiral drug bioanalytical methods which are capable of dealing with large sample numbers, what is needed is simple methods that are easy to develop. Liquid–liquid extraction followed by achiral derivatisation with a π-donor-containing group and then LC on a Pirkle column will fit the bill very well, as has been noted. Also suitable is the use of solid phase extraction followed by direct injection of a large volume of the resulting 'cleaned-up' solution from SPE on to a microbore Hypercarb LC column on which chiral separation is brought about by a cyclodextrin mobile phase additive.

The success of this on-column sample focusing approach depends on the very high retentivity of the porous graphitic carbon Hypercarb phase compared with all the reversed-phase materials used in SPE cartridges, and that its planar surface in conjunction with a chiral selector is useful in bringing about enantioselectivity. The feasibility of this approach has been demonstrated (Prangle *et al.* 1998) and it would seem to have good potential for general use in chiral drug bioanalysis.

The challenge presented by the preparative scale chromatographic isolation of enantiomers is clearly a demanding one, because the chiral resolution on the analytical scale must be sufficient to withstand the loss of resolution that will accompany the loss of chromatographic efficiency that will occur when the column is overloaded to achieve high throughput. It is, therefore, in this area that it might still be worthwhile looking for new chiral selectors that have higher enantioselectivity towards certain classes of compound.

The isolation of enantiomerically pure samples of drug candidates is so important in pharmaceutical R&D that often the most economical approach overall is simply to accept the high cost of scaling up a separation performed on a very expensive commercial CSP. This need not be such a problem, because among the most expensive commercial CSP are in fact the most easy to prepare. In fact costs can be reduced even further given that the enantioselectivity obtainable on derivatised cellulose or amylose CSP is occasionally great enough for the preparative scale isolation to be performed on 'flash chromatography' grade silica (Matlin *et al.* 1995). In pharmaceutical R&D, however, such 'self-help' can be a diversion from core activities so that the cost of scaling up separations achieved on derivatised polysaccharide CSP will be reduced only when such CSP come off patent.

The alternative approaches to scale-up with some overload involve systems which can make the most of limited enantioselectivity. Since the earliest days of preparative chiral chromatography this has been achieved by recycling the sample components round through the pump and column again in a closed loop. The use of such methodology persisted even into the mid-nineteen-nineties (Dingenen & Kinkel 1994) and it is still an option worth considering. Using this approach, if the extra-column volume, specially in the pumping system is not excessive the resolution will improve with every cycle through the pump until the peaks are resolved to baseline. The throughput of such a system can be improved by skimming off, during each cycle, the part of the front end of the first peak and the tail of the second peak which contain pure enantiomer. If this is done the column is less overloaded on each consecutive cycle so that the improvement in resolution is greater than it would otherwise have been.

Displacement chromatography is another means of achieving higher sample throughput from limited enantioselectivity. The major practitioners of this approach to preparative chiral HPLC (Fig. 7.11) were Vigh *et al.* (1990) who worked with systems in which the chiral selector was a cyclodextrin. This approach has not been widely adopted but it should, in principle, to attractive to synthetic organic chemists, because essentially it is little different from the flash chromatography they use regularly.

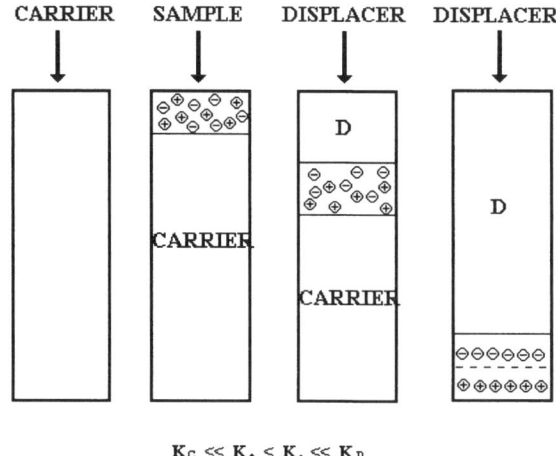

$K_C \ll K_* < K_* \ll K_D$

Fig. 7.11. Chiral displacement chromatography. A cyclodextrin CSP might be used for the chromatography illustrated.

Although displacement chromatography, recycling, *etc.*, have their place, in recent years simulated moving bed (SMB) chromatography has had the most significant impact in industry. This is based on a simulation, as the name suggests, of a true moving bed system which is a counter-current absorption process made of four zones that involves actual circulation of solid-phase adsorbent. When a sample is applied, the component of the sample that has the greatest affinity for the liquid phase will be transported against the flow of the solid, if the flow of liquid is great enough. In contrast, the component of the sample with the lowest affinity for the liquid phase will be carried along with the flow of the solid.

By balancing the flow rates of the solid and liquid phases it is possible to separate compounds with extremely similar affinities for the solid/liquid phases *e.g.* very low enantioselectivity. Two inlet lines enable continuous introduction of mobile phase and solution containing the mixture of enantiomers to be separated. Two outlet lines enable continuous extraction of the pure products. In reality such a true moving bed system is not practical because it involves the circulation of a solid adsorbent. To simulate the counter-current flow the two inlet and two outlet lines are, therefore, all moved one column (or more) forward at fixed time intervals (Fig. 7.12). Preparative simulated moving bed separations are now well established—they have, for example, been used to isolate enantiomers of a chiral epoxide (Nicoud *et al.* 1993) and have been succinctly described and critically reviewed in 1996 (*anon*, Merck commercial magazine 1996).

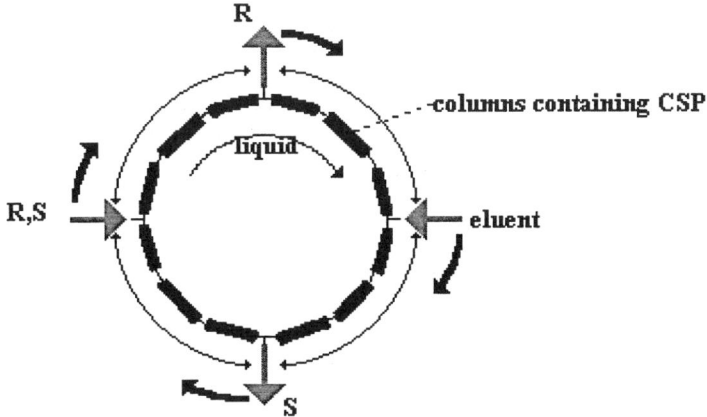

Fig. 7.12. Schematic diagram of a simulated moving-bed system for preparative chiral separations. Fixed bed chiral columns are connected in series and counter current flow is simulated by moving the racemate feed, eluent and collection points for purified or enriched *R* and *S* forward in the fluid flow direction.

The most significant future developments in chiral resolution might well be in SMB, or in other preparative systems in which there is still demonstrably a need for further improvements. Even in an apparently mature area such as chiral resolution, however, there is no accounting for what might come next. For example, the ever-innovative Armstrong has recently made progress in processing large amounts of material relatively inexpensively by using adsorptive bubble separation methods (Armstrong *et al.* 1994b, 1998). Given the possibility of ingenuity such as this from future 'giants' in separation science it is not yet 100% certain that all the important developments in chiral resolution are now behind us.

REFERENCES

Ahnoff, M. & Einarsson, S. (1989) Chiral Derivatisation. In: *Chiral Liquid Chromatography*. Lough, W. J. (Ed.) Blackie Academic & Professional, Glasgow. 39.

Amini, A., Wiersma, B., Westerlund, D. & Paulsen-Sorman, U. (1999) *Eur. J. Pharm. Sci.,* **9**, 17.

Angelaud, R., Matsumoto, Y., Korenaga, T., Kudo, K, Senda, M. & Mikami, K. (2000) *Chirality*, **12**, 544.

Anon. (1996) Simulated Moving Bed Chromatography—The Answer to Chiral Purification Problems? In: *Chromatography Insights.* 7, 6. Merck Chromatography, Poole, Dorset, UK.
Armstrong, D. W. (1980) *J. Liquid Chromatogr.,* **3**, 895.
Armstrong, D. W., DeMond, W. & Czech, B. P. (1985) *Anal. Chem.,* **57**, 481.
Armstrong, D. W., He, F.-Y. & Han, S. M. (1988) *J. Chromatogr.,* **448**, 345.
Armstrong, D. W., Tang, Y., Chen, S., Zhou, E. Y. & Bagwill, C. (1994) *Anal. Chem.,* **66**, 1473.
Armstrong, D. W. & Zhou, Y. (1994a) *J. Liq. Chromatogr.,* **12**, 1695.
Armstrong, D. W., Zhou, E. Y., Chen, S. Le, K. & Tang, Y. (1994b) *Anal. Chem.,* **66**, 4278.
Armstrong, D. W., Schneiderheinze, J. M., Hwang, Y.-S. & Sellergren, B. (1998) *Anal. Chem.,* **70**, 3717.
de Biasi, V., Evans, M. B. & Lough, W. . J. (1990) Chiral Ion-Pair Chromatography of Basic Novel Anti-Hypertensive Agents. In: *Recent Advances in Chiral Separations* Stevenson, D. & Wilson, I. D. (Eds) Plenum Publishing, 93.
Bourque, A. J. & Krull, I. S. (1991) *J. Chromatogr.* **537**, 123.
de Camp, W. H. (1989) The Importance of Enantiomer Separations. In: *Chiral Liquid Chromatography.* Lough, W. J. (Ed.) Blackie Academic & Professional, Glasgow. 14.
Casy, A. F. & Mercer, A. D. (1998) *Mag. Res. Chem.,* **26**, 765.
Casy, A. F., Cooper, A. D., Jefferies, T. M., Gaskell, R. M., Greatbanks, D. & Pickford, R. (1991) *J. Pharm. Biomed. Anal.,* **9**, 787.
Cram, D. J., Helgeson, R. C., Sousa, L. R., Timko, J. M., Newcomb, M., Moreau, P., De Jong, F., Gokel, G. W., Hoffman, D. H., Domeier, L. A., Peacock, S. C., Madan, K. & Kaplan, L. (1975) *Pure Appl. Chem.,* **43**, 327.
Cram, M., Lough, W. J. (1998) *J. Pharm. Pharmacol.*
Czugler, M, Korkas, P. P., Bombicz, P., Seichter, W. & Weber, E. (1997) *Chirality,* **9**, 203.
Dalgliesh, C. E. (1952) *J. Chem. Soc.* 3490.
Davankov, V. A. & Rogozhin, S. V. (1971) *J. Chromatogr.,* **60**, 280.
Davankov, V. A., Rogozhin, S. V., Semechkin, A. V. & Sachkova, T. P. (1973) *J. Chromatogr.,* **82**, 359.
Dawson, B. A. & Black, D. B. (1995) *J. Pharm. Biomed. Anal.,* **13**, 39.
Dingenen, J. & Kinkel, J. N. (1994) *J. Chromatogr.,* **666**, 627.
Doyle, T. D. (1989) Synthetic Multiple-Interaction Chiral Bonded Phases. In: *Chiral Liquid Chromatography.* Lough, W. J. (Ed.) Blackie Academic & Professional, Glasgow. 102.
Ekborg-Ott, K. H., Kullman, J. P., Wang, X., Gahm, K., He, L. & Armstrong, D. W. (1998) *Chirality,* **10**, 627.
Fanali, S. (1996) *J. Chromatogr. A.,* **735**, 77.
Gil-Av, E., Feibush, B. & Charles-Sigler, R. (1966) *Tetrahedron Lett.,* 1009.
Gil-Av, E. (1975) *J. Mol. Evol.,* **6**, 131.
Henderson, G. M. & Rule, H. G. (1939) *J. Chem. Soc.,* 1568.
Hesse, G. & Hagel, R. (1973) *Chromatographia,* **6**, 277.
Ingersoll, A. W. & Adams, R. (1922) *J. Amer. Chem. Soc.,* **44**, 2930.
Jacques, J., Collet, A. & Wilen, S. H. (1981) *Enantiomers, Racemates and Resolutions.* John Wiley & Sons, New York.
Karagunis, G. & Coumoulos, G. (1938) *Nature,* **142**, 162.

Kirkland, K. M., Neilson, K. L. & McCombs, D. A. (1991) *J. Chromatogr.* **545**, 43.
Klemm, L. H. & Reed, D. (1960) *J. Chromatogr.* **3**, 364.
Knox, J. H. & Jurand, J. (1982) *J. Chromatogr.* **234**, 222.
König, W. A. (1992) *Gas Chromatographic Enantiomer Separation with Modified Cyclodextrins*. Hüthig, Heidelberg.
Krstulovic, A. M. (Ed.) (1989) *Chiral Separation by HPLC: Applications to Pharmaceutical Compounds*. Ellis Horwood Ltd., Chichester.
Kurganov, A. A., Facklan, C. & Davankov, V. A. (1987) Abstracts *11th International Symposium of Column Liquid Chromatography*, Amsterdam Th-P-32, 51.
Liu, L. & Nussbaum, M. (1999) *J. Pharm. Biomed. Anal.*, **19**, 679.
Lochmuller, C. H. & Ryall. R. R. (1978) *J. Chromatogr.*, **150**, 511.
Lough, W. J. (Ed.) (1989) *Chiral Liquid Chromatography*. Blackie Academic & Professional, Glasgow.
Lough, W. J. & Noctor, T. A. G. (1994) In: *Biomedical and Pharmaceutical Applications of Liquid Chromatography*. Riley, C. M., Lough, W. J. & Wainer, I. W. (Eds.) Pergamon Press, Oxford.
Lough, W. J., Groves, S-J., Law, B., Maltas, J., Mills, M. J. & Saeed, M. (1996) *Methodological Surveys in Bioanalysis of Drugs* **24**, 142.
MacNicol, D. D. & Rycroft, D. S. (1977) *Tetrahedron Lett.*, **25**, 2173.
Matlin, S. A., Grieb, S. J. & Belenguer, A. M. (1995) *J. Chem. Soc. Chem. Commun.*, 301.
Mikes, F., Boshart, G. & Gil-Av, E. (1976) *J. Chromatogr.*, **122**, 205.
Nicoud, R.-M., Fuchs, G., Adam, P., Baily, M., Kusters, E., Antia, F. D., Reuille, R. & Schmid, E. (1993) *Chirality*, **5**, 267.
Pasteur, L. (1848) *C. R. Acad. Sci.*, **26**, 535.
Pasteur, L. (1853) *C. R. Acad. Sci.*, **37**, 162.
Pasteur, L. (1858) in *Oevres*, **II**, 25, 129.
Pettersson, C. & Schill, G. (1982) *Chromatographia*, **16**, 192.
Pettersson, C. & Josefsson, M. (1986) *Chromatographia*, **21**, 321.
Pirkle, W. H. & Sikkenga, D. L. (1977) *J. Org. Chem.*, **42**, 1370.
Pirkle, W. H. & Hauske, J. R. (1977) *J. Org. Chem.*, **42**, 1839.
Pirkle, W. H. & House, D. W. (1979) *J. Org. Chem.*, **44**, 1957.
Pirkle, W. H., House, D. W. & Finn, J. M. (1980) *J. Chromatogr.*, **192**, 143.
Prangle, A. S., Noctor, T. A. G. & Lough, W. J. (1998) *J. Pharm. Biomed. Anal.*, **16**, 1205.
Prelog, V. & Wieland, P. (1944) *Helv. Chim. Acta*, **27**, 1127.
Pryde, A. (1989) Chiral Liquid Chromatography: Past and Present. In: *Chiral Liquid Chromatography*. Lough, W. J. (Ed.) Blackie Academic & Professional, Glasgow. 23.
Rogozhin, S. V. & Davankov, V. A. (1970) *Dokl. Akad. Nauk SSR,* 192.
Schill, G., Wainer, I. W. & Barkan, S. A. (1986) *J. Chromatogr.* **365**, 73.
Soo, E. C. & Lough, W. J. (1999a) *Chem. Ind.*, **17**, 202.
Soo, E. C., Lough, W. J. & de Biasi, V. (1999b) *Pharm. Sci. & Tech. Today*, **2**, 418.
Stewart, K. K. & Doherty, R. F. (1973) *Proc. Nat. Acad. Sci.*, **70**, 2850.
Suedee, R. & Heard, C. M. (1997) *Chirality*, **9**, 139.
Tamura, R., Ushio, T., Takahashi, H., Nakamura, K., Azuma, N., Toda, F. & Endo, K. (1997) *Chirality*, **9**, 220.
Van Overbeke, A., Sandra, P., Medvedovici, A., Baeyens, W. & Aboul-Enein, H. Y. (1997) *Chirality*, **9**, 126.

Vigh, G., Quintero, G. & Farkas, G. (1990) *J. Chromatogr.*, **506**, 481.
Wainer, I. W. (1987) *Trends in Anal. Chem.*, **6**, 125.
Welch, C. J. (1993) *Chemistry in New Zealand*, 9.
Wu, Z., Goodall, D. M. & Lloyd, D. K. (1990) *J. Pharm. & Biomed. Anal.*, **8**, 357.
Zukowski, J., Pawlowska, M. & Armstrong, D. W. (1992) *J. Chromatogr.* **623**, 33.

Chapter 8
Electronic Circular Dichroism—Fundamentals, Methods and Applications

Piero Salvadori and Lorenzo Di Bari
Centro di Studio del C.N.R. per le Macromolecole Stereordinate ed Otticamente Attive, Dipartimento di Chimica e Chimica Industriale, Università di Pisa, I-56126 Pisa (Italy)
Carlo Rosini
Dipartimento di Chimica Università della Basilicata a Potenza, I-85100 Potenza, Italy.

INTRODUCTION

Chiral molecules (*i.e.* those devoid of rotoreflection axes) interact peculiarly with plane polarized radiation; this interaction can be studied by means of specific spectroscopic techniques. The first peculiarity of a chiral structure is the ability to rotate plane polarized light by a characteristic angle α, a phenomenon discovered as early as in 1812. It was also observed that α is a function of the wavelength (the measurement of which gives rise to optical rotatory dispersion, ORD) and that the absorption bands of chiral molecules are characterized by anomalous bisignated ORD curves; these 'Cotton effects' were first studied in 1895.

A related phenomenon is the differential absorption of the left and right components of circularly polarized light, denoted circular dichroism, CD. ORD and CD were originally studied in the UV–Vis region, *i.e.* in the region of electronic transitions. Later it was realised that these phenomena are also apparent at wavelengths >800 nm and this led to the introduction of vibrational (Diem 1994; Nafie et al. 1994), Raman (Barron & Hecht 1994; Nafie et al. 1994) and NIR-CD allied to the f–f transition in some lanthanide(III) ions (Salvadori *et al.* 1984; Shoene *et al.* 1991) and to overtones of vibrational excitations (Abbate *et al.* 1996). In addition, CD can be observed indirectly by exciting a sample with circularly polarized light and measuring the intensity of fluorescence (Tinoco & Turner 1976). The differential spontaneous emission of left and right circularly polarized radiation by luminescent molecules can also be measured (Dekkers 1994; Richardson & Metcalf 1994; Riehl 1994).

In this chapter discussion is limited to electronic CD in the UV–Vis region. Readers interested in the other applications should refer to the general literature listed at the end of this chapter.

The aim of this chapter is to discuss the basic aspects of this field, to give an introduction to the essence of the phenomenon, and to provide an overview of the methods of interpretation which correlate UV–Vis CD with the molecular structure of organic molecules. Metal complexes, organometallic compounds, synthetic polymers and biological macromolecules will not be treated here.

PHENOMENOLOGICAL ASPECTS

Optical rotation and circular dichroism, are two different aspects of the same physical phenomenon: the interaction of plane polarized electromagnetic radiation with a collection of chiral non-racemic molecules. Left and right circularly polarized radiation having the same frequency and intensity (Figs. 8.1a and b) combine to yield plane polarized radiation (Fig. 8.1c).

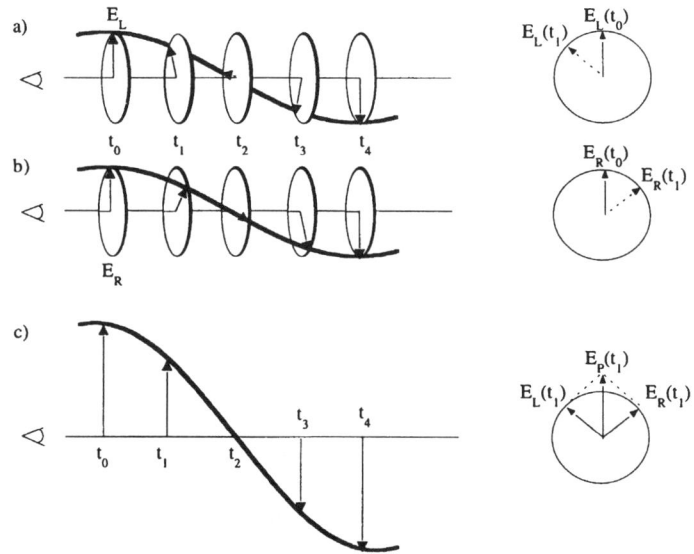

Fig. 8.1 Left (a) and right (b) circularly polarized light. Plane polarized light (c).

In Fig. 8.2 a source S produces UV–Vis radiation over the range 200–800 nm; a monochromator M (*e.g.* a prism) selects one wavelength λ_1 (more properly a narrow band around it) and the radiation, travelling in the z direction, encounters a polarizer P, which transmits only the component having the electric field **E** oscillating in the plane x,z (P_0). This radiation enters a cell of length l (dm) containing a solution of a chiral compound at concentration c (g/100 mL solution).

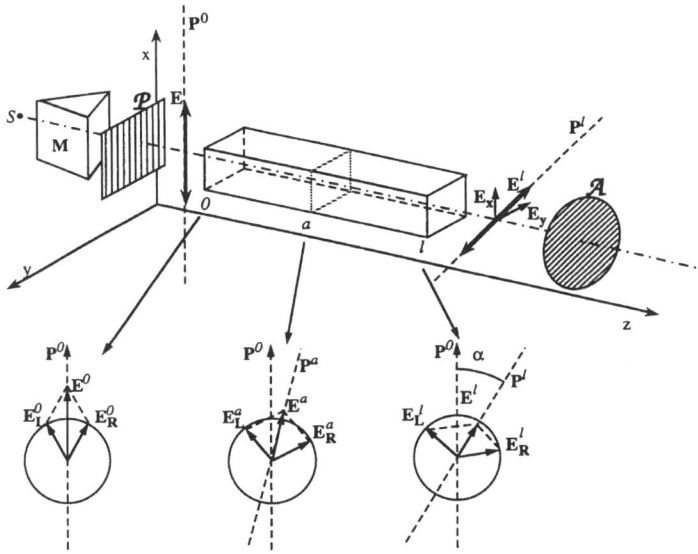

Fig. 8.2 Idealized optical rotation experiment.

At the point 0 the electric field $\mathbf{E}^0$ is the resultant of two circularly polarized components $\mathbf{E}_R^0$ and $\mathbf{E}_L^0$. In the optically active medium the two circularly polarized components have different refractive indices (Mathieu 1946; Mason 1963, 1982) i.e. $n_L \neq n_R$. Assuming, for instance, that $n_L < n_R$, then:

$$\frac{c}{v_L} < \frac{c}{v_R} \text{ and } v_L > v_R \qquad (1)$$

Thus in the optically active medium the velocity of the left-polarized component is greater than that of the right-polarized component. When the radiation leaves the cell, the situation will be described as in Fig. 8.2, point *l*—the vector $\mathbf{E}^l$ will no longer oscillate in the plane P^0 but rather in a plane P^l rotated through an angle α relative to P^0. This can be experimentally observed by means of a second polarizer filter A (analyser). When A is kept parallel to P, the observed radiation is attenuated, but as soon as A is rotated through an angle α the entire $\mathbf{E}^l$ vector is transmitted and maximum intensity is observed. This is the principle on which the measurement of optical rotatory power is based. The angle of rotation α at the wavelength λ is related to the left and right refraction indices, n_L and n_R, through the Fresnel equation (Mathieu 1946; Mason 1963; Snatzke 1967):

$$\alpha = (n_L - n_R)\frac{\pi}{\lambda} l \qquad (2)$$

The specific rotatory power is given by:

$$[\alpha]_\lambda^T = \frac{\alpha}{cl} 100 \tag{3a}$$

where α is the measured angle of rotation, l is the path length in dm, c is the concentration of solute (g/100 mL), T is the temperature, and λ the wavelength of the radiation used, often the sodium D-lines at 589.0 and 589.6 nm, referred to as D. For pure liquids, rather than solutions, the definition of specific rotatory power is modified as follows:

$$[\alpha]_\lambda^T = \frac{\alpha}{\delta l} \tag{3b}$$

where δ is the density of the substance at the temperature T.

This treatment completely neglects that the two circularly polarized components in the optically active medium not only propagate at different velocities, but also are absorbed differently. In other words, at points $a, b, ..., l$ of Fig. 8.2 the moduli E_L and E_R are different.

This more general situation is depicted in Fig. 8.3. On leaving the sample cell, the two components have different phases and amplitudes and thus combine into elliptically polarized radiation, *i.e.* the point of the emerging electric field vector traces an elliptical path.

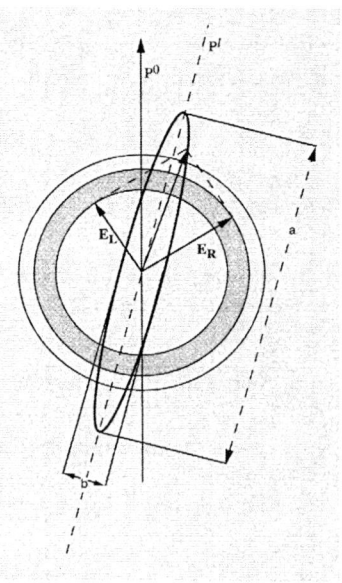

Fig. 8.3 Elliptically polarized light.

The *ellipticity*, ψ, is defined as:

$$\text{tg}\Psi = \frac{b}{a} \tag{4}$$

where b and a are the lengths of the minor and major axes of the ellipse. Normalizing for the pathlength and concentration, one can define the *specific* and *molar* ellipticities, respectively as:

$$[\Psi] = \frac{\Psi}{cl} \text{ and } [\theta] = \frac{[\Psi]M}{100} \tag{5}$$

where M is the molecular weight.

The differential absorption of the two circularly polarized components by an optically active medium implies two different extinction coefficients, ε_L and ε_R. The quantity:

$$\Delta\varepsilon = \varepsilon_L - \varepsilon_R \tag{6}$$

is the *circular dichroism*, CD, and is directly related to the molar ellipticity $[\theta]$ by:

$$\Delta\varepsilon = 3300[\theta] \tag{7}$$

Both $n_L - n_R$, determining $[\alpha]$, and $\varepsilon_L - \varepsilon_R$, related to $\Delta\varepsilon$, depend on λ (Fig. 8.4), thus giving rise to ORD and CD spectra, respectively.

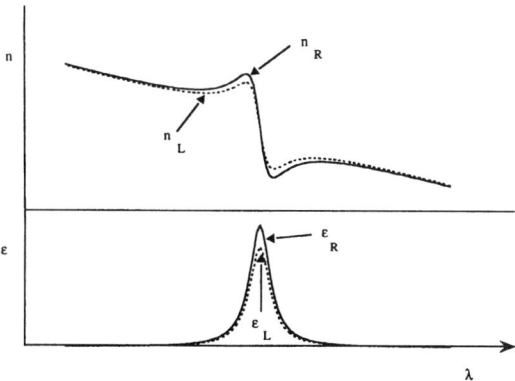

Fig. 8.4 Differential refractive indices and extinction coefficients for left and right circularly polarized light.

Two antipodes of the same molecule give, while having the same isotropic absorption spectra, rise to ORD and CD curves which are mirror images of each other. A hypothetical example for a specific electronic transition is depicted in Fig. 8.5.

ORD and CD are linked by a mathematical transformation (the dispersion and absorption components of the light–matter interaction are linked by Kramers–Kronig transforms, which can be found elsewhere, *e.g.* Caldwell & Eyring 1971) and give equivalent structural information; because CD can be observed only in the vicinity of absorption bands, however, its interpretation is easier than that of its dispersive counterpart, where contributions arising from different transitions often merge.

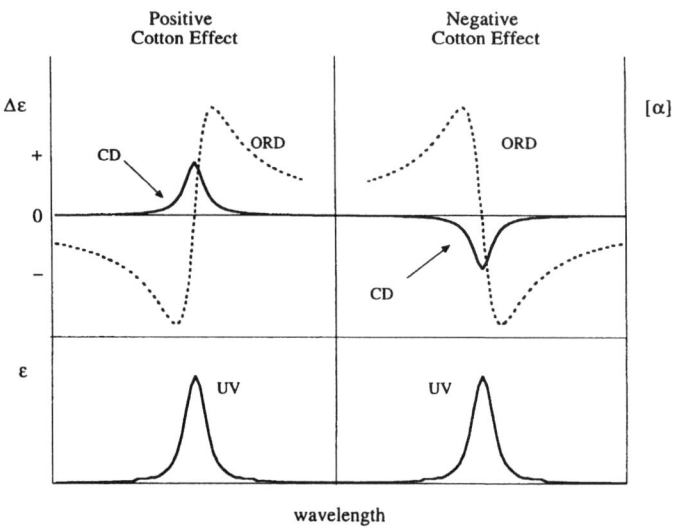

Fig. 8.5 The Cotton effect associated with an absorption peak.

INSTRUMENTATION—ESSENTIAL DESCRIPTION OF A DICHROGRAPH
(Mathieu 1946; Snatzke 1967; Mason 1982; Drake 1989; Purdie & Brittain 1994)

A dichrograph measures the difference between the absorbance of left- and right-circularly polarized light. It can be described as a differential spectrometer in which two oppositely polarized CPL beams are used at different times and the results subtracted. Thus it is somewhat similar to an ordinary spectrometer but contains a suitable polarizing filter.

It can be shown that to produce CPL it is necessary to superimpose two plane polarized beams at square angles with a phase shift of $\pi/2$. This requirement amounts to saying that one of the two PPLs must lag $\lambda/4$ behind the other. The key feature of a dichrograph is, therefore, a linear polarizer, usually placed between the source and the sample. The light source is usually a xenon arc lamp, covering wavelengths from 200 to 1000 nm, followed by a monochromator which selects a small range of wavelengths.

The quarter-wave retardation is ensured by a photoelastic modulator (PEM), a piezoelectric device the effect of which is to produce left and right CPL alternately, according to an electric current oscillating at 50–100 MHz or above. The beam then goes through the sample, usually contained in a quartz cell, and is finally analysed by means of a photomultiplier.

For optically inactive samples the absorption coefficients for the two CPL beams are equal and the output current of the detector is constant, whereas, for an optically active solution, the current is modulated at the frequency of the PEM. The AC and DC components are separated by means of a lock-in-amplifier, the latter being associated with the average of the two extinction coefficients (and hence with the absorbance) the former with the circular dichroism. In some instruments the architecture is more sophisticated and, for example, enables simultaneous detection of several wavelengths or coupling of the dichrograph to other devices, *e.g.* a stopped-flow apparatus or HPLC. Also, especially for IR measurements, the various parts can be arranged differently.

TOWARDS AN UNDERSTANDING OF OPTICAL ACTIVITY AND RELATED PHENOMENA

PRELIMINARY CONSIDERATIONS

Macroscopic characteristics of an optically active medium (Atkins 1983)

So far the phenomena occurring when plane-polarized radiation interacts with a solution of a non-racemic chiral medium have been described. The origin of these effects must be investigated and the question "Why do chiral molecules respond differently to left and right circularly polarized light?" must be answered.

The fundamental reason lies in the spatial extension of a molecule and in the variation of the external electromagnetic field over it. Thus the molecular dimensions with regard to the wavelength of incident radiation must not be neglected. Maxwell equations show that the spatial variation of the electric field of a light wave is proportional to the time variation of the magnetic field $\left(\frac{\partial \mathbf{E}}{\partial x} \propto \frac{\partial \mathbf{H}}{\partial t}\right)$. The electric polarization **P** resulting from the interaction with the electromagnetic field can thus be expressed as the sum of two terms, one related to **E**, the second proportional to its spatial variation, in turn proportional to $\frac{\partial \mathbf{H}}{\partial t}$:

$$\mathbf{P} = \alpha \mathbf{E} + \beta \frac{\partial \mathbf{H}}{\partial t} \tag{8}$$

By analogy:

$$\mathbf{M} = \kappa \mathbf{H} + \beta \frac{\partial \mathbf{E}}{\partial t} \tag{9}$$

where $\kappa = 0$ for diamagnetic systems.

In Eqs (8) and (9) a non-vanishing β implies rotation of the plane of polarized light. In fact, if **E** and **H** are the vectors of a plane-polarized radiation they oscillate in two perpendicular directions and **P** thus consists of two components, one parallel to the incident electric field **E**, the other, proportional to β, perpendicular to it. The resultant polarization is thus rotated from the initial orientation. Now, for a given chiral structure, β has a definite signed value, which becomes its opposite if the coordinates are inverted, *i.e.* if the enantiomer is used. For a racemic mixture, and for an ensemble of non-chiral molecules, β averages to 0, and thus optical rotation vanishes.

The helix model (Feynman *et al.* 1963)

The prototype of a chiral structure is a helix and one of the simplest models accounting for optical rotation consists in an electron constrained to move along a helical path.

A plane-polarized light beam propagating along z with the electric dipole parallel to x is described by:

$$E_x(z,t) = E_0 e^{\left(i\omega\left(t-\frac{z}{c}\right)\right)} \tag{10}$$

This field interacts with a right-handed helix of diameter d with the axis parallel to x. Let the two extreme points along z be A and B, according to Fig. 8.6.

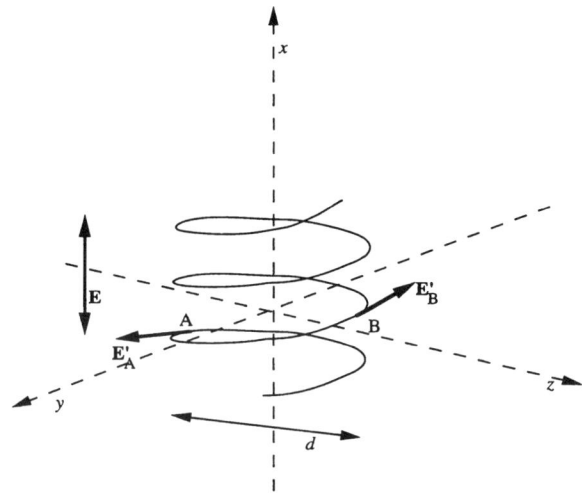

Fig. 8.6 Helix model. The radiation propagates along z; the induced fields at points A and B are shown.

Moving between A and B in the time:

$$\Delta t_{AB} = \frac{d}{c} \tag{11}$$

the electrons move on the helix under the action of electric field of the radiation $E_x(z,t)$ and produce, in turn, an induced field $\mathbf{E}'(z)$, necessarily parallel to the helical segment around z. To make the situation explicit at the two extremes, A and B, a coefficient

$c(v)$ is introduced which takes into account the specific characteristics of the interaction. The y component of $\mathbf{E}'$ in A can be expressed as:

$$E'_y(A) = c(v)\exp\left[i\omega\left(t - \frac{z_A}{c}\right)\right] \tag{12}$$

At B the helix is oriented with a y component opposite to that at A, thus:

$$E'_y(B) = -c(v)\exp\left[i\omega\left(t - \frac{z_B}{c}\right)\right] \tag{13}$$

which uses the fact that the orientation of the helix with respect to y is opposite for the two points A and B. It is apparent that $E'_y(B) - E'_y(A)$ does not vanish and is equal to:

$$E'_y(B) - E'_y(A) = 2c(v)\exp\left[i\omega\left(t - \frac{m}{c}\right)\cos\left(\omega\frac{d}{2c}\right)\right] \tag{14}$$

where $m = (z_B + z_A)/2$. This equation not only indicates that there is non-vanishing polarization in the y direction after interaction with the helix, but also that it is frequency-dependent, even when $c(v)$ is constant, notably outside all absorption phenomena.

In particular, if the other terms can be considered quasi-constant, then:

$$E'_y(B) - E'_y(A) \propto \cos\left(2\pi\frac{d}{\lambda}\right) \tag{15}$$

which, incidentally, gives a $1/\lambda^2$ expansion as in one of the first equations formulated for ordinary optical rotatory dispersion, that of Drude (Mason 1982).

THE NATURE OF ELECTRONIC TRANSITIONS (Snatzke 1979; Michl & Thulstrup 1986; Eliel *et al.* 1994)

Because this discussion is limited to the chiroptical properties of the electronic transitions of organic molecules in the UV–vis range, it will be useful to give here a brief account of the main characteristics of absorption. In chemistry the behaviour of a molecule is ascribed to the presence of a group (the functional group) responsible for reactivity. Similarly, the concept of a chromophore can be introduced as the molecular fragment giving rise (*i.e.* where the electronic transition is mostly localized) to characteristic absorption in the UV or visible spectrum, *i.e.* the group which determines the spectroscopic behaviour of the molecule.

Absorption of a photon by a molecule is accompanied by passage from the ground $|0\rangle$ state to an excited state $|i\rangle$. At a very simple level electronic transitions from an occupied (σ, π) or a nonbonding (n) MO to an empty (π^*, σ^*) MO can be considered, giving n→π^* and π–π^* transitions, etc. The distribution of electronic charge is different in the two states and, during the transition, redistribution of electronic charge occurs, giving rise to a transient moment. An electric dipole transition moment is generated if such redistribution correspond to a linear displacement of charge, whereas a magnetic transition moment is generated by a rotation of charge.

The transition charge density can be defined as:

$$\rho_{0i}(x,y,z) = -e\psi_o(x,y,z)\psi_i(x,y,z), \tag{16}$$

where ψ_o is the starting MO and ψ_i the final MO. That is, it can be calculated by picking a point in space and determining the signed product of two MO values for that point.

The $\pi \rightarrow \pi^*$ transition of the olefin chromophore is presented in Fig. 8.7.

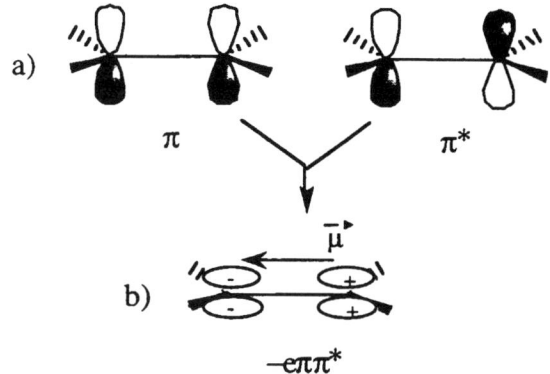

Fig. 8.7 Bonding and antibonding π orbitals in the olefin chromophore (a) and transition charge density for the $\pi \rightarrow \pi^*$ transition (b).

The transition charge density is given by the product $-e\pi\pi^*$ which corresponds to an electric transition dipole moment $\mu_{\pi\pi^*}$, directed from the atom with the deficient negative charge to the atom with excess negative charge. This transition is electric dipole-allowed and so has a high extinction coefficient (ε 10^4). Other electrically allowed transitions present in some organic chromophores are illustrated in Fig. 8.8 with their polarization directions, which can, in principle, be established theoretically (MO calculations; Snatzke 1979; Michl & Thulstrup 1986) or experimentally (linear dichroism measurements; Michl & Thulstrup 1986).

Although electrically allowed transitions usually dominate ordinary absorption spectra, magnetic transition dipoles play an essential role in optical activity. The $n \rightarrow \pi^*$ transition of the carbonyl chromophore can be chosen as an example. With the choice of axes represented in Fig. 8.9 the n orbital is an almost pure $p_y(O)$ orbital located on the oxygen atom, whereas the π^* MO is a linear combination of $p_x(C)$ and $p_x(O)$ with suitable coefficients c' and c'' respectively, so that:

$$\rho_{n\pi^*}(x,y,z) = -en\pi^* = -ep_y(O)[c'p_x(C) + c''p_x(O)] \approx -ec''p_y(O)p_x(O) \tag{17}$$

Indeed, if it is taken into account that $p_x(O)$ is different from zero only in proximity of the oxygen atom, the transition charge density will be non vanishing only about this point.

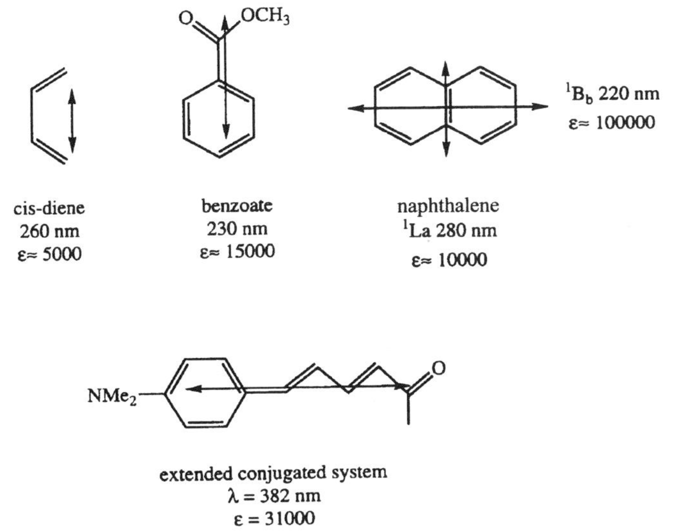

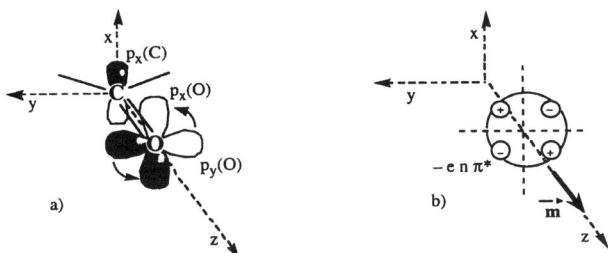

O
OCH3
1Bb 220 nm
ε≈ 100000
cis-diene
260 nm
ε≈ 5000
benzoate
230 nm
ε≈ 15000
naphthalene
1La 280 nm
ε≈ 10000
NMe2
O
extended conjugated system
λ = 382 nm
ε = 31000

Fig. 8.8 Electrically allowed transitions in some organic chromophores.

x
px(C)
y
C
px(O)
O
py(O)
a)
z
x
y
−e n π*
b)
m
z

Fig. 8.9 (a) Atomic orbitals involved in the n→π* transition of the carbonyl chromophore. (b) The corresponding magnetic moment and electric quadrupole.

The transition charge density has a vanishing electric dipole and, to a good approximation, corresponds to a $p_y \rightarrow p_x$ excitation on the oxygen atom. The charge rotates around the z axis and a magnetic transition dipole moment $\boldsymbol{m}_{n\pi^*}$ is developed along the positive z axis (according to the choice of phases of the p orbitals shown in the figure). Such a transition is classified as electrically (dipole) forbidden magnetically allowed.

Furthermore, Eq. (17) describes a pattern of alternating positive and negative charges around the oxygen atom, defining an electric quadrupole in the plane parallel to xy as represented in Fig. 8.9b.

Dipolar strength (Mathieu 1946; Mason 1963, 1982; Snatzke 1967, 1979; Charney 1979)

The extinction coefficient about an absorption peak, or, more precisely, its integral through a band, can be related to a microscopic quantity, the *dipolar strength*. If $|0\rangle$ and $|i\rangle$ are the ground and excited states the dipolar strength, D_{0i}, is defined as:

$$D_{0i} = |\mu_{0i}|^2 \tag{18}$$

where μ_{0i} is the electric dipole moment and contributions from higher order electric and magnetic multipoles have been neglected.

Experimentally, this quantity is given by the area under the corresponding absorption band and can be approximated (in c.g.s. units) as:

$$D_{0i} = 9.2 \times 10^{-39} \varepsilon_{max} \frac{\Delta \tilde{\nu}}{\tilde{\nu}_{max}} \tag{19}$$

where $\Delta \tilde{\nu}$ is the width of the band at half-height.

Rotational strength (Mathieu 1946; Mason 1963, 1982; Snatzke 1967, 1979; Charney 1979)

Discussion so far has been limited to purely electrically or magnetically allowed transitions. The two are not, however, mutually exclusive. When the electron describes a helical path, it defines at the same time a rotation and a linear displacement, and the electric and the magnetic dipole moments are parallel. As already discussed, such a trajectory is the prototype of chirality, thus it is not surprising that the prerequisite for a transition to be optically active is that it has non-vanishing and not mutually orthogonal μ and $\mathbf{m}$. The parameter measuring this property is called the *rotational strength* and is defined, for the transition $|0\rangle \to |i\rangle$, as:

$$R_{0i} = \text{Im}\{\mu_{0i} \cdot \mathbf{m}_{0i}\} = |\mu_{0i}||\mathbf{m}_{0i}|\cos\theta \tag{20}$$

where θ is the angle between the two vectors. With this definition, a right-handed helical charge displacement is associated with a *positive* R and a left-handed displacement with a *negative* R, as shown in Fig. 8.10

It can be demonstrated that the rotational strength is proportional to the integral of the CD spectrum over a band, exactly as for the dipolar strength in UV. As for the isotropic absorption discussed above, the rotational strength can be related to the integral of $\Delta\varepsilon/\nu$ over a CD band by:

$$R_{0i} = 23 \times 10^{-40} \Delta\varepsilon_{max} \frac{\Delta\tilde{\nu}}{\tilde{\nu}_{max}} \tag{21}$$

again in c.g.s. units.

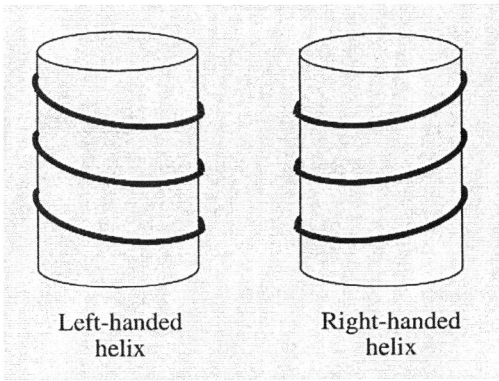

Fig. 8.10 Left- and right-handed helices.

Symmetry properties of the rotational strength—spectroscopic foundation of optical activity (Mason 1963; Charney 1979; Atkins 1983)

The definition of rotational strength given in Eq. (20) satisfies the symmetry requirements of chirality: it is necessarily zero for molecules with an improper rotation axis of any order and is opposite for a pair of enantiomers. Both these properties are derived from the nature of the factors of the scalar product defining R: **μ** corresponds to a charge translation and **m** to a rotation; consequently, the former transforms as a vector, the latter as a pseudovector.

The transformation properties of these two objects under reflection operations are complementary: a vector is transformed into itself by reflection onto a plane containing it and changes sign when reflected on a plane perpendicular to it, whereas a pseudovector changes sign in the former instance and remains unchanged in the latter. Consequently, because the rotational strength, R, is the product of the two terms, it changes sign as a result of any reflection operation, while remaining the same as a result of a rotation. This implies that whenever a molecule has a rotoreflection axis, $\hat{S}_n$, its transitions can have electric- or magnetic-dipole character (or even a combination of the two), but they must transform according to $\hat{S}_n$ and R should be equal to its opposite, which implies that is zero. In the same way, for a pair of enantiomers, *i.e.* mirror image structures, the two Rs must be equal and opposite.

These rules correspond to the macroscopic criterion for identification of chiral structures (Eliel *et al.* 1994).

Another important relationship concerning R is that ΣR_{nm}, the sum over *all transitions* $|n\rangle \rightarrow |m\rangle$ is given by:

$$\sum_{nm} R_{nm} = 0 \qquad (22)$$

Justification of this equation requires quantum mechanical treatment (Caldwell & Eyring 1971) and is beyond the scope of this chapter.

CLASSIFICATION OF OPTICALLY ACTIVE MOLECULES (Moscowitz 1961)

Optically active molecules can be divided into two types: those containing an intrinsically chiral chromophore and those in which a symmetric chromophore is inserted into a chiral environment. In the former type each transition generates non-orthogonal electric and magnetic dipole moments; this results in non-vanishing rotational strength. In symmetric chromophores each transition has either the electric or magnetic moment only; the non-zero R arises because the missing dipole (magnetic or electric) is induced by the presence of the chirally arranged perturber. A few examples are shown in Fig. 8.11.

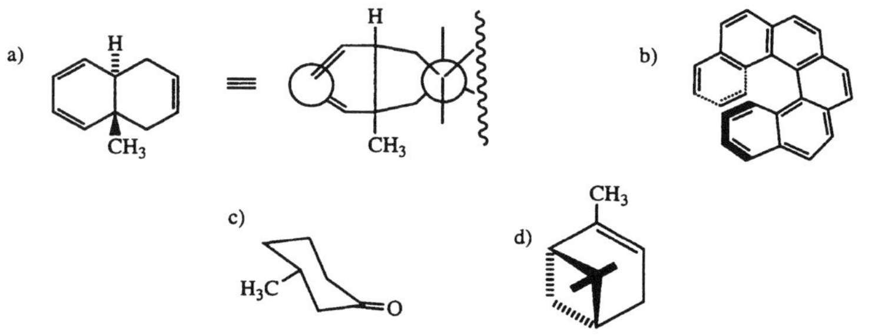

Fig. 8.11 The distorted *cis*-diene in a ring (a) and hexahelicene (b) are two examples of intrinsically chiral chromophores. In 3-methylcyclohexanone (c) and in α-pinene (d) the chromophores (saturated carbonyl and olefin, respectively) are locally symmetric but perturbed by the chiral environment.

Optical activity of dissymmetric chromophores

It will be shown here that an electronic transition occurring in a twisted chromophore is characterized by non-orthogonal electric and magnetic transition moments and is therefore optically active. To this end let us consider glucal (Fig. 8.12), a molecule containing a twisted enol ether chromophore (Snatzke 1979).

The lowest energy $\pi \rightarrow \pi^*$ transition of this compound is at 205 nm; this can be assigned, using the simple MO scheme of Fig. 8.12a as the $\pi^0 \rightarrow \pi^-$ transition. The transition charge density, represented in Fig. 8.12b, corresponds to an electric dipole oriented from C1 to O, and is accompanied by a rotation, which by the right-hand rule provides a magnetic moment directed from O to C1. The two dipoles are antiparallel, giving rise to a negative rotatory strength, as shown in Fig. 8.12c. For the negative torsional angle θ of the chromophore, therefore, a negative CD is expected.

This discussion is a qualitative application of LCAO–MO theory. MO calculations of different sophistication have been used to estimate rotational strengths. The first example was that of Moscowitz (Moscowitz 1961; Moscowitz *et al.* 1961), who introduced the use of Hückel wave functions to treat intrinsically dissymmetric chromophores such as hexahelicenes and skewed dienes. Since then there have been many reports of the use of molecular wave functions derived from semi-empirical MO methods—Hückel, extended Hückel, PPP, CNDO and INDO. In addition, *ab initio* calculations of rotational strength have appeared since 1968 (Yaris *et al.* 1968).

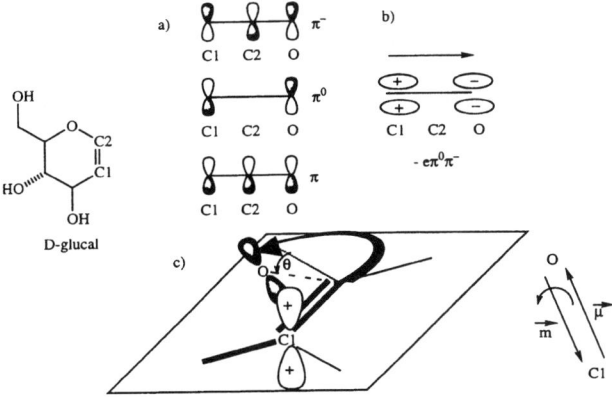

Fig. 8.12 Lowest-energy $\pi \rightarrow \pi^*$ transition in glucal, a twisted enol ether chromophore.

Among the most recent applications of MO methods, several papers must be cited:
(1) PPP calculations of atropisomeric biflavones (Harada *et al.* 1992), troponoids (Harada *et al.* 1987) and spiro compounds (Harada *et al.* 1989);
(2) CNDO/S–CI calculations of optically active dipyridine (Rashidi-Ranjbar *et al.* 1992), tetrahydropyridine (Nilsson *et al.* 1991), coumarins (Hargitai *et al.* 1991) and biaryl derivatives (Bringmann *et al.* 1997); and
(3) *ab initio* SCF–MO–CI calculations on simple chromophores (*e.g.* oxirane (Carnell *et al.* 1991), aziridine (Shustov et al. 1992a, b; Shustov & Rauk 1995) and peroxide (Huang & Suter 1994)) and even on larger systems (*e.g.* Grimme *et al.* 1995). The reliability of this method depends on the quality of the wavefunctions of the ground and excited states.

Use of MO methods for the calculation of natural electronic optical activity has been reviewed by Hansen & Bouman (1980, 1989) and by Volosov & Woody (1994) and will not be further discussed here. More recently new computational approaches have appeared (Bak *et al.* 1995, Pedersen & Hansen 1995; Grimme 1996).

An alternative method, the independent systems approach, originally devised for calculation of the CD of chirally perturbed achiral chromophores, has been proposed.

THE INDEPENDENT SYSTEMS APPROACH

In the independent systems (IS) approach (Caldwell & Eyring 1971; Mason 1978, 1982; Charney 1979) a molecule is divided into one or more chromophores, in which the electronic transition is located, and one or more perturbers, 'modifying' the transition. Optical activity arises from the interaction(s) between chromophore(s) and perturber(s). Such interaction is assumed to arise solely as a result of electrostatic interaction between charge distributed on chromophore and perturber. No electron exchange and negligible differential overlap are also assumed.

At this level of approximation, the historical development of interpretative models of optical rotation can be followed, and two separate treatments can be distinguished:

(a) for magnetically allowed electrically forbidden transitions (*e.g.* the n→π* transition of the ketone chromophore); and

(b) for electrically allowed magnetically forbidden excitations (*e.g.* the π→π* transition of the same chromophore).

Problem (a), *i.e.* how can a magnetically allowed transition acquire the necessary non-orthogonal electric dipole moment, was initially approached in the 1930s by the Princeton group (Kauzmann *et al.* 1940), who developed the so-called one-electron theory. They considered a simple symmetric chromophore provided with only two excited states, |a⟩ and |b⟩, transition |0⟩ →|a⟩ being magnetically allowed and transition |0⟩ →|b⟩ electrically allowed. Because the presence of an external perturber reduces the overall molecular symmetry, mixing the two states, the magnetically allowed transition acquires the necessary electric dipole moment and *vice versa*. In this respect, this approach is reminiscent of the crystal field theory of the electronic absorption spectra of transition metal complexes. This mechanism can be more easily understood with reference to Fig. 8.13.

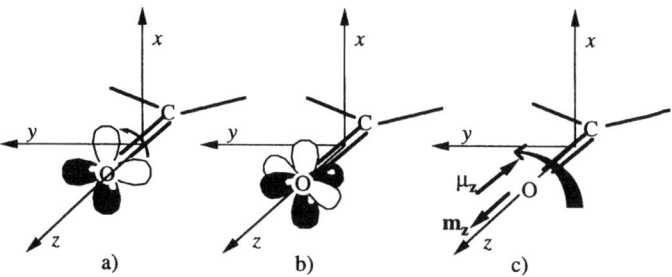

Fig. 8.13 One-electron model interpretation of the optical activity of the n→π* transition of the carbonyl chromophore.

As shown in Fig. 8.9 and discussed in the section *The Nature of Electronic Transitions*, in the symmetric carbonyl chromophore (Fig. 8.13a) the n→π* transition corresponds to $p_y(O) \to p_x(O)$ excitation, *i.e.* rotation of charge around z, and provides a magnetic dipole moment m_z with no collinear electric dipole. The symmetry reduction as a result of the presence of the perturber causes the aforementioned mixing of states; as a consequence the new π* orbital could result from contamination of a d_{yz} function on the oxygen atom (Fig. 8.13b). As is apparent from Fig. 8.13c the new n→π* transition is no longer confined to the *x,y* plane but involves displacement of charge along z, *i.e.* an electric transition dipole moment along the axis of the magnetic polarization.

Such a situation gives rise to non-vanishing rotational strength. In other words, the introduction of even a small fraction of d_{yz} in the π* orbital makes the original rotational charge distribution helical. It must be clear at present that this simple picture, although giving an understandable description of the one-electron model, gives no clue about either the sense of the transitional helix (hence the sign of the rotational strength) or the magnitude of R. This can be obtained only after defining the reciprocal arrangement of the chromophore and the perturber, specifying the nature of the interaction potential, and so on.

In 1968 a different mechanism was introduced to account for optical activity of the n→π* transition of saturated ketones (Weigang & Hohn 1968). As was shown above, this excitation is associated with a magnetic dipole m_z and an electric quadrupole θ_{xy}. In Fig. 8.14 these two terms are represented for a dissymmetrically perturbed carbonyl group (here on a 3-substituted cyclohexanone). Assuming that P is an isotropically polarizable perturber, at coordinates ($+x$, $+y$, $-z$) an electric dipole μ_z is induced on P by θ_{xy} (here only the effect of the quadrupolar charge nearest to P is considered): m_z and μ_z are thus parallel, which shows that a substituent at $+x$, $+y$, $-z$ (upper left rear octant) gives rise to a positive rotational strength.

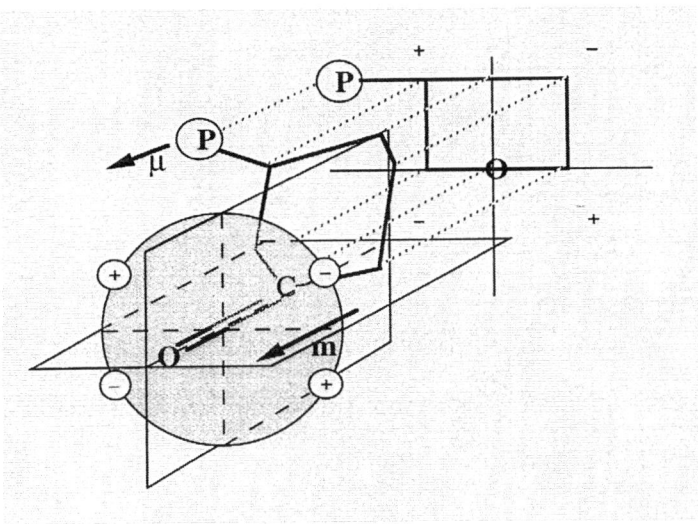

Fig. 8.14 Octant representation of a 3-substituted cyclohexanone.

The essential difference between the two mechanisms described here for the saturated ketone n→π* transition is that in the former the perturber simply reduces the symmetry of the chromophore, without directly interacting with light; in the latter, there is a reciprocal interaction between chromophore and perturber. For this reason, the first case is called *static coupling*, the second *dynamic coupling*.

COUPLED OSCILLATORS

Let us consider a simple (hypothetical) molecule containing two chromophores, each having a single electrically allowed transition, polarized along non-coplanar directions, depicted in Fig. 8.15a.

The two oscillating dipoles couple with one another and give rise to two modes, one *in-phase*, shown in Fig. 8.15b, the other *in anti-phase*, Fig. 8.15c, according to their *reciprocal* orientation. These two modes are not equivalent; because the former brings charges of the same sign nearer than the latter, energy splitting as a result of electrostatic interaction must be expected. The oscillating charge of μ_1 generates in P_2 a magnetic field given by:

$$\mathbf{m}(P_2) \propto \mu_1 \times \mathbf{r}_{12} \tag{23}$$

The vectors μ_1, $\mathbf{r}_{12}$, $\mathbf{m}(P_2)$ and μ_2 for the in-phase mode are represented on the left of Fig. 8.15b. It is apparent that $\mathbf{m}(P_2)$ and μ_2 are not orthogonal. Therefore, for the transition localized in P_2, the following non-vanishing rotational strength is obtained:

$$R \propto \mu_2 \cdot \mu_1 \times \mathbf{r}_{12} = \mu_2 \times \mu_1 \cdot \mathbf{r}_{12} \tag{24}$$

A similar expression can be obtained for the transition in P_1, on the right of Fig. 8.15b, and the overall rotational strength for the in-phase mode is given by:

$$R \propto 2\mu_2 \times \mu_1 \cdot \mathbf{r}_{12} \tag{25}$$

Similarly, for the antiphase mode, depicted in Fig. 8.15c, a value of R is obtained which is opposite to that of the in-phase given in Eq. (25).

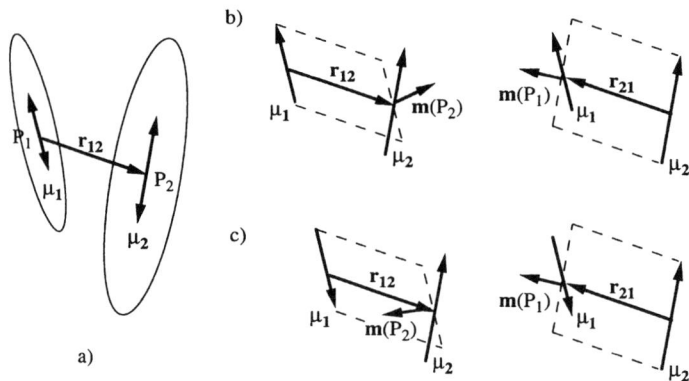

Fig. 8.15 Coupled oscillators: (a) two non-coplanar transition dipole moments. (b) The *in-phase coupling* mode. (c) The *anti-phase* coupling mode.

This is a very short description of the coupled oscillator model formulated classically by Kuhn (1930) in the early 1930s and quantum mechanically by Kirkwood (1937). The model for two equal dipoles (exciton coupling) is that of Moffitt (1956).

APPLICATIONS TO STRUCTURAL CHEMISTRY

EMPIRICAL CORRELATIONS

This section describes how structural information is obtained from a CD spectrum. Because the energy, position, sign, and intensity of a single Cotton effect are dependent on the overall chemical structure, CD and, more generally, chiroptical data, are potential reservoirs of structural information. The extraction of stereochemical information from CD spectra is, however, not always immediate. CD data can be interpreted, accepting the classification of Snatzke (1968), by following empirical, semi-empirical, or non-empirical paths.

Empirical analysis of CD spectra implies proposing spectra–structure relationships with reference to similar molecules of known structure. This method can be illustrated here by the assignment of the absolute configuration of some 1,4-benzodiazepinones (Salvadori *et al.* 1991; Fig. 8.16). The *S* antipode of 7-chloro-1,3-dihydro-3-*i*-propyl-5-phenyl-2*H*-1,4-benzodiazepin-2-one has positive CD between 250 and 260 nm.

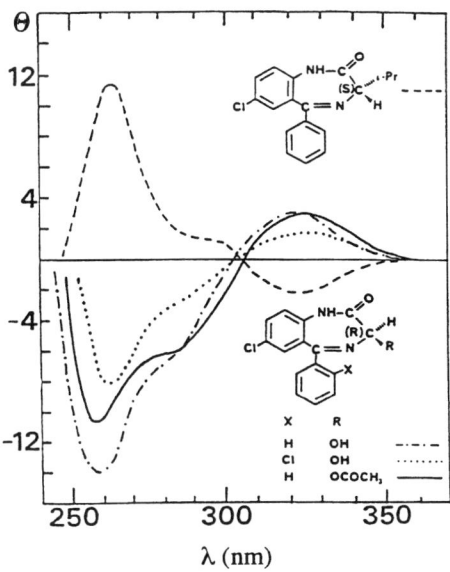

Fig. 8.16 Empirical correlation of the CD spectra and absolute configurations of substituted benzodiazepines.

Because compounds with negative CD in the same spectral range are structurally very similar to the reference substance, the *R* absolute configuration can be assigned to them. Nowadays empirical analysis of CD data (*i.e.* simple comparison of spectra) is still widely employed (Drake *et al.* 1996; Mukakami *et al.* 1997; Tejero *et al.* 1997).

The popularity of this old approach can be attributed to its rapidity and simplicity. It requires no spectroscopic or quantum mechanical knowledge—configurational assignment is achieved solely on the basis of comparison of the CD spectra of two compounds.

The recent configurational assignment of some dihydroisocoumarins (Salvadori *et al.* 1996) is highly illustrative of the proper use of the method. Compound **1** (of known 3*S*,1*S* absolute configuration) has the CD spectrum reported in Fig. 8.17.

Compounds **2**, **3**, and **4** have CD spectra which are nearly superimposable on each other and on that of **1**. The same absolute configuration at C-3 can be assigned to these compounds, in fact the features of the CD spectrum are determined only by the absolute configuration of the C-3 stereocentre, as is shown by **2** and **1** (which have the opposite absolute configuration at the exocyclic stereocentre) having the same CD; the same is true for **2** and **3** also, which have different substituents at the same position. The conclusion which can be drawn from this example is that the main features of CD depend only on the core of these systems (*i.e.* the cyclic part) and the nature of the exocyclic substituent does not significantly affect the spectrum. Therefore the absolute configuration of the core can be correctly assigned by comparing CD data.

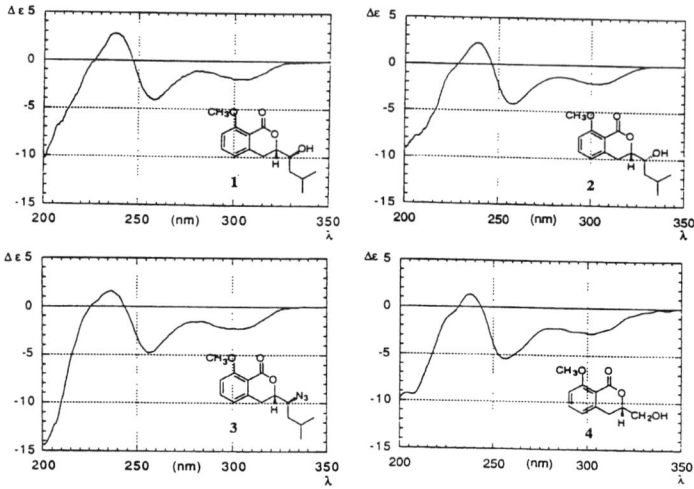

Fig. 8.17 Comparison of the CD spectra of compounds **1–4**. An illustrative example of the correct approach to configurational assignment.

Generalizing this procedure can, however, hide difficulties and risks, and it is certainly worth discussing this approach further, to obtain a better idea of its applications and limitations. Firstly, assignment of absolute configuration must be performed with extreme caution—totally analogous chromophoric systems must be compared and the chromophoric environments of the reference substance and of the compound under examination must by strictly analogous, as in the examples shown above. Thus correlations of this kind are generally not possible between molecules of similar structure but with different chromophores. For instance, colchicine, **5**, cannot be correlated with known biphenyl systems such as glaucine, **6**, because the presence of the tropolone chromophore introduces profound differences to the CD spectra, making configurational correlation impossible (Hrbek *et al.* 1982).

It must also be remarked that sometimes even very small structural differences make the application of this method completely unreliable. The compounds below (Rosini *et al.* 1994, 1995) are different in structure only in the substitution of a naphthalene ring (α rather than β substitution). The CD spectrum of the ββ compound has a negative Cotton effect in the ^1L_a spectral region followed by strong positive–negative Cotton effects which correspond to the ^1B transition. The αβ derivative has a positive Cotton effect for ^1L_a and a two strong negative–positive Cotton effects in the ^1B spectral region: *i.e.* the CD spectra of these compounds look like mirror images.

Naïve application of the empirical method of comparison would suggest the opposite absolute configuration for the two compounds. A more accurate analysis of the CD spectra, with particular reference to the strong high-energy Cotton effects, reveals (Rosini *et al.* 1994, 1995) that the two compounds have the same absolute configuration. The spectra allied to the transition considered above are opposite in sign for geometric reasons only—the most stable conformation is almost the same but the different linkage of the naphthalene rings leads to a different reciprocal orientation of the transition dipole moments.

This situation is not unique. A recent paper by Sandström *et al.* (1997) confirms that compounds with very similar structure and the same absolute configuration, *e.g.* (+)-**7** and (−)-**8**, below, can have mirror-image CD spectra. CD and molecular mechanics calculations reveal that (+)-**7** and (−)-**8** have the same absolute configuration; the CD spectra have opposite signs for corresponding transitions because the different constraints of two- and three-carbon bridges lead to significantly different relative orientations of the chromophores, The general conclusion of the authors is that this investigation shows "*…a caveat against deducing identity in absolute stereochemistry from strong similarity in CD spectra.*"

SEMI-EMPIRICAL RULES

Since the beginning of the sixties, attempts have been made to formulate rules (mainly on empirical grounds, although a theoretical basis has been found for some) for spectrum–structure relationships. These rules are classified as described below.

Helicity rules

These correlate the sense of helicity of an intrinsically dissymmetric chromophore with the CD sign of a given Cotton effect. An example is provided by the *cis* diene rule (Moscowitz *et al.* 1961; Fig. 8.18).

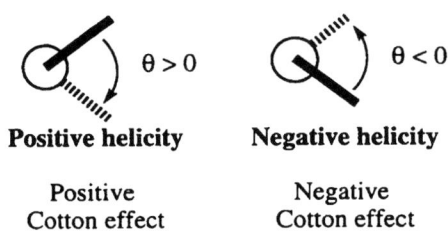

Positive helicity   **Negative helicity**

Positive   Negative
Cotton effect   Cotton effect

Fig. 8.18 1,3-Diene helicity rule for the lowest-lying $\pi \to \pi^*$ transition.

In its most simple form (Fig. 8.18), the helicity rule, like the diene rule, establishes that a positive twist of double bonds gives rise to positive rotational strength for the lowest energy $\pi \to \pi^*$ transition. It was, however, noticed some time ago (Snatzke 1978) that the situation is not always so clear—if the helical segment is short, it can appear left-handed when viewed along an axis and right-handed when viewed from a direction orthogonal to the first. Indeed, Buß and coworkers (Buß & Kolster 1996; Eggers *et al.* 1996, 1997) have recently reported an example of the apparent violation of the helicity rule—they demonstrated negative rotational strength for the lowest energy $\pi \to \pi^*$ transition of a C_2 symmetric chiral monomethine dye endowed with a chromophore of P helicity. When homologous systems with longer chromophores are considered, however, the expected sign of R is found to be in agreement with the helicity rule if the helix is sufficiently long.

Other rules have been formulated for the disulphide (Neubert & Carmack 1974), the α,β-unsaturated ketone (Gawronski 1982) and the biphenyl (Ringdahl *et al.* 1981) chromophores. The last is very useful both for determining the absolute configuration of metabolites of polycyclic aromatic hydrocarbons (Balani *et al.* 1987) and in the field of natural product chemistry (*e.g.* aporphine alkaloids) (Ringdahl *et al.* 1981).

Sector rules

These provide the sign of the contribution of a substituent to a given Cotton effect when the position of the substituent is changed relative to the chromophore itself. The oldest and most reliable sector rule is the *Octant Rule* (Moffit *et al.* 1961; Lightner 1994) for the saturated carbonyl chromophore (Fig. 8.19).

The space around the carbonyl chromophore is divided by the three planes A, B and C into eight sectors (octants) and a substituent falling into a given sector makes a contribution to CD of the n→π* transition as reported in Fig. 8.19. Groups falling on the nodal planes do not contribute to CD.

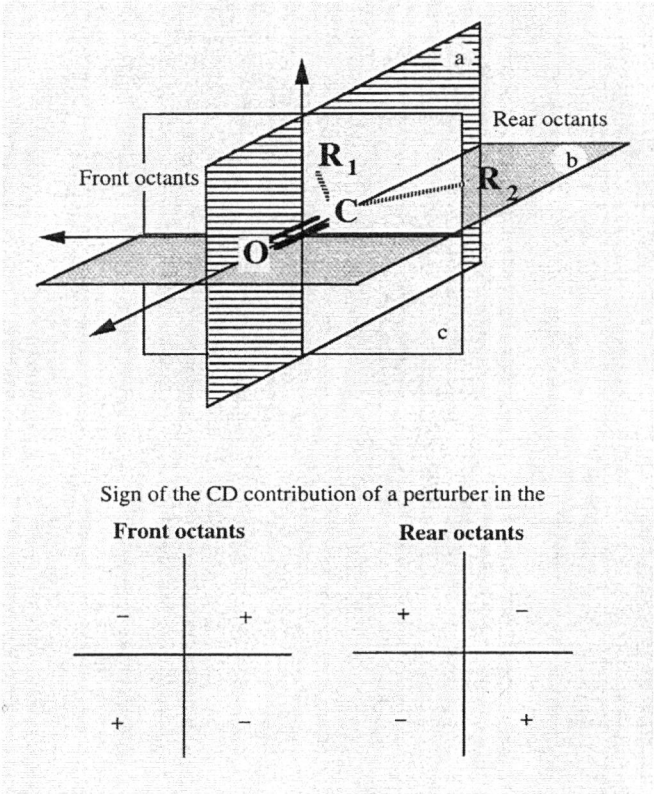

Fig. 8.19 Nodal planes and octants for the carbonyl rule.

An example of usefulness of this rule can be found in the field of natural product synthesis (Partridge *et al.* 1973). Loganine, an important building block in much of the plant world, can be prepared by the sequence presented in Fig. 8.20. The key intermediates in the synthetic strategy are the optically active *trans*-2-methylcyclopent-3-en-1-ols, obtained by asymmetric hydroboration of 5-methylcyclopentadiene using (+)- or (−)-di-3-pinanylborane—only the alcohol with the 2*R* absolute configuration can be employed to complete the sequence.

Application of the Octant Rule to the corresponding methyl cyclopentanones enables the 2*R* absolute configuration to be assigned to the (−) antipode, indicating that it is the correct isomer for the synthesis. This example indicates that such a complex synthetic sequence resides completely in assignment of the absolute configuration of (−)-*trans*-3-methylcyclopenten-4-ol.

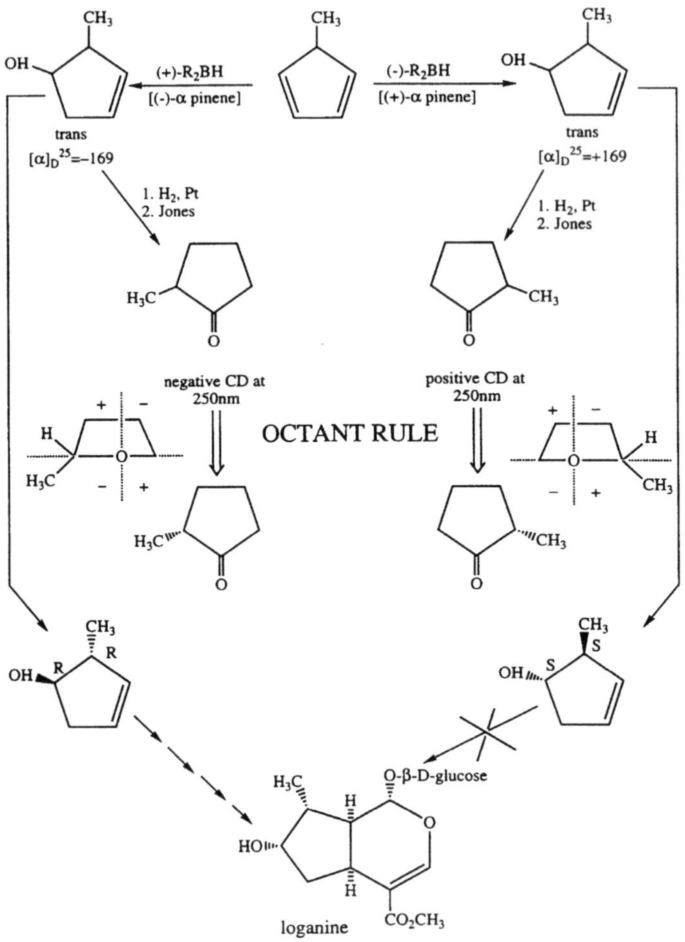

Fig. 8.20 The stereoselective synthesis of loganine.

Chirality rules

These correlate geometric characteristics of the chiral molecule with the CD sign of a given Cotton effect. Examples of chirality rules are the allylic axial rule (for 1,2-disubstituted olefins and for 1,3-dienes; Burgstahler & Barkhurst 1970), the salicylideneamino rule for amines (Smith 1983), and the C_2-symmetric chromophore chirality rule (Hug & Wagniere 1972). The Lowe (1965) rule relating optical rotation at the sodium D-line of chiral allenes (determined by the sign of the 175 nm $^1A_1 \rightarrow {}^1B_2$ Cotton effect of the allene chromophore) with absolute configuration will be described briefly. This rule states (Fig. 8.21) that if R_1 is more polarizable than R_2 and R_3 is more polarizable than R_4, a clockwise screw pattern of polarizability is described and dextrorotation at the sodium D-line must be expected.

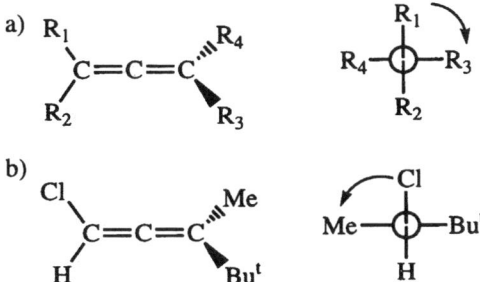

Fig. 8.21 (a) The Brewster Rule for chiral allenes. (b) The correct order of polarization of the substituents is Cl > Bu^t > Me > H.

This rule has been formulated on an empirical basis, *i.e.* taking into account the absolute configuration of some known allenes. In 1-chloro-4,4-dimethylpenta-1,2-diene the configuration had been misassigned as (−)-*S*; this forced analysts, to enable them to make a self-consistent prediction, to assume that the Me group is more polarizable than the Bu^t group.

NON-EMPIRICAL METHODS OF ANALYSIS (Mason 1978, 1982; Charney 1979; Harada & Nakanishi 1983; Berova *et al.* 1994; Eliel *et al.* 1994; Rodger & Norden 1997)

The knowledge of the mechanisms by which a certain transition can attain both electric and magnetic dipole moments, non-orthogonal and non-vanishing, and thus non-zero rotational strength, constitute the basis for non-empirical analysis of CD spectra. This approach is particularly important because it enables the interpretation of CD data avoiding all the uncertainties, discussed above, connected with empirical analysis.

Some of the sector or helicity rules have theoretical foundation. The octant rule, for instance, can be based on calculations in the one-electron frame and, as shown in a previous section, dynamic Höhn–Weigang coupling (Weigang & Höhn 1968) even affords pictorial justification. Within the independent systems approach, those dedicated to electrically allowed transitions and based on coupled oscillators or polarizability models have received considerable attention. By means of a coupled oscillators approach, moreover, simple rules have been formulated which correlate the spectrum and structure qualitatively (Harada & Nakanishi 1983; Berova *et al.* 1994).

In the section on coupled oscillators it was seen that an aggregate of two chirally arranged electric dipoles can give rise to a non-vanishing rotational strength. To illustrate this method, one of the first examples of its application, to (+)-*trans*-stilbene oxide (Gottarelli *et al.* 1970), will be discussed.

The 210–185 nm region of the spectrum of this compound shows the electrically allowed $^1A_{1g} \rightarrow {}^1E_{1u}$ transition of the benzene chromophore, polarized in the plane of the phenyl ring (200 nm, ε 90000). In the CD spectrum, a succession of a positive and a negative band is observed. This particular feature, typical of the CD of 'dimeric' chiral molecules, is called a CD *couplet* (it is defined as positive or negative according to the sign of the lower-energy component). The two excited states of the benzene chromophores can interact such that in-phase and anti-phase oscillations are separated in energy by a term proportional to the point-dipolar interaction term G_{12}:

$$G_{12} = \frac{\mathbf{e}_1 \cdot \mathbf{e}_2}{r_{12}^3} - \frac{3(\mathbf{e}_1 \cdot \mathbf{r}_{12})(\mathbf{e}_2 \cdot \mathbf{r}_{12})}{r_{12}^5} \tag{26}$$

where $\mathbf{e}_1$ and $\mathbf{e}_2$ are the directions of polarization of the two dipoles.

Considering the *R,R* enantiomer of *trans*-stilbene oxide as depicted in Fig. 8.22 results in a left-handed helix arrangement of the two electric dipoles for the high-energy coupling mode, corresponding to a negative rotational strength.

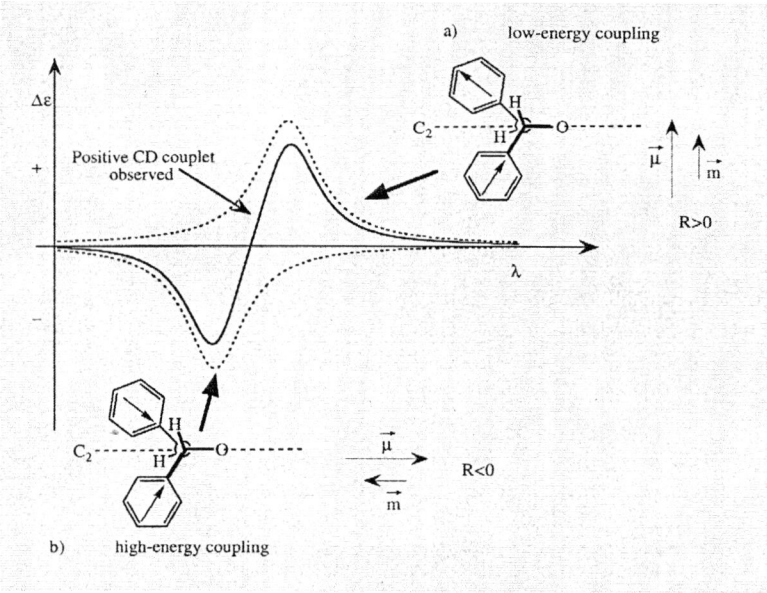

Fig. 8.22 The CD couplet of *trans*-stilbene oxide.

Conversely, the low-energy coupling gives rise to a right-handed helix and thus to a positive rotational strength. The superposition of the two branches gives rise to an *exciton couplet*, similar to the experimental CD spectrum. The sign and magnitude of the CD couplet depend, through Eqs. (25) and (26), on the geometry (and in particular on the relative orientations) of the two transition dipole moments. Thus, once their position in the chromophore is known, information on the molecular structure can be derived.

This treatment has been applied to numerous 'dimeric' molecules including such important groups as the binaphthyl derivatives (Mason *et al.* 1974), the paracyclophanes (Weigang & Nugent 1969) and 1,2-diarylethane-1,2-diols (Rosini *et al.* 1994, 1995). In the last example conformational distribution as a result of rotation about the C–C bond would make any assignment questionable, and it seems much safer to immobilize the molecule by the straightforward formation of a cyclic ketal. As

will be shown below, the real importance of the exciton analysis of CD spectra resides in two points:
(1) This approach is a non-empirical method for assigning absolute configuration in solution which is totally reliable and comparable with X-ray analysis in the solid state. The answer provided by the exciton model is always in agreement with results from X-ray analysis. In the only examples in which the two were in apparent contradiction it was shown that the wrong locations of the transition dipoles had been used for exciton couplet calculations (Mason 1973).
(2) Exciton analysis is not limited to dimeric molecules—the approach can be successfully extended to transparent molecules also, simply by derivatizing the functional groups present in the structure to introduce suitable chromophores (Harada & Nakanishi 1983). Thus the simple diol 1R,2R-cyclohexanediol, shown in Fig. 8.23, can be transformed into the corresponding p-dibromobenzoate (Harada & Nakanishi 1983).

In this example the 246 nm band of the p-bromobenzoate gives rise to a couplet with extremes at 236 and 256 nm. The sign of the Cotton effect at the longer wavelength represents the chirality of the systems and enables assignment of the absolute configuration of the two stereocentres.

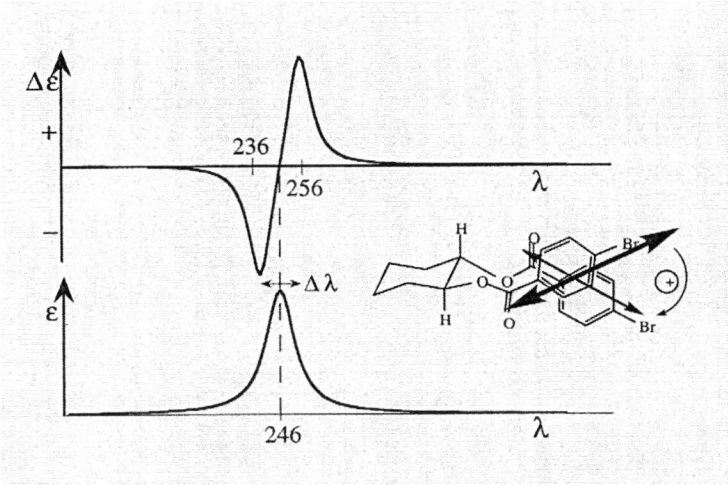

Fig. 8.23 Use of dibenzoates for determination of configuration by the exciton chirality method.

This approach can also be applied to the coupling of different chromophores. In this example, as demonstrated both theoretically and experimentally by Harada & Nakanishi (1983), the two branches of the couplet appear split in correspondence with the absorption wavelengths of the two oscillators. The sign of the chirality defined by the transition dipoles is reflected in the sign of the first (lower energy) Cotton effect.
One such example is the determination of the absolute stereochemistry of (+)-cis-2-hydroxy-2-phenylcyclohexanecarboxylic acid (cycloxilic acid, Fig. 8.24a), an important choleretic agent and intermediate in the synthesis of spasmolytic agents (Pini et al. 1994).

The molecule contains an alkyl-substituted benzene chromophore which has an intense absorption band at *ca* 185 nm as a result of the electrically allowed degenerate $^1A_{1g} \to {}^1E_{1u}$ transition polarized in the plane of the benzene ring. By introducing a second chromophore with a well defined electrically allowed transition, exciton coupling effects can be observed in the CD spectrum, enabling assignment of the absolute configuration.

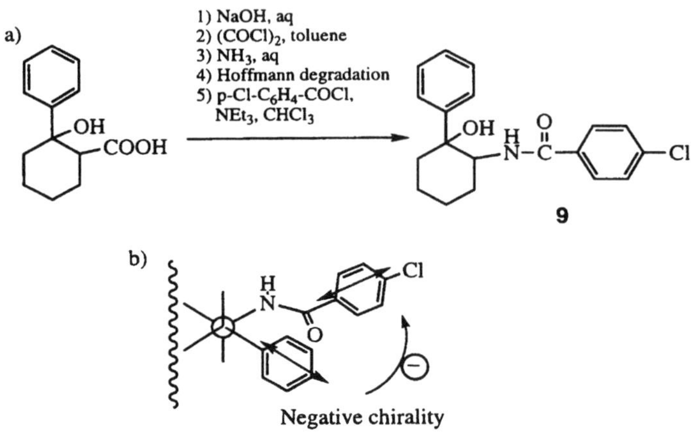

Fig. 8.24 Determination of the absolute configuration of cycloxilic acid, derivatized according to the sequence in (a). The negative chirality of the amide **9**, depicted in (b), justifies the observed negative couplet.

The transformation (Fig. 8.24a) of COOH to NH_2 by Hofmann degradation of the primary amide, affords the corresponding aminoalcohol, which, in turn, can be transformed in the *p*-chlorobenzamide, **9**, which has the same absolute stereochemistry as the starting acid, because the Hofmann reaction occurs with complete retention of configuration. The *p*-chlorobenzamide chromophore has an electrically allowed $\pi \to \pi^*$ transition at 235 nm, polarized along the C*–N axis.

The coupling between this and the above mentioned transitions of the alkylbenzene chromophore affords the CD spectral feature suitable for stereochemical assignment. Indeed, the absorption spectrum of **9** has a band at 235 nm (ε_{max} 12900) assignable to the *p*-chlorobenzamide $\pi \to \pi^*$ transition. Correspondingly, the CD spectrum contains a negative Cotton effect ($\Delta\varepsilon = -4.9$); this is attributable to the exciton coupling between the benzamide transition and the allowed transition of the alkylbenzene chromophore at C2.

Assuming *R,R* absolute configuration for the two stereogenic centres and assuming also that the cyclohexane ring adopts the more stable chair conformation with the largest substituents (phenyl and benzamide groups) both equatorial, the relative position of the two chromophores can be represented by the Newman projection of Fig. 8.24b . The two transition moments define a negative exciton chirality, in agreement with the negative sign of the observed CD Cotton effect at 235 nm. The absolute configuration at the two stereogenic centres of **9** is, therefore, 1*R*,2*R*, and so the 1R,2S configuration can be assigned to the dextrorotatory antipode of cycloxilic acid.

Such considerations constitute the basis of the so-called exciton chirality method, developed by Nakanishi and coworkers (Harada & Nakanishi 1983; Berova *et al.* 1994). It has found many applications in studies related to assignment of the absolute configuration of natural products and in the structural analysis of carbohydrates. In addition, these methods have also been applied to the determination of the absolute configuration of acyclic polyols.

The efficiency and versatility of the exciton chirality method was recently emphasized by Tan *et al.* (1997), who determined the absolute sense of twist of the C2–C3 bond of the retinal chromophore in bovine rhodopsin. This is important work because the conformation of the chromophore in the binding pocket of a large protein can be determined from the sign of the bisignated exciton coupled split CD spectra—in other words, subtle conformational aspects of ligand–receptor interactions can be probed by use of this CD method.

It has been observed that stereochemical assignment resides in the measurement of a couplet or, at least, of a CD band. Because (refer to Eqs. (25) and (26)) the intensity of the observed Cotton effect is, among other terms, given by:

$$\Delta\varepsilon \propto \mathbf{\mu}_1 \times \mathbf{\mu}_2 \cdot \mathbf{r}_{12} G_{12} \propto \varepsilon_1 \varepsilon_2 \frac{1}{r_{12}^2} \sin\theta \cos\theta \qquad (27)$$

where θ is the angle between the two transition dipoles $\mathbf{\mu}_1$ and $\mathbf{\mu}_2$, corresponding to the absorption bands, with extinction coefficients ε_1 and ε_2. The Cotton effects will be easily measurable when strongly absorbing chromophores are placed near each other (*e.g.* the benzoates of cyclohexanediol of Fig. 8.23).

It is of interest to point out that the lack of interaction between chromophores, because of an unfavorable intensity/distance ratio, can sometimes be an advantage in the determination of the absolute configuration of one stereocentre. This is so for ciguatoxin (Satake *et al.* 1997), a compound with several hydroxyl groups in addition to a conjugated diene chromophore. When the molecule was perbenzoylated, the configuration of the stereocentre near the 1,3-diene moiety could be determined, because the benzoate–diene couplet was not perturbed by the other benzoate chromophores, which were far apart from each other.

In general, however, the problem is rather that of having strong Cotton effects even when the chromophores are distant or the quantity of compound under examination is minute. Thus chromophores with larger extinction coefficients have been used successfully. Using porphyrins, for example, with ε_{max} 350000 (at *ca* λ_{max} = 400 nm) the method has been extended to distances as large as 50 Å and less than 1 mg starting material (Matile *et al.* 1996b). Among the most recent methods is also the determination of the configuration of a single stereogenic centre, by virtue of the strong stacking properties of porphyrins (Matile *et al.* 1996a).

Among the coupled-oscillator methods for CD calculations, the all-order polarizability model of DeVoe (1964) can be profitably employed. In this model a molecule is considered to comprise a set of sub-systems, the chromophores. These are polarized by the external electromagnetic radiation and are coupled to each other by their own dipolar oscillating fields. The optical properties (absorption, refraction, optical rotatory dispersion and circular dichroism) of the molecule under study can be calculated by taking into account the above interaction of the subsystems. Each group is represented in terms of one (or more) classical oscillator(s)—each oscillator represents an electric dipole-allowed transition, defined by the polarization direction e_i and the complex polarizability $\alpha_i(\tilde{v}) = R_i(\tilde{v}) + iI_i(\tilde{v})$.

$I_i(\tilde{v})$ is obtainable experimentally, *i.e.* from the absorption spectra of compounds, which can be considered good models of the subsystems, $R_i(\tilde{v})$ can be calculated from $I_i(\tilde{v})$ by means of a Kronig–Kramers transform (Caldwell & Eyring 1971). The main applications of these calculations to the solution of stereochemical problems are described in references listed at the end of this chapter. The absolute configuration of chiral acetylenic alcohols has, for example, been established by studying the CD of the 230-nm transition of the corresponding benzoates. The derivatives for which positive CD is observed for this band can be assigned the *S* absolute configuration (Rosini *et al.* 1984).

This rule has been also extended (Caporusso *et al.* 1986) to tertiary acetylenic alcohols. Because the benzoate of the laevorotatory antipode of 3,4,4-trimethyl-1-pentyn-3-ol **10** has negative CD it must be assigned the *R* absolute configuration, and thus the *R* absolute configuration must also be assigned to (−)-1-chloro-4,4-dimethylpenta-1,2-diene, obtained from (−)-*R*-**10** by treatment with $SOCl_2$.

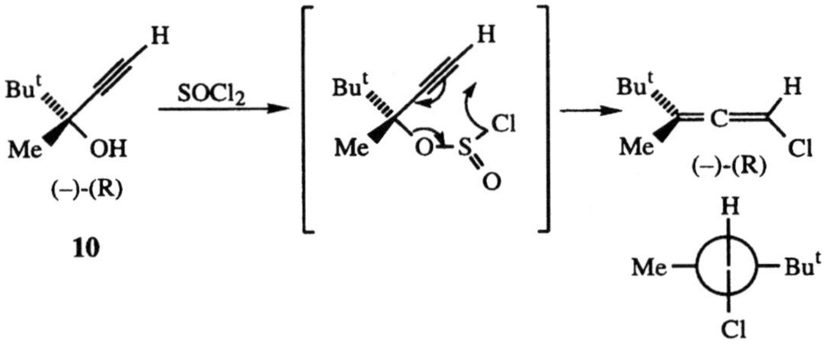

As a consequence, the first configurational correlation between a centrochiral molecule and an axially chiral system must be corrected, as reported in the scheme above (Caporusso *et al.* 1986). Furthermore, order can be put in the Lowe rule, without requiring that the Bu^t group is less polarizable than the Me group (*Chirality rules* section).

Coupling effects of this type are responsible for the optical activity of electrically allowed excitations in simple organic molecules such as those containing a symmetric chromophore chirally perturbed by alkyl groups only.

The chiral diene compound illustrated below is a good example (Rosini *et al.* 1993).

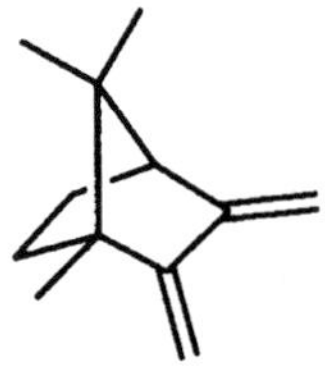

The spectrum of this molecule contains a band at 260 nm, $\Delta\varepsilon = -0.6$. Because the chromophore is planar and because the molecular chirality arises solely as a result of the presence of the C*–CH$_3$ fragment, the optical activity is derived from the coupling effect between the C1–C4 polarized 260 nm transition of the diene chromophore and the excitations of the C*–CH$_3$ fragment, polarized along the same axis. Indeed, when a DeVoe calculation is performed within this simple two oscillators scheme, employing suitable input data (Rosini *et al.* 1993), a $\Delta\varepsilon$ value of -0.57 is predicted, in excellent agreement with the experimental -0.63.

These considerations demonstrate how the coupled oscillator interaction is responsible for the CD of electrically allowed transitions not only in 'dimeric' molecules, but also in the numerous molecules in which the chromophore is perturbed by alkyl groups only, *i.e.* by groups that are usually not regarded as chromophores. DeVoe calculations have, furthermore, been shown to cope very well with many-body effects (Rosini *et al.* 1993).

CONCLUSIONS AND FUTURE PERSPECTIVES

As shown in this discussion of the principles and applications of CD spectroscopy, it seems clear that important advances have been achieved in the interpretation of CD data, enabling easier clarification of spectra–structure relationships. One of the most relevant examples of this progress has been the development of the coupled oscillators model, a very simple tool for interpreting CD spectra, *i.e.* the exciton chirality rules enable provision of an absolute stereochemical answer, simply by inspection of a molecular model.

Much, however, remains to be done in the formulation of reliable correlations between spectra and structure, even for simple systems. The example of the diene chromophore is quite illustrative—for this system conclusive rules have not yet been formulated (Salvadori *et al.* 1997); even for the deeply studied benzene chromophore empirical correlations only have been reported (Smith 1994). Despite this, CD is one of the most successful non-destructive techniques for determining the molecular structure of chiral substrates in solution and an effort must be made to make its interpretation more accessible to scientists of different cultural origin.

In the last few years CD spectroscopy has received a renewed attention and two fields of application, at least, deserve mention. The first is the coupling of HPLC on chiral stationary phases with CD detection (Salvadori *et al.* 1991, 1994). This system has several advantages in the liquid chromatographic analysis of chiral compounds, because of the selective monitoring of optically active molecules. Its use enables reliable determination of enantiomeric excess and orders of elution.

The on-line recording of CD spectra and the evaluation of dissymmetry factors also makes CD detection very powerful in the characterization of the stereochemistry of chiral eluates. This enables the saving of much time in the testing of the relationships between stereochemistry and activity of enantiomers and in improving separation methodologies and asymmetric synthesis.

A second topic not discussed here, but of great importance is the use of CD in determining the structure of polypeptides (Johnson 1990) and nucleic acids (Johnson 1994). This spectroscopy is a straightforward means of determining the relative amounts of ordered structures, even on relatively small amounts of biomacromolecules.

The sensitivity of CD to overall conformational variations can also be exploited in the study of supramolecular structures, for example the drug–protein adducts formed by non-covalent interactions (stereoselective binding at target site; Bertucci *et al.* 1997) which are of fundamental importance in the understanding of the mechanism of drug action. The technique can also be used to investigate molecular recognition problems (Gallivan & Schuster 1995), photoresponsive polymers (Pieroni & Ciardelli 1995), chiral molecular switches (Feringa *et al.* 1996, Eggers & Buss 1997), and induced cholesteric phases (Solladié & Zimmermann 1984).

Finally, attention must be drawn to the regulatory laws that are being introduced in the use and distribution of enantiomeric molecules in medicine, agriculture, and food chemistry. Determination of enantiomeric purity and better understanding of the relationship between the stereochemistry and biological activity of such products will certainly be challenges in the near future and CD is one of the tools that will help meet them.

REFERENCES

Abbate, S., Longhi, G., Givens III, J. W., Boiadjiev, S. E., Lightner, D. A. & Moscowitz, A. (1996) *Appl. Spectrosc.*, **50**, 642.

Atkins, P. W. (1983) *Molecular Quantum Mechanics*, Oxford University Press, Oxford.

Bak, K. L., Hansen, Aa. E. Ruud, K., Helgaker, T., Olsen, J. & Jørgensen, P. (1995) *Theor. Chim. Acta*, **90**, 441.

Balani, S. K., van Bladeren, P. J., Cassidy, S. E., Boyd, D. R. & Jerina, D. M. (1987) *J. Org. Chem.*, **52**, 137.

Barron, L. D. & Hecht, L. (1994). In: *Circular Dichroism: Application and Interpretation* (Ed. by N. Berova, R. W. Woody & K. Nakanishi). VCH, New York.

Berova, N., Woody, R. W. & Nakanishi, K. (Eds.) (1994) *Circular Dichroism: Application and Interpretation*, VCH, New York.

Bertucci, C., Salvadori, P. & Domenici, E. (1997). In: *The Impact of Stereochemistry in Drug Development* (Ed. by H. Y. Aboul-Enein & I. W. Wainer). Chemical Analysis Series **142**, 521.

Bringmann, G., Stahl, M. & Gulden K. P. (1997), *Tetrahedron*, **53**, 2817.

Burgstahler, A. W. & Barkhurst, R. C. (1970) *J. Am. Chem. Soc.*, **92**, 7602.

Buß, V. & Kolster, K. (1996) *Chem. Phys.*, **203**, 309.

Caldwell, D. J. & Eyring, H. (1971) *The Theory of Optical Activity*, Wiley-Interscience, New York.

Caporusso, A. M., Rosini, C., Lardicci, L., Polizzi, C. & Salvadori, P. (1986) *Gazz. Chim. Ital.*, **116**, 467.

Carnell, M., Peyerimhoff, S. D., Breest, A., Goeddels, K. H., Hochmann, P. & Hormes, J. (1991) *Chem. Phys. Lett.*, **180**, 477.

Charney, E. (1979) *The Molecular Basis of Optical Activity—Optical Rotatory Dispersion and Circular Dichroism*, J. Wiley & Sons, New York.

Dekkers, H. P. J. M. (1994). In: *Circular Dichroism: Application and Interpretation* (Ed. by N. Berova, R. W. Woody & K. Nakanishi). VCH, New York.

DeVoe, H. (1964) *J. Chem. Phys.*, **43**, 3199.

Diem, M. (1994). In: *Analytical Applications of Circular Dichroism* (Ed. by N. Purdie & H. G. Brittain). Elsevier, Amsterdam.

Drake, A. F. (1989). In *Physical Methods of Chemistry*, Vol. IIIB, 2nd edn. (Ed. by B. W. Rossiter & J. F. Hamilton). J. Wiley & Sons, New York.
Drake, A. F., Garofalo, A. Hillman, J. M. L., Merlo, V., McCague, R. & Roberts, S. M. (1996) *J. Chem. Soc. Perkin 1*, 2739.
Eggers, L. & Buss, V. (1997) *Angew. Chem. Int. Ed. Eng.*, 36, 881.
Eggers, L., Buß, V. & Henkel, G. (1996) *Angew. Chem. Int. Ed. Engl.*, 35, 870.
Eggers, L., Kolster, K. & Buß, V. (1997) *Chirality*, 9, 243.
Eliel, E., Wilen, S., & Mander (1994) *Stereochemistry of Organic Compounds*, Chapter 13. J. Wiley & Sons, New York.
Feringa, B. L., Schoevaars, A. M., Jager, W. F., De Lange, B. & Huck, N. P. M. (1996) *Enantiomer*, 1, 325.
Feynman, R. P., Leighton, R. B. & Sands, M. (1963) *The Feynman Lectures on Physics*, Vol. I. Addison–Wesley, Reading, USA.
Gallivan, J. P. & Schuster, G. B. (1995) *J. Org. Chem.*, 60, 2423.
Gawronski, J. (1982) *Tetrahedron*, 38, 3.
Gottarelli, G., Mason, S. F. & Torre, G. (1970) *J. Chem. Soc. B*, 1349.
Grimme, S. (1996), *Chem. Phys. Lett.*, 259, 128.
Hansen, Aa. E. & Bouman, T. D. (1980) *Adv. Chem. Phys.*, 44, 545.
Hansen, Aa. E. & Bouman, T. D. (1989) *Croat. Chim. Acta*, 62, 227.
Harada, N. & Nakanishi, K. (1983) *Circularly Dichroic Spectroscopy—Exciton Coupling in Organic Stereochemistry*. University Science Books, Mill Valley.
Harada, N., Uda, H., Nozoe, T., Hokamoto, Y., Wakabayashi, H. & Ishikawa, S. (1987) *J. Am. Chem. Soc.*, 109, 1661.
Harada, N., Iwabuchi, J., Yokota, Y. & Uda, H. (1989) *Croat. Chem. Acta.*, 62, 267.
Harada, N., Ono, H., Uda, H., Parveen, M., Kahn, N. U., Achari, B. & Dutta, P. K. (1992) *J. Am. Chem. Soc.*, 114, 7687.
Hargitai, T., Reinholdsson, P. & Sandström, J. (1991) *Acta Chem. Scand.*, 45, 1076.
Hrbek, J., Hruben, L., Simanek, L., Santavy, F., Snatzke, G. & Yemul, S. (1982) *Collect. Chech. Chem. Commun.*, 47, 2258.
Huang, M. B., & Suter, H. U. (1994), *J. Mol. Struct. (Theochem)*, 337, 173.
Hug, W. & Wagniere, G. (1972) *Tetrahedron*, 28, 1241.
Johnson Jr, W. C. (1990) *Protein: Structure, Function and Genetics*, 7, 205.
Johnson Jr, W. C. (1994). In: *Circular Dichroism: Application and Interpretation* (Ed. by N. Berova, R. W. Woody & K. Nakanishi). VCH, New York.
Kauzmann, W., Walter J., & Eyring, H. (1940) *Chem. Rev.*, 26, 339.
Kirkwood, J. G. (1937) *J. Chem. Phys.*, 5, 479.
Kuhn, W. (1930) *Trans. Farad. Soc.*, 26, 293.
Lightner, D. A. (1994). In: *Circular Dichroism: Application and interpretation* (Ed. by N. Berova, R. W. Woody & K. Nakanishi). VCH, New York.
Lowe, G. (1965) *Chem. Commun.*, 411
Mason, S. F. (1963) *Quart. Rev.*, 17, 20.
Mason, S. F. (1973) *Chem. Commun.*, 239 and references cited therein.
Mason, S. F. (Ed.) (1978) *Optical Activity and Chiral Discrimination*. D. Riedel Publishing Co., Dordrecht, Holland.
Mason, S. F. (1982) *Molecular Optical Activity and Chiral Discrimination*, Cambridge University Press, Cambridge.
Mason, S. F., Seal, R. H. & Roberts, D. R. (1974) *Tetrahedron*, 30, 1671.
Mathieu, J. P. (1946) *Les Theories Moleculaires du Pouvoir Rotatoire Naturel*, Gauthier–Villars, Paris.
Matile, S. Berova, N. & Nakanishi, K. (1996a) *Enantiomer*, 1, 1.

Matile, S., Berova, N., Nakanishi, K., Fleischhauer, J. & Woody, R. (1996b) *J. Am. Chem. Soc.*, **118**, 5198.

Michl, J. & Thulstrup, E. W. (1986) *Spectroscopy with Polarized Light*, VCH, New York

Moffitt, W. (1956) *J. Chem. Phys.*, **25**, 467.

Moffit, W., Woodward, R. B., Moscowitz, A., Klyne, W. & Djerassi, C. (1961) *J. Am. Chem. Soc.*, **83**, 4013.

Moscowitz, A. (1961) *Tetrahedron*, **13**, 48.

Moscowitz, A., Charney, E., Weiss, U. & Ziffer, H. (1961) *J. Am. Chem. Soc.*, **83**, 4661.

Mukakami, N., Wang, W., Aoki, M., Tsutsui, Y., Higuki, K., Aoki, S. & Kobayashi, M. (1997) *Tetrahedron Lett.*, **38**, 5533.

Nafie, L. A, Citra, M., Ragunathan, N., Yu, G.-S. & Che, D. (1994). In: *Analytical Applications of Circular Dichroism* (Ed. by N. Purdie & H. G. Brittain). Elsevier, Amsterdam.

Neubert, L. A. & Carmack, M. (1974) *J. Am. Chem. Soc.*, **96**, 943.

Nilsson, K., Hallberg, A., Issacson, R. & Sandström, J. (1991) *Acta Chem. Scand.*, **45**, 716.

Partridge, J. J., Chadha, N. K. & Uskokovic, M. R. (1973) *J. Am. Chem. Soc.*, **95**, 532.

Pedersen, T. B. & Hansen, Aa. E. (1995) *Chem Phys. Lett.*, **246**, 1.

Pieroni, O. & Ciardelli, F. (1995) *Trends Polym. Sci.*, **3**, 282.

Pini, D., Petri, A., Rosini, C., Salvadori, P., Giorgi, R., Di Bugno, C., Turbanti, L. & Marchetti, F. (1994) *Tetrahedron*, **50**, 205.

Purdie, N. & Brittain, H. G. (Eds.) (1994) *Analytical Applications of Circular Dichroism*. Elsevier, Amsterdam.

Rashidi-Ranjbar, P., Sandström, J., Wong, H. N. C. & Wang, X. C. (1992) *J. Chem. Soc. Perkin Trans. II*, 1625.

Richardson, F. S. & Metcalf, D. H. (1994). In: *Circular Dichroism: Application and Interpretation* (Ed. by N. Berova, R. W. Woody & K. Nakanishi). VCH, New York.

Riehl, J. P. (1994). In: *Analytical Applications of Circular Dichroism* (Ed. by N. Purdie & H. G. Brittain). Elsevier, Amsterdam.

Ringdahl, B., Chan, R. P. K., Cymerman C. J., Cava, M. P. & Shamma, M. (1981) *J. Nat. Prod.*, **44**, 80.

Rodger A. & Norden, B. (1997) *Circular Dichroism and Linear Dichroism*, Oxford University Press, Oxford.

Rosini, C., Giacomelli, G. & Salvadori, P. (1984) *J. Org. Chem.*, **49**, 3394.

Rosini, C., Bertucci, C., Salvadori, P. & Zandomeneghi, M. (1985) *J. Am. Chem. Soc.*, **107**, 17

Rosini, C., Zandomeneghi, M. & Salvadori, P. (1993) *Tetrahedron: Asymmetry*, **4**, 545

Rosini, C., Scamuzzi, S., Uccello-Barretta, G. & Salvadori, P. (1994) *J. Org. Chem.*, **59**, 7395.

Rosini, C., Scamuzzi, S., Pisani-Focati, M. & Salvadori, P. (1995) *J. Org. Chem.*, **60**, 8289.

Salvadori, P., Rosini, C. & Bertucci, C. (1984) *J. Am. Chem. Soc.*, **106**, 2439.

Salvadori, P., Bertucci, C. & Rosini, C. (1991) *Chirality*, **3**, 376.

Salvadori, P., Bertucci, C.& Rosini, C. (1994). In: *Circular Dichroism: Application and Interpretation* (Ed. by N. Berova, R. W. Woody & K. Nakanishi). VCH, New York.

Salvadori, P., Minutolo, F. & Superchi, S. (1996) *J. Org. Chem.*, **61**, 4190.

Salvadori, P., Rosini, C. & Di Bari, L. (1997). In: *The Chemistry of Dienes and Polyenes* (Ed. by Z. Rappoport). J. Wiley & Sons, New York.

Sandström, J., Ripa, L. & Allberg, A. (1997) *J. Am. Chem. Soc.*, **119**, 5701.

Satake, M., Morohashi, A., Oguri, H., Oishi, T., Hirama, M., Harada, N. & Yasumodo, T. (1997) *J. Am. Chem. Soc.*, **119**, 11325.

See for instance: Grimme, S., Pischel, I., Vögtle, F., & Nieger, M. (1995), *J. Am. Chem. Soc.*, **117**, 157.

Shoene, K. A., Quagliano, J. R. & Richardson, F. S. (1991) *Inorg. Chem.*, **30**, 3803.

Shustov, G. V. & Rauk, A. (1995) *J. Org. Chem.*, **60**, 5891 (1995).

Shustov, G. V., Kachanov, A. V., Kadorkina, G. K., Kostyanovsky, R. G. & Rauk, A. (1992a) *J. Am. Chem. Soc.*, **114**, 8257.

Shustov, G. V., Kadorkina, G. K., Varlamov, S. V., Kachanov, A. V., Kostyanovsky, R. G. & Rauk, A. (1992b) *J. Am. Chem. Soc.*, **114**, 1616.

Smith, H. E. (1983) *Chem. Rev.*, **83**, 359.

Smith, H. E. (1994). In: *Circular Dichroism: Application and Interpretation* (Ed. by N. Berova, R. W. Woody & K. Nakanishi). VCH, New York.

Snatzke, G. (Ed.) (1967) *Optical Rotatory Dispersion and Circular Dichroism in Organic Chemistry*. Heyden & Son, London, UK.

Snatzke, G. (1968) *Angew. Chem. Int. Ed. Engl.*, **7**, 14.

Snatzke, G. (1978). In: *Optical Activity and Chiral Discrimination* (Ed. by S.F. Mason). D. Riedel Publishing Co., Dordrecht, Holland

Snatzke, G. (1979) *Angew. Chem. Int. Ed. Engl.*, **18**, 363.

Solladié, G. & Zimmermann, R. G. (1984) *Angew. Chem. Int. Ed. Eng.*, **23**, 348.

Tan, Q., Lou, J., Borhan, B., Karnaukhova, E., Berova, N. & Nakanishi, K. (1997) *Angew. Chem. Int. Ed. Engl.*, **36**, 2089.

Tejero, T., Dondoni, A. Rojo, I. Merchon, F. L. & Merino, P. (1997) *Tetrahedron*, **53**, 3301.

Tinoco Jr, I. & Turner, D. H. (1976) *J. Am. Chem. Soc.*, **98**, 6453.

Volosov, A. & Woody, R. W. (1994). In: *Circular Dichroism: Application and Interpretation* (Ed. by N. Berova, R. W. Woody & K. Nakanishi). VCH, New York.

Weigang, O. E. (1979) *J. Am. Chem. Soc.*, **101**, 1965.

Weigang, O. E. & Höhn, E. G. (1968) *J. Chem. Phys.*, **48**, 1127.

Weigang, O. E. & Nugent, M. J. (1969) *J. Am. Chem. Soc.*, **91**, 4555

Yaris, M., Moscowitz, A. & Barry, R. S. (1968) *J. Chem. Phys.*, **49**, 3150.

General references

Barrett, G. C. (1972) *Tech. Chem.* (NY), **4**, 515.

Barron, L. D. (1982) *Molecular Light Scattering and Optical Activity*. Cambridge University Press, Cambridge.

Berova, N., Woody, R. W. & Nakanishi, K. (Eds.) (1994) *Circular Dichroism: Application and Interpretation*, VCH, New York.

Caldwell, D. J. & Eyring, H. (1971) *The Theory of Optical Activity*, Wiley–Interscience, New York.

Charney, E. (1979) *The Molecular Basis of Optical Activity—Optical Rotatory Dispersion and Circular Dichroism*, J. Wiley & Sons, New York.

Ciardelli, F. & Salvadori, P. (Eds.) (1973) *Fundamental Aspects and Recent Developments of Optical Rotatory Dispersion and Circular Dichroism*. Heyden & Son Ltd., London, UK.

Drake, A. F. (1989). In *Physical Methods of Chemistry*, Vol. IIIB, 2nd edn. (Ed. by B. W. Rossiter & J. F. Hamilton). J. Wiley & Sons, New York.

Eliel, E., Wilen, S., & Mander (1994) *Stereochemistry of Organic Compounds*, Chapter 13. J. Wiley & Sons, New York.

Harada, N. & Nakanishi, K. (1983) *Circularly Dichroic Spectroscopy—Exciton Coupling in Organic Stereochemistry*. University Science Books, Mill Valley.

Legrand, M. & Rougier, M. J. (1977). In: *Stereochemistry: Fundamentals and Methods* (Ed. by H. B. Kagan), Vol. 2, p. 33. G. Thieme, Stuttgart.

Mason, S. F. (1963) *Quart. Rev.*, **17**, 20.

Mason, S. F. (Ed.) (1978) *Optical Activity and Chiral Discrimination*. D. Riedel Publishing Co., Dordrecht, Holland.

Mason, S. F. (1982) *Molecular Optical Activity and Chiral Discrimination*, Cambridge University Press, Cambridge.

Mathieu, J. P. (1946) *Les Theories Moleculaires du Pouvoir Rotatoire Naturel*, Gauthier–Villars, Paris.

Michl, J. & Thulstrup, E. W. (1986) *Spectroscopy with Polarized Light*, VCH, New York

Purdie, N. & Brittain, H. G. (Eds.) (1994) *Analytical Applications of Circular Dichroism*. Elsevier, Amsterdam.

Rodger, A. & Norden, B. (1997) *Circular Dichroism and Linear Dichroism*, Oxford University Press, Oxford.

Snatzke, G. (Ed.) (1967)*Optical Rotatory Dispersion and Circular Dichroism in Organic Chemistry*. Heyden & Son, London, UK.

Snatzke, G. (1968) *Angew. Chem. Int. Ed. Engl.*, **7**, 14.

References to the application of DeVoe calculations to structural problems

Applequist, J., Sundberg, K. R. & Olson, M. L. (19??) *J. Chem. Phys.*, **70**, 1240.

Caporusso, A. M., Rosini, C., Lardicci, L., Polizzi, C. & Salvadori, P. (1986) *Gazz. Chim. Ital.*, **116**, 467.

Clericuzio, M., Rosini, C., Persico, M. & Salvadori, P. (1991) *J. Org. Chem.*, **56**, 4343.

Hug, W., Ciardelli, F. & Tinoco, I. (1974) *J. Am. Chem. Soc.*, **96**, 3407.

Rosini, C. & Zandomeneghi, M. (1981) *Gazz. Chim. Ital.*, **111**, 493.

Rosini, C., Salvadori, P. & Zandomeneghi, M. (1978) *J. Chem. Soc. Dalton Trans.*, 822.

Rosini, C., Giacomelli, G. & Salvadori, P. (1984) *J. Org. Chem.*, **49**, 3394.

Rosini, C., Bertucci, C., Salvadori, P. & Zandomeneghi, M. (1985) *J. Am. Chem. Soc.*, **107**, 17.

Rosini, C., Zandomeneghi, M. & Salvadori, P. (1993) *Tetrahedron: Asymmetry*, **4**, 545.

Salvadori, P., Bertucci, C., Rosini, C., Zandomeneghi, M., Gallo, G. G., Martinelli, E. & Ferrari, P. (1981) *J. Am. Chem. Soc.*, **103**, 5553.

Zandomeneghi, M. (1979) *J. Phys. Chem.*, **83**, 2926.

Zandomeneghi, M., Rosini, C. & Salvadori, P. (1976) *Chem. Phys. Lett.*, **44**, 533.

Zandomeneghi, M., Rosini, C. & Drake, A. F. (1981) *J. Chem. Soc. Faraday Trans. 2*, **77** 567.

Chapter 9
Chirality in the Natural World: Chemical Communications

Kenji Mori
Department of Chemistry, Science University of Tokyo, Japan

INTRODUCTION

Because chemical communications among organisms is a vast subject, this review is restricted to insect pheromones, compounds which have been studied in great depth in the latter half of the twentieth century. The author has been engaged in enantioselective synthesis of pheromones since 1973 and has, therefore, witnessed the process by which all the pheromone scientists have become convinced of the importance of chirality in pheromone perception. Although his position as the pioneer of this field influences the perspective of this chapter, there is absolutely no doubt that his tendency to regard organic synthesis as very important is fully justified.

Pasteur remarked "L'univers est dissymétrique." Pheromone science as it is now fully supports his view. It began, however, with a study by Butenandt *et al.* (1959) of an achiral molecule named bombykol (**1**, Fig. 9.1), which was isolated as the sex attractant produced by the female silkworm moth, *Bombyx mori* (Hecker & Butenandt 1984). The term 'pheromone' was coined in the same year by Karlson & Lüscher (1959). The name is derived from the Greek *pherein* to transfer, and *hormon*, to excite. Pheromones are substances that are secreted by an individual and received by a second individual of the same species, in which they induce a specific reaction such as special behaviour or a developmental process.

The chiral nature of pheromone molecules became apparent in the late 1960s, when Silverstein and co-workers identified several chiral pheromones, for example **2** and **3**, from beetles. To understand fully the action of pheromones chemists must first determine the absolute configuration of naturally occurring chiral pheromones.

In 1973 when the author began his studies on pheromone synthesis, almost nothing was known about the absolute configuration of chiral pheromones. Difficulties are often encountered in stereochemical studies of pheromones, because the compounds are usually obtained in small quantities only (several micrograms to several milligrams) as volatile oils. Stereochemical studies of pheromones are therefore beyond the scope of conventional methods of stereochemical assignment such as degradation to a simple compound of known absolute configuration or X-ray crystallographic analysis. The best way to circumvent these difficulties is to perform enantioselective synthesis of the target pheromone starting from a compound of known absolute configuration. This approach generates the target molecule of known absolute configuration. If the chiroptical properties, for example specific rotation or ORD/CD spectrum, of the natural pheromone are recorded, these can be compared with those of the synthetic material. The absolute configuration of the natural pheromone will thus be clarified.

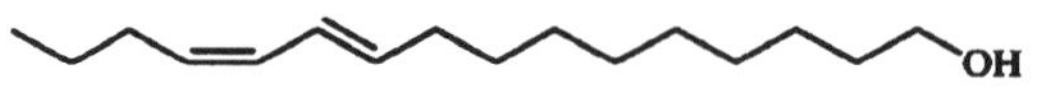

Bombykol (1)

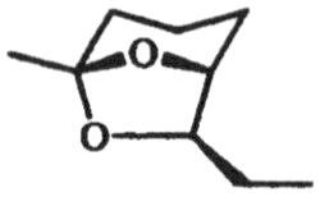

exo-Brevicomin (2)

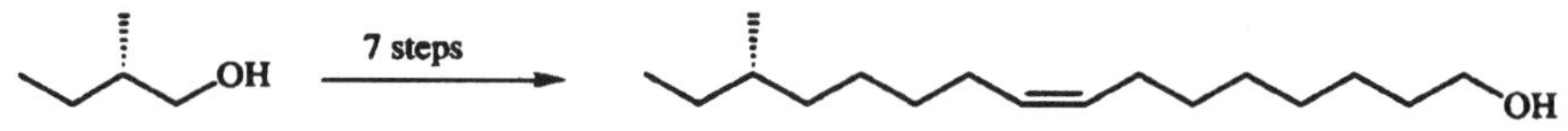

(S)-2-Methyl-1-butanol → 7 steps → (S)-3 = dextrorotatory

Natural 3 = levorotatory

∴ Natural 3 = (R)-3

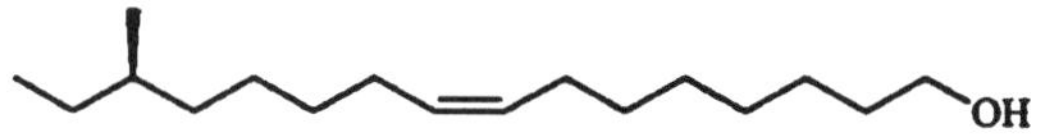

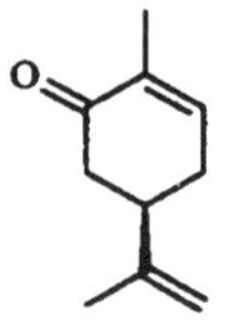

(R)-(−)-Carvone (4)

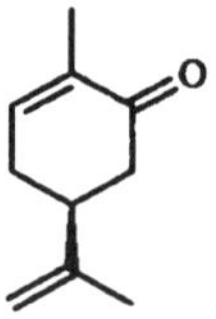

(S)-(+)-Carvone (4)

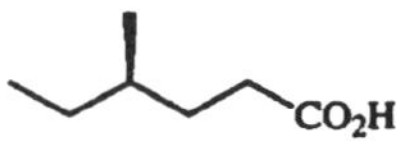

(R)-4-Methylhexanoic acid (5)

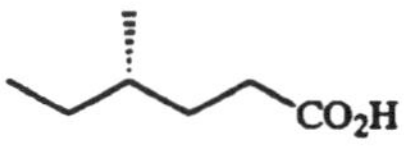

(S)-4-Methylhexanoic acid (5)

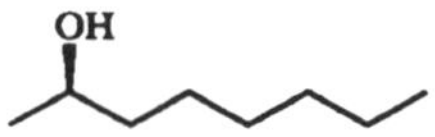

(R)-6

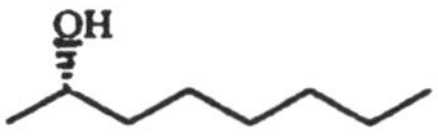

(S)-6

Fig. 9.1 Structures of some pheromones and odorous compounds.

The usefulness of this approach was first demonstrated by the author in 1973 (Fig. 9.1). The synthesis (Mori, K. 1973, 1974a) of the *S* enantiomer of 14-methyl-8-

hexadecen-1-ol (**3**), the dermestied beetle pheromone artefact (Rodin *et al.* 1969), from *S*-2-methylbutanol showed *S*-**3** to be dextrorotatory. Because natural **3** isolated from the insect was laevorotatory, its absolute configuration was unambiguously shown to be *R*. (Later, the genuine pheromone of the dermestied beetle, *Trogoderma inclusum*, was identified as *R,E*-14-methyl-8-hexadecenal.) Since 1973 synthesis starting from a compound of known absolute configuration has become the standard method for determining the absolute configuration of a chiral pheromone with known chiroptical properties or gas chromatographic properties on chiral stationary phases.

Although determination of the absolute configuration of a pheromone might be the only task for chemists working in the field of chemical communications, it is merely a part of traditional natural products chemistry. The most interesting and exciting discovery in this area was the diversity and complexity of the relationships between absolute configuration and pheromone activity as will be detailed in the rest of this chapter.

WHEN WAS CHIRALITY FOUND TO BE IMPORTANT IN CHEMICAL COMMUNICATIONS?

BACKGROUND—THEORY OF OLFACTION IN THE 1960s AND EARLY 1970s

Before the advent of molecular biology and gene manipulation, there were two theories about the mechanism of olfaction. Wright's vibration theory of olfaction (1964) predicted the unimportance of enantiomerism in pheromone perception. From the generalization that no example was known of the human nose detecting an odour for one of a pair of the enantiomers but not for the other, Wright (1963) inferred that the primary process of olfaction must be a physical rather than chemical interaction. Reports of slight differences between the odours of some enantiomers might, he thought, result from purity differences. According to Wright's theory, the vibrational frequencies of an odorous molecule in the far infrared region (500–50 cm^{-1}) determine the quality of an odour. Amoore (1970), on the other hand, emphasized the importance of molecular shape in determining odour quality. His stereochemical theory is a version of Emil Fischer's lock-and-key concept so well known in enzyme and drug theory. An odorous molecule must have a stereostructure complementary to that of the receptors site. In fact, a highly significant correlation existed between molecular shape and ant alarm-pheromone activity (Amoore *et al.* 1969). In Amoore's theory two enantiomers should have different odours, because they are not superimposable.

In 1971 differences between the odours of highly purified *R*-(−)-carvone (**4**, Fig. 9.1) and its *S*-(+) isomer were reported by three independent groups (Friedman & Miller 1971; Leitereg *et al.* 1971; Russell & Hills 1971). (*R*)-Carvone had the odour of spearmint, whereas the odour of the antipode was that of caraway—both the enantiomers of carvone were odorous to the human nose. The author thought at the time that study of chiral insect pheromones might furnish a clearer result, for example bioactivity residing in a single enantiomer only.

PIONEERING WORK WITH INSECTS

Even before 1974, when synthetic enantiomers of some insect pheromones became available, pioneers tested both enantiomers of odorous compounds to determine whether or not insects could discriminate between enantiomers.

Schneider, who invented electroantennography to record electrical signals from antennae of insects in response to bombykol and other pheromones, tested the enantiomers of 4-methylhexanoic acid (**5**) prepared by Ohloff (Kafka *et al.* 1973). They found that single olfactory receptor cells of the migratory locust (*Locusta migratoria migratorioides*) and honeybee drones (*Apis mellifica*) were stimulated by both *R*-(−)-**5** and *S*-(+)-**5**, although the respective stimulus effectiveness of the enantiomers differed from one cell to another, and even between adjacent cells in the same sensillum. For two-thirds of cells tested *R*-(−)-**5** was the more effective molecule; discrimination between the two enantiomers of **5** by honeybee drones, after conditioning, was also shown to be significant. It was concluded that the discrimination between the enantiomers of **5** suggested the existence of at least two sorts, or variants, of receptor in the olfactory cell membrane of the insects.

Lensky & Blum (1974) subsequently found that conditioned honeybee workers *Apis mellifera ligustica*) can discriminate between the enantiomers of carvone (**4**) and 2-octanol (**6**); the odours of the enantiomers of the latter compound cannot be distinguished by the human nose. It was concluded that the presence of chiral chemoreceptors for discrimination between enantiomers, coupled with the ability to memorize information encoded in these specific signals, has enabled honeybee workers to exploit chirality-dependent cues emanating from food plants. They also concluded that their data did not support Wright's vibrational theory of odour.

THE BIOACTIVITY OF PHEROMONES DEPENDS ON THEIR CHIRALITY

In 1974 three groups synthesized enantiomerically highly enriched pheromones independently; their bioassay definitely proved that the bioactivity depends on the chirality of pheromones.

Silverstein and co-workers (Riley *et al.* 1974) prepared the enantiomers of 4-methyl-3-heptanone (**7**, Fig. 9.2). The dextrorotatory *S* isomer of this ketone is the principal alarm pheromone of the leaf cutting ant (*Atta texana*). The enantiomers of **7** were synthesized from the resolved enantiomers of 2-methyl-4-pentenoic acid as the key intermediates. *S*-(+)-**7** was *ca* 100 times more active than *R*-**7** workers of *Atta texana* and the unnatural *R*-**7** did not inhibit response to *S*-**7**. Thus (±)-**7** was also active as a pheromone.

Marumo and co-workers (Iwaki *et al.* 1974) synthesized the enantiomers of disparlure (**8**), the sex pheromone produced by the female gypsy moth (*Lymantria dispar*). They prepared both 7*R*,8*S*-**8** and 7*S*,8*R*-**8** from glutamic acid (with recourse to diastereomer separation in the course of the synthesis). Behavioural response to the enantiomers of **8** in the laboratory showed 7*R*,8*S*-(+)-**8** (active at 10^{-10} g mL^{-1}) to be more active than (±)-**8** (active at 10^{-7} g mL^{-1}) whereas 7*S*,8*R*-(−)-**8** was active at 10^{-4} g mL^{-1} only. In electroantennographic (EAG) studies, the threshold concentration of (+)-**8** was 10^{-7} g mL^{-1}), slightly less than that of (±)-**8** (10^{-6} g mL^{-1}), whereas a much higher concentration of (−)-**8** (10^{-4} g mL^{-1}) was required. These data implicated the involvement of a chiral receptor system in pheromone perception by the male gypsy moth and the leaf cutting ant.

At this time the author wondered whether the enantiomers of highly dissymmetric compounds might evoke totally different olfactory reactions, and thus undertook the synthesis of the enantiomers of *exo*-brevicomin (**2**) (Mori, K. 1974b). *exo*-Brevicomin and frontalin (**9**) were known to be the components of the aggregation pheromone of the western pine beetle (*Dendroctonus brevicomis*), although the absolute configuration of the natural products had not yet been determined. In 1973, when the synthesis was planned, the author, a young associate professor, was teaching introductory organic stereochemistry to undergraduates. This, of course, entailed covering Pasteur's first optical resolution of racemic tartaric acid (*acide recémique*). At this time recognition of the coincidence of the configuration of tartaric acid with the configuration at C-1 and C-7 of brevicomin came as a revelation. Thus 1*R*,5*S*,7*R*-(+)-**2** was synthesized from unnatural 2*S*,3*S*-tartaric acid—first obtained by Pasteur in 1848. Natural 2*R*,3*R*-tartaric acid was converted to 1*S*,5*R*,7*S*-(+)-**2**.

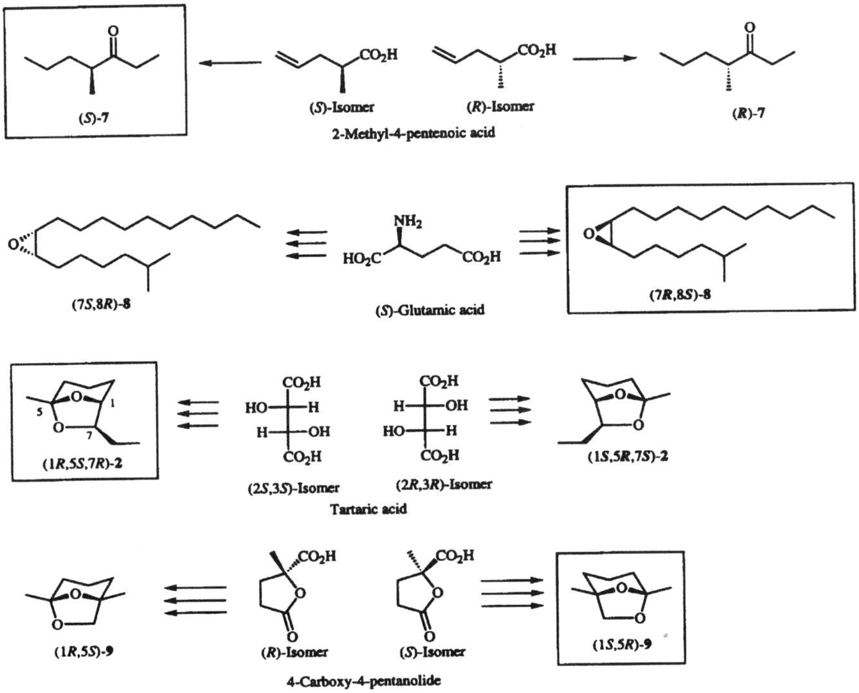

Fig. 9.2 Early syntheses of the enantiomers of pheromones (1).

These two enantiomers of brevicomin were sent to Wood at Berkeley, California, in early 1974. When the author visited Berkeley in June 1974, Wood's co-worker Browne took him to the laboratory, and showed him a spectacular bioassay experiment with western pine beetles. Only 1*R*,5*S*,7*R*-(+)-**2** was bioactive when mixed with (±)-frontalin (**9**) and a monoterpene myrcene. The antipodal 1*S*,5*R*,7*S*-(+)-**2** had no activity. It was an unforgettable moment to watch the beetles attracted by (+)-**2** only. This result was reported at the 9th IUPAC International Symposium on the Chemistry of Natural Products in Ottawa on June 27 1974 (*cf.* ref. 2 of Mori, K. 1975a).

Because frontalin (**9**) is also the pheromone component, the enantiomers of **9** were then synthesized starting from the enantiomers of a lactonic acid (Fig. 9.2) (Mori, K. 1975a). Only 1*S*,5*R*-(−)-frontalin (**9**) was bioactive against the western pine beetle. The biological details of this work were published later by Wood *et al.* (1976).

In all the studies discussed above, a single enantiomer only of the pheromone was found to be highly bioactive. It thus became clear that bioactivity depends on the chirality of the pheromones, as for other bioactive substances such as steroid hormones. A common belief among bioorganic chemists at that time was that bioactivity resides in the naturally occurring enantiomer and that bioactive and chiral substances must usually be enantiomerically pure. These 1974 findings in pheromone science seemed to support the above general view, and it became apparent that the lock and key concept was also applicable to pheromone perception.

HOW WAS THE UTMOST IMPORTANCE OF CHIRALITY IN CHEMICAL COMMUNICATIONS RECOGNIZED?

SYNERGISTIC RESPONSE BASED ON ENANTIOMERS—SULCATOL

Sulcatol (**10**, Fig. 9.3) is the aggregation pheromone produced by males of *Gnathotrichus sulcatus*, an economically important ambrosia beetle in the Pacific coast of North America (Byrne *et al.* 1974). The natural pheromone was shown to be a 35:65 mixture of *R*-**10** and *S*-**10** by ^{1}H NMR analysis of its Mösher ester (α-methoxy-α-trifluoromethylphenylacetate). The reason the beetle produced a mixture of the enantiomers of **10** was unclear at the time of its discovery. The enantiomers of sulcatol were synthesized starting from the enantiomers of glutamic acid (Mori, K. 1975b) and when they were assayed in Canada, Borden *et al.* (1976) found neither *R*-**10** nor *S*-**10** to be bioactive. The maximum response of the beetle was to a racemic mixture (50:50) of the enantiomers and the response to (±)-**10** was significantly greater than that to a 35:65 mixture.

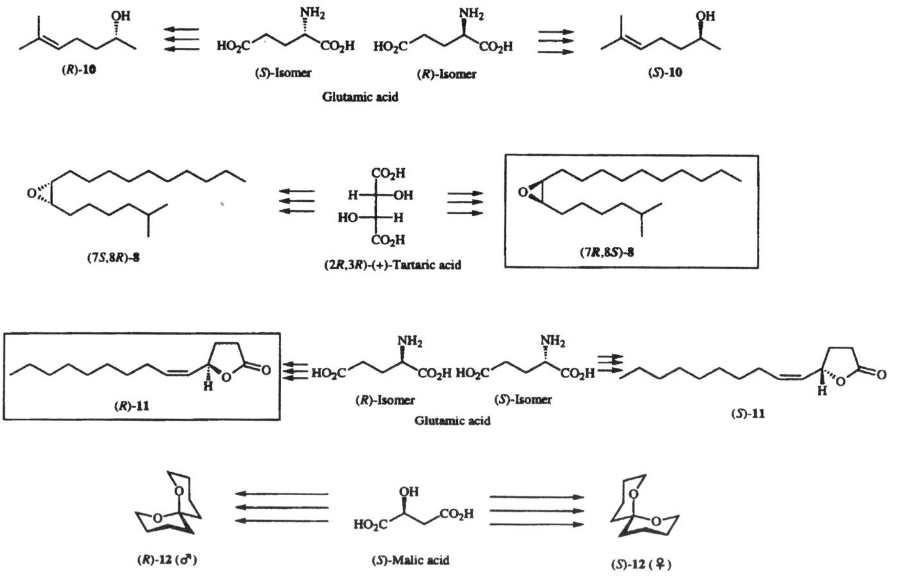

Fig. 9.3 Early syntheses of the enantiomers of pheromones (2).

It thus became clear that the beetles must produce an enantiomeric mixture of **10** if they are to communicate with each other. This discovery was the first example of a synergistic response based on enantiomers.

INHIBITION BY THE WRONG ENANTIOMER—DISPARLURE AND JAPONILURE

As already described above, (±)-disparlure (**8**) is not as active as 7*R*,8*S*-(+)-**8** and 7*S*,8*R*-(−)-**8** is almost inactive (Iwaki *et al.* 1974). Dosage–response effects of the inactive 7*S*,8*R*-(−)-**8** were therefore evaluated by Vité *et al.* (1976) on the gypsy moth by employing the enantiomers of **8** synthesized by Mori, K. *et al.* (1976). Although the response to low concentrations (10^{-3} to 10^{-5} dilution) of 7*R*,8*S*-(+)-**8** was not affected substantially by addition of equal or lower concentrations of 7*S*,8*R*-(−)-**8**, concentrations of 7*S*,8*R*-(−)-**8** higher than those of the antipode drastically reduced the response of the moths.

It seems that the inactive 7*S*,8*R*-**8** must be present in larger concentrations than 7*R*,8*S*-**8** if it is to saturate the receptor site and inhibit the effects of the latter. Under field conditions males of the gypsy moth (*Lymantria dispar*) and males of the nun moth (*Lymantria monacha*) responded to 7*R*,8*S*-**8** and addition of 7*S*,8*R*-**8** significantly suppressed the response of *L. dispar* but had less effect on the response of *L. monacha*. EAG studies on *L. dispar* by Miller *et al.* (1977) by use of the differential receptor saturation technique suggest the existence of one receptor type with the greatest affinity for 7*R*,8*S*-**8** and another type with greater affinity for 7*S*,8*R*-**8**. It should be added that the enantiomers of disparlure (Mori, K. *et al.* 1976, 1979) were synthesized from (+)-tartaric acid, the acid so closely associated with Pasteur.

Japonilure (*R*-(−)-**11**) is the sex pheromone produced by the female of the Japanese beetle (*Popillia japonica*). Because (±)-**11** was inactive, Tumlinson *et al.* (1977) carefully studied the relationship between the enantiomeric purity of **11** and its bioactivity. They synthesized both enantiomers of **11** starting from the enantiomers of glutamic acid. The bioactive enantiomer is *R*-(−)-**11**; *S*-(+)-**11** severely inhibits the action of *R*-**11**. Accordingly, (−)-**11** of 99% e.e. is *ca* two-thirds as active as pure *R*-**11**; that of 90% e.e. is *ca* one-third as active, that of 80% e.e. is ca one-fifth as active, and both (−)-**11** of 60% e.e. and (±)-**11** were inactive. These results illustrate dramatically the utmost importance of chirality in chemical communications. Leal (1996) recently, found that the sex pheromone of the female scarab beetle, *Anomala osakana*, is *S*-**11**, and that *R*-**11** interrupts the attraction caused by *S*-**11**.

ONE ENANTIOMER IS ACTIVE AGAINST MALES WHEREAS THE OPPOSITE ENANTIOMER AFFECTS FEMALES—OLEAN

An unusual example of the relationship between stereochemistry and pheromone activity is that of olean (**12**), the sex pheromone produced by the female olive fruit fly, *Bactrocera oleae* (Haniotakis *et al.* 1986). Both enantiomers of **12** were synthesized from *S*-malic acid (Mori, K. *et al.* 1985a, b) and bioassayed in Greece by Haniotakis *et al.* (1986). Surprisingly, *R*-**12** was active against males whereas *S*-**12** was active against females. Chiral GC analysis of natural olean by Schurig revealed it to be (±)-**12** (Haniotakis *et al.* 1986). Thus the female-produced pheromone activates male olive fruit flies and the female herself.

It is evident that the availability of the pure enantiomers of pheromones as a result of enantioselective synthesis has enabled understanding of the diverse chiral world of chemical communications.

CHIRALITY PLAYS A ROLE EVEN WITH AN ACHIRAL PHEROMONE

The European corn borer (*Ostrinia nubilalis*) and the redbanded leaf roller (*Argyrotaenia velutinana*) use Z-11-tetradecenyl acetate (13, Fig. 9.4) both as a sex attractant and as a precopulatory behaviour pheromone. The two pheromone receptor systems are different. The sex attraction system requires specific ratios of Z- to E-11-tetradecenyl acetate for each insect, and the precopulatory behaviour system is relatively insensitive to the presence or absence of E-11-tetradecenyl acetate. Chapman *et al.* (1978) demonstrated that within the precopulatory behaviour pheromone system there are at least two different receptors for **13**, that the receptors for achiral **13** are chiral, and that **13** is coiled differently (**13A** and **13B**) in the two receptors. Chapman's unique strategy for clarifying the situation was to synthesize the pure enantiomers of **14** which were regarded as conformationally fixed mimics of the two conformers of **13**.

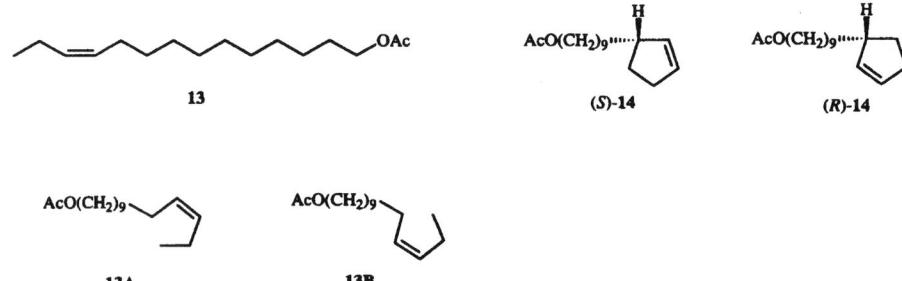

Fig. 9.4 The two conformers **13A** and **13B** of Z-11-tetradecenyl acetate, which correspond to the chiral pheromone mimics S- and R-**14**.

Although the European corn borer responds to S-**14** (which mimics conformer **13A**) as strongly as it does to the natural pheromone **13**, it responds only weakly to R-**14** (which mimics conformer **13B**). The response to (±)-**14** is intermediate between the responses to the pure enantiomers. These data are consistent with the presence of a single stereoselective pheromone receptor. The redbanded leaf roller, on the other hand, responds equally to R- and S-**14**, but responds much more strongly to (±)-**14** than to the either enantiomer. The greater activity of (±)-**14** in the redbanded leaf roller requires two stereospecific receptors, one sensitive to R-**14** and the other sensitive to S-**14**. The redbanded leaf roller has thus evolved two receptors which sense different conformations (**13A** and **13B**) of the achiral (but prochiral) pheromone **13**.

It has become clear from the work of Chapman *et al.* (1978) that the insect detection system for this particular precopulatory behaviour pheromone system makes very clever use of the prochiral character of the achiral olefinic pheromone. Thus chirality is important even among achiral olefinic pheromones.

CURRENT UNDERSTANDING OF THE IMPORTANCE OF CHIRALITY IN CHEMICAL COMMUNICATIONS—STEREOCHEMISTRY–BIOACTIVITY RELATIONSHIPS AMONG PHEROMONES

The most notable advance in two decades of pheromone science is clear understanding of the significance of chirality in pheromone perception. The results obtained from work on 27 chiral pheromones are summarized in Figs 9.5–9.14; the examples are selective rather than comprehensive. As is apparent from the figures, stereochemistry–pheromone activity relationships are quite complicated. The relationships are divided into ten categories as detailed below.

ONLY ONE ENANTIOMER IS BIOACTIVE, AND ITS ANTIPODE DOES NOT INHIBIT THE ACTION OF THE ACTIVE STEREOISOMER

This is the most common relationship; most (*ca* 60%) chiral pheromones belong to this category. Although many workers therefore believed this relationship must be the only one, emphasizing the importance of a single bioactive enantiomer, it is merely one of the diverse relationships found for pheromones.

Three pheromones belonging to this category are shown in Fig. 9.5. The alarm pheromone *S*-**7** and the aggregation pheromone 1*R*,5*S*,7*R*-**2** have already been discussed. 1*R*,5*S*,7*R*-Dehydro-*exo*-brevicomin (**15**) is a chemical signal produced by the male house mouse (*Mus musculus*) to show his male state to others; its enantiomers have been synthesized by Mori, K. & Seu (1986). This mammalian chemical communication is now known to be an enantioselective process (Novotny *et al.* 1995). The only bioactive isomer, 1*R*,5*S*,7*R*-**15**, has the same absolute stereochemistry as the western pine beetle pheromone, *exo*-brevicomin (**2**). It is interesting to note that such different animals as mouse and pine beetle biosynthesize heterocycles with the same absolute configuration for use in chemical communication.

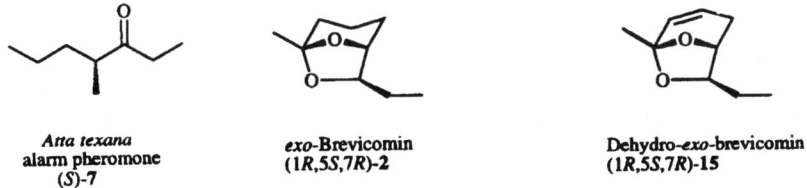

Atta texana
alarm pheromone
(*S*)-**7**

exo-Brevicomin
(1*R*,5*S*,7*R*)-**2**

Dehydro-*exo*-brevicomin
(1*R*,5*S*,7*R*)-**15**

Fig. 9.5 Stereochemistry and pheromone activity. 1. Only one enantiomer is bioactive and its antipode does not inhibit the action of the active stereoisomer.

ONLY ONE ENANTIOMER IS BIOACTIVE, AND ITS ANTIPODE INHIBITS THE ACTION OF THE PHEROMONE

Disparlure (**8**) and japonilure (**11**) have already been discussed. Lardolure (1*R*,3*R*,5*R*,7*R*-**16**, Fig. 9.6) is the aggregation pheromone of the acarid mite *Lardoglyphus konoi*. Again both the enantiomers have been synthesized (Mori, K. & Kuwahara 1986); 1*R*,3*R*,5*R*,7*R*-**16** is active whereas 1*S*,3*S*,5*S*,7*S*-**16** is inhibitory. A mixture of all possible stereoisomers of **16** is, therefore, only marginally active (Kuwahara *et al.* 1991).

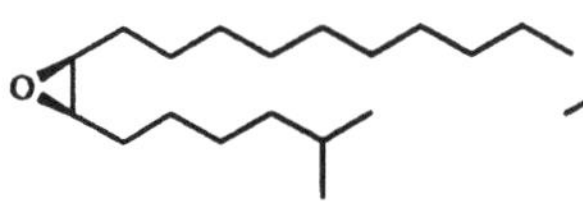

Disparlure
(7R,8S)-8

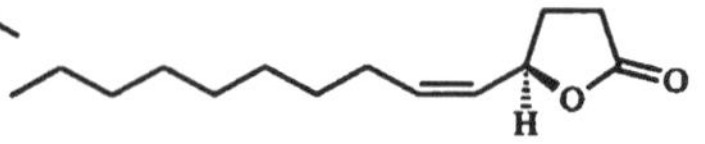

Japonilure
(R)-11

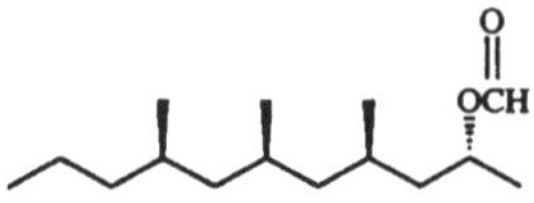

Lardolure
(1R,3R,5R,7R)-16

Fig. 9.6 Stereochemistry and pheromone activity. 2. Only one enantiomer is bioactive and the antipode inhibits the action of the pheromone.

ONLY ONE ENANTIOMER IS BIOACTIVE, AND ITS DIASTEREOMER INHIBITS THE ACTION OF THE PHEROMONE

Serricornin (4*S*,6*S*,7*S*-**17**, Fig. 9.7) is the sex pheromone produced by the female cigarette beetle, *Lasioderma serricorne*. The bioactivity of the stereoisomers of **17** was studied carefully by Chuman and co-workers in the course of the development of practical pheromone traps (Mori, M. *et al.* 1986). Although only 4*S*,6*S*,7*S*-**17** was bioactive, and the antipode 4*R*,6*R*,7*R*-**17** did not inhibit the action of the pheromone, the isomer 4*S*,6*S*,7*R*-**17** was quite inhibitory against the action of 4*S*,6*S*,7*S*-**17**. Accordingly, the commercial pheromone lure must be manufactured free from contamination by the 4*S*,6*S*,7*R* isomer.

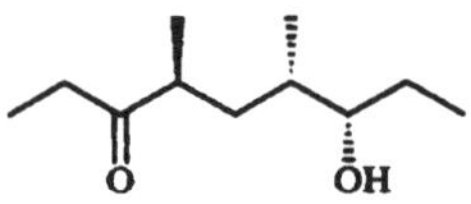

Serricornin
(4*S*,6*S*,7*S*)-**17**

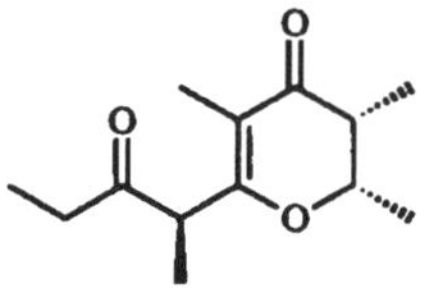

Stegobinone
(2*S*,3*R*,1′*R*)-**18**

Fig. 9.7 Stereochemistry and pheromone activity. 3. Only one enantiomer is bioactive and its diastereomer inhibits the action of the pheromone.

Stegobinone (2*S*,3*R*,1′*R*-**18**) is one of two components of the pheromone produced by the female drugstore beetle, *Stegobium paniceum*. Very low activity of the racemic and diastereomeric mixture of stegobinone was indicative of the presence of inhibitor(s) in the synthetic products. It was later shown that addition of 2*S*,3*R*,1′*S*-epistegobinone to **18** significantly reduced the response of male drugstore beetles (Kodama *et al.* 1987).

THE NATURAL PHEROMONE IS A SINGLE ENANTIOMER, AND ITS ANTIPODE OR DIASTEREOMER IS ALSO ACTIVE

Dominicalure 1 (*S*-**19**, Fig. 9.8) and dominicalure 2 (*S*-**20**) are components of the aggregation pheromone produced by the male lesser grain borer, *Rhyzopertha dominica*. They attract both sexes of the insect. Field testing showed that the natural *S*-

19 and *S*-**20** are approximately twice as active as the unnatural isomers *S*-**19** and *S*-**20**, respectively (Williams *et al.* 1981).

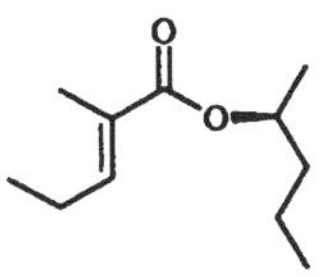

Dominicalure 1
(*S*)-**19**

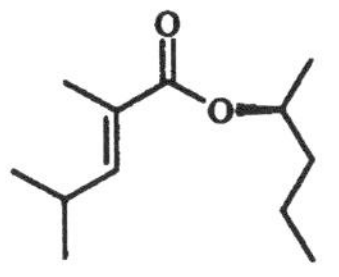

Dominicalure 2
(*S*)-**20**

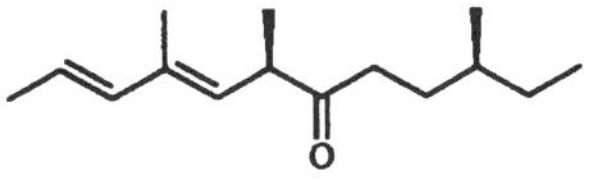

Matsucoccus feytaudi pheromone
(3*S*,7*R*)-**21**

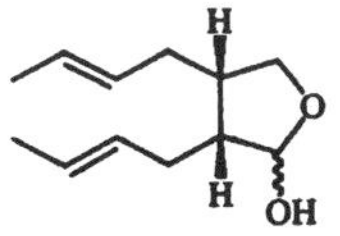

Biprorulus bibax pheromone
(3*R*,4*S*)-**22**

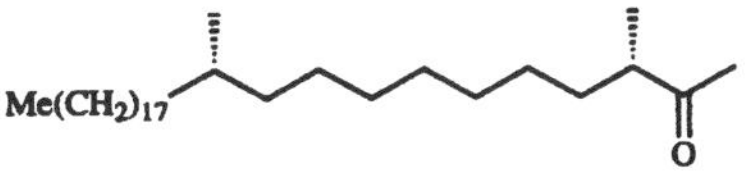

Blattella germanica pheromone
(3*S*,11*S*)-**23**

Fig. 9.8 Stereochemistry and pheromone activity. 4. The natural pheromone is a single enantiomer, and its antipode or diastereomer is also active.

Females of the maritime pine scale (*Matsucoccus feytaudi*) produce 3*S*,7*R*-**21** as a sex pheromone. Although the bioactivity of the 3*R*,7*R* isomer is similar to that of the natural pheromone, *M. feytaudi* males respond very weakly to the two other stereoisomers (Jactel *et al.* 1994). It therefore seems that the stereochemistry at C-3 is not important for the expression of bioactivity.

The male spined citrus bug (*Biprorulus bibax*) produces 3*R*,4*S*-**22** as aggregation pheromone. When synthetic enantiomers of **22** were bioassayed (Mori, K. *et al.* 1993) it was found that the 3*S*,4*R* isomer of the pheromone was also bioactive, indicating that *B. bibax* does not discriminate between the enantiomers (James & Mori 1995).

The female German cockroach (*Blattella germanica*) produces 3*S*,11*S*-**23** as her contact sex pheromone (Mori, K. *et al.* 1981). Male German cockroaches do not, however, discriminate the four stereoisomers of **23** from one another (Nishida & Fukami 1983). *B. germanica* males seem, therefore, to have no stereoselectivity with regard to perception of the pheromone **23**.

THE NATURAL PHEROMONE IS AN ENANTIOMERIC MIXTURE, AND BOTH ENANTIOMERS ARE SEPARATELY ACTIVE

Female Douglas fir beetles (*Dendroctonus pseudotsugae*) produce a 55:45 mixture (on average) of *R*- and *S*-**24** (Fig. 9.9) (Lindgren *et al.* 1992). The combined effect of the enantiomers was additive, rather than synergistic, and both enantiomers are required for maximum response.

Male southern pine beetles (*Dendroctonus frontalis*) produce an 85:15 mixture of 1*S*,5*R*-frontalin (**9**) and its 1*R*,5*S* isomer. In laboratory and field bioassays, the response of *D. frontalis* to a mixture of 1*S*,5*R*-**9** and α-pinene was significantly greater than that to 1*R*,5*S*-**9** and α-pinene (Payne *et al.* 1982). EAG studies showed that antennal olfactory receptor cells were significantly more responsive to 1*S*,5*R*-**9** than to

1*R*,5*S*-9. Both enantiomers of **9** stimulated the same olfactory cells, suggesting that each cell has at least two types of enantioselective receptor.

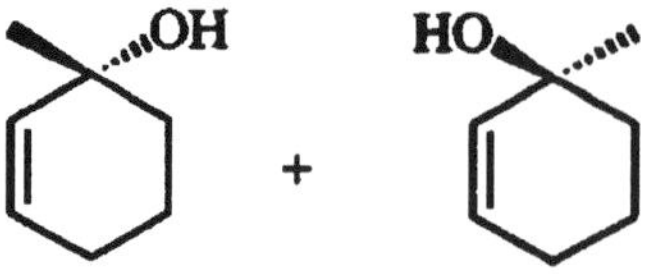

(55:45)

Dendroctonus pseudotsugae
pheromone
(*R*)-24   (*S*)-24

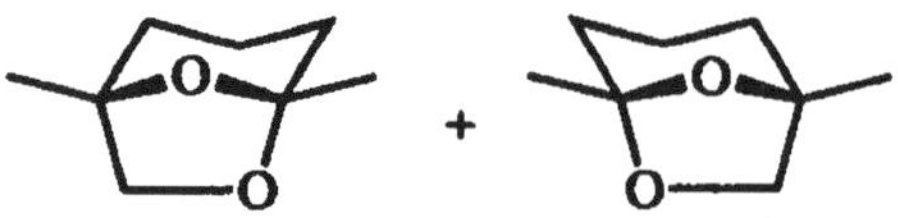

(85:15)

Frontalin
(1*S*,5*R*)-9   (1*R*,5*S*)-9

Fig. 9.9 Stereochemistry and pheromone activity. 5. The natural pheromone is a mixture of enantiomers, and both are separately active.

DIFFERENT ENANTIOMERS OR DIASTEREOMERS ARE EMPLOYED BY DIFFERENT SPECIES

S-Ipsdienol (**25**, Fig. 9.10) is the pheromone component of the California five-spined ips (*Ips paraconfusus*). The bark beetles *I. calligraphus* and *I. avulsus* both, however, respond to its antipode, *R*-**25** (Vité *et al.* 1978). The pine engraver *I. pini* in the USA responds to a mixture of the enantiomers of **25**. Variation of the enantiomeric purity of **25** in *I. pini* was studied in detail by Seybold *et al.* (1995).

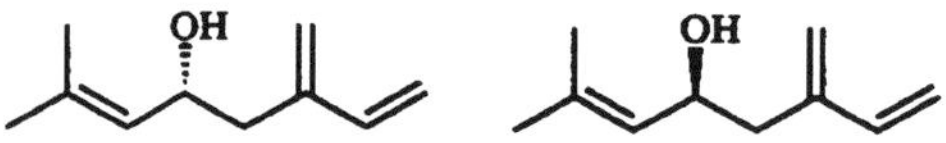

Ipsdienol
(*R*)-25   (*S*)-25

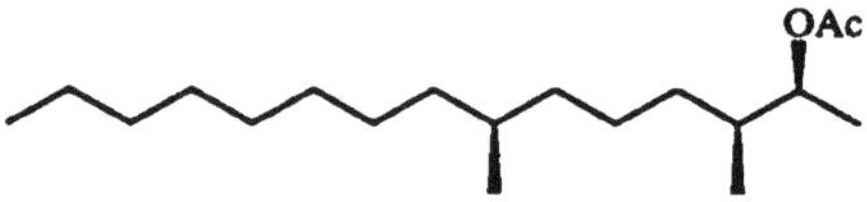

Neodiprion pinetum pheromone
(1*S*,2*S*,6*S*)-26

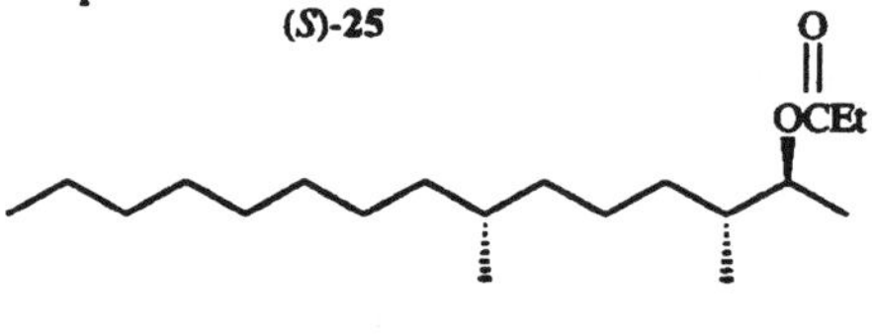

Diprion similis pheromone
(1*S*,2*R*,6*R*)-27

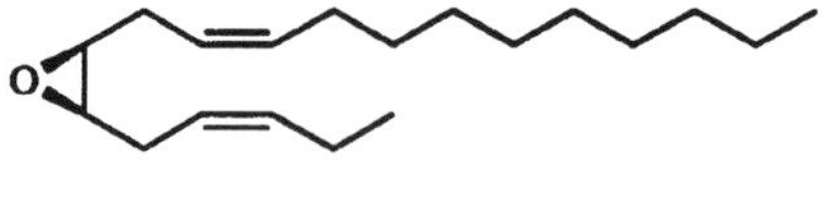

Colotois pennaria pheromone
(6*R*,7*S*)-28

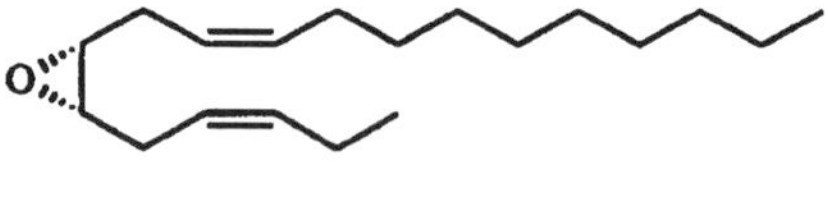

Erannis defoliaria pheromone
(6*S*,7*R*)-28

Fig. 9.10 Stereochemistry and pheromone activity. 6. Different enantiomers or diastereomers are used by different species.

The complicated relationships between the stereochemistry and bioactivity of pine sawfly pheromones have been studied extensively (Norin 1996). For example, in the USA the white pine sawfly (*Neodiprion pinetum*) uses 1*S*,2*S*,6*S*-**26** as its sex pheromone (Kraemer *et al.* 1979; Olaifa *et al.* 1986) whereas 1*S*,3*R*,6*R*-**27** is the pheromone of the introduced pine sawfly (*Diprion similis*) (Olaifa *et al.* 1986).

Use of the chirality of pheromones is an important means of discriminating between two species of the winter flying geometrid moths in Middle Europe. Thus 6*R*,7*S*-**28** is the pheromone of *Colotois pennaria*, whereas that of *Erannis defoliaria* is 6*S*,7*R*-**28** (Szöcs *et al.* 1993).

In these examples, insects use chirality to segregate different species.

BOTH ENANTIOMERS ARE NECESSARY FOR BIOACTIVITY

The synergistic response of the beetle *Gnathotrichus sulcatus* to the enantiomers of sulcatol (**10**, Fig. 9.3) has already been discussed. Similarly, the aggregation pheromone of the grain beetle *Cryptolestes turcicus* is an 85:15 mixture of *R*- and *S*-**29** (Fig. 9.11), and neither is separately bioactive (Millar *et al.* 1985).

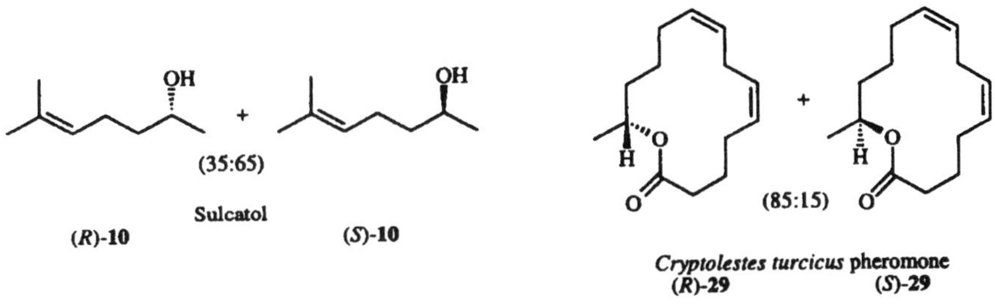

Fig. 9.11 Stereochemistry and pheromone activity. 7. Both enantiomers are necessary for bioactivity.

ONE ENANTIOMER IS MORE ACTIVE THAN THE OTHER STEREOISOMER(S), BUT AN ENANTIOMERIC OR DIASTEREOMERIC MIXTURE IS MORE ACTIVE THAN THAT ENANTIOMER ALONE

The smaller tea tortrix moth (*Adoxophyes* sp.) uses **30** (Fig. 9.12) as a minor component of its pheromone bouquet, and *R*-**30** was found to be slightly more active than *S*-**30**. Field tests suggested 95:5 was the optimum *R*:*S* ratio for trapping males (Tamaki *et al.* 1980).

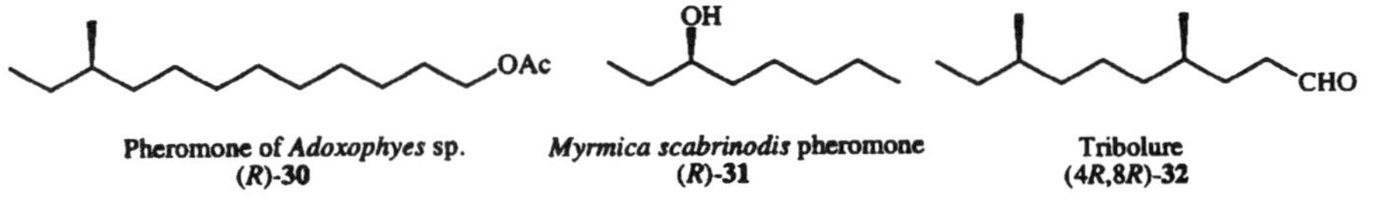

Fig. 9.12 Stereochemistry and pheromone activity. 8. One enantiomer is more active than the other stereoisomer(s), but an enantiomeric or diastereomeric mixture is more active than that enantiomer alone.

In the ant *Myrmica scabrinodis* the naturally occurring 9:1 mixture of *R*- and *S*-**31** was more attractive than pure *R*-**31** or (±)-**31**; *S*-**31** was inactive (Cammaerts & Mori 1987).

Tribolure (4*R*,8*R*-**32**) is the aggregation pheromone of the male red flour beetle, *Tribolium castaneum*. Suzuki *et al.* (1984) found that 4*R*,8*R*-**32** was as active as the natural pheromone and that an 8:2 mixture of 4*R*,8*R*-**32** and its 4*R*,8*S* isomer was *ca* ten times more active than 4*R*,8*R*-**32** alone.

ONE ENANTIOMER IS ACTIVE ON MALES, THE OTHER ON FEMALES

As described above, the stereochemistry–bioactivity relationship for olean (**12**) is unique (Haniotakis *et al.* 1986).

5-Methyl-3-heptanone (**33**, Fig. 9.13) was isolated as a pheromone in the coelomic fluid of gravid specimens of nereid marine polychaetes by Zeeck *et al.* (1988). It is responsible for the induction of the nuptial dance behaviour prior to the release of gametes in *Platynereis dumerilii*. Interestingly, *S*-(+)-**33**, produced by the female, activates the males and *R*-(−)-**33**, produced by the male, is active on females (Hardege *et al.* 1996; Zeeck *et al.* 1992).

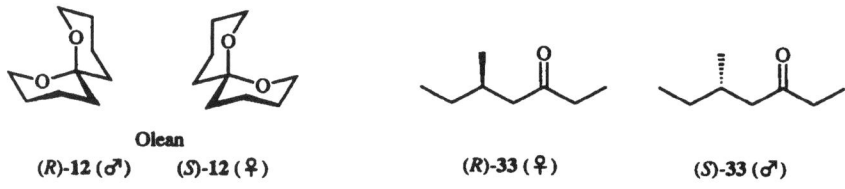

(*R*)-**12** (♂)   (*S*)-**12** (♀)   (*R*)-**33** (♀)   (*S*)-**33** (♂)

Olean

Fig. 9.13 Stereochemistry and pheromone activity. 9. One enantiomer is active on males, the other on females.

ONLY THE MESO ISOMER IS ACTIVE

In the tsetse fly, *meso* alkanes seem to be the bioactive sex pheromones. Thus 13*R*,23*S*-**34** (Fig. 9.14) was active as the sex stimulant pheromone of the female tsetse fly, *Glossina palidipes* but neither 13*R*,23*S*-**34** nor 13*R*,23*S*-**34** was bioactive (McDowell *et al.* 1985).

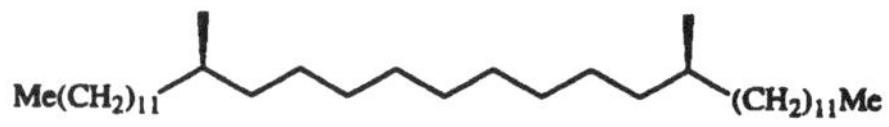

Me(CH$_2$)$_{11}$ ~~~~~~~~~~ (CH$_2$)$_{11}$Me

Glossina pallidipes pheromone
(13*R*,23*S*)-**34**

Fig. 9.14 Stereochemistry and pheromone activity. 10. Only the *meso* isomer is active.

The ten categories described above were discovered solely as a result of experimentation. In all instances synthesis of pure enantiomers and stereoisomers was followed by bioassay to clarify the situation.

THE PRACTICAL IMPORTANCE OF CHIRALITY IN PHEROMONE-BASED PEST CONTROL

If pheromones are to be applied practically their stereochemical properties must be solved before any attempt at large-scale use. It should be remembered that stereoisomers of pheromones can inhibit pheromone action.

Trapping of male gypsy moths in traps baited with 7R,8S-disparlure (**8**) has been used for several years to monitor the population of the gypsy moth in the USA. It has been shown that only highly enantiomerically pure 7R,8S-**8** works as a powerful trapping agent. For the mass trapping of the Japanese beetle in the USA and Japan, only enantiomerically pure R-japonilure (**11**) works ideally. Both 7R,8S-**8** and R-**11** are commercially available as the pure enantiomers.

If serricornin (4S,6S,7S-**17**), the cigarette beetle pheromone, is to be used practically it is essential to prepare the commercial product without any contamination by the 4S,6S,7R isomer. The commercial lure is a racemic and diastereomeric mixture of **17** with no contamination by the 4S,6S,7R isomer. Accordingly, the 4*,6*,7* isomer of **17** should not be generated in the course of the industrial synthesis of the lure.

Highly sophisticated stereoselective synthetic technology as thus needed to prepare pheromones to be used practically. Synthesis of pheromones was thoroughly reviewed by Mori, K. (1992), and results recently obtained by the author's group have also been reviewed (Mori, K. 1997)

CONCLUSION

The importance of chirality in chemical communications is now well established. Much of the progress in this field can be directly attributed to advances in stereoselective organic chemistry. The advance in the last quarter of the 20th century is most evident in that pheromones with enantiomeric purity even higher than that of the naturally occurring compounds can now be synthesized.

Biodiversity can clearly be recognized in pheromone perception. Insects employ chirality to enrich their communication systems and to secure greater specificity in perception. In this way they have acquired a strong adaptive advantage.

The precise enantioselective mechanisms of pheromone perception are now under active investigation and it is possible to look forward to further advances. As in the days of Pasteur, experimental work will continue to tell us more about chirality in the natural world.

ACKNOWLEDGEMENT

The author thanks his co-workers, especially Dr H. Takikawa, for their help with the preparation of the manuscript.

REFERENCES

Amoore, J. E. (1970) *Molecular Basis of Odor*, Charles C. Thomas, Springfield, Illinois.

Amoore, J. E, Palmieri, G., Wanke, E. & Blum, M. S. (1969) Ant alarm pheromone activity: correlation with molecular shape by scanning computer. *Science*, **165**, 1266–9.

Borden, J. H., Chong, L., McLean, J. A., Slessor, K. N. & Mori, K. (1976) *Gnathotrichus sulcatus*; synergistic response to enantiomers of the aggregation pheromone sulcatol. *Science*, **192**, 894–6.

Butenandt, A., Beckmann, R., Stamm, D. & Hecker, E. (1959) Ober den Sexuallockstoff dos Seidenspinners *Bombyx mori*. Reindarstellung und Konstitutionsermittlung. *Zeitschrift für Naturforschung*, **14b**, 283–4.

Byrne, K. W., Swigar, A., Silverstein, R. M., Borden, J. H. & Stokkink, E. (1974) Sulcatol: population aggregation pheromone in *Gnathotrichus sulcatus* (Coleoptera: Scolytidae). *Journal of Insect Physiology*, **20**, 1895–900.

Cammaerts, M.-C. & Mori, K. (1987) Behavioral activity of pure chiral 3-octanol for the ants *Mynnica scabrinodis* Nyl. and *Mynnica rubra* L. *Physiological Entomology*, **12**, 381–5.

Chapman, O. L., Mattes, K. C., Sheridan, R. S. & Klum, J. A. (1978) Stereochemical evidence of dual chemoreceptors for an achiral sex pheromone in Lepidoptera. *Journal of the American Chemical Society*, **100**, 4878–84.

Friedman, L. & Miller, J. G. (1971) Odor incongruity and chirality. *Science*, **172**, 1044–6.

Haniotakis, G., Francke, W., Mori, K., Redlich, H. & Schurig, V. (1986) Sex-specific activity of (R)-(–)- and (S)-(+)-1,7-dioxaspiro[5.5]undecane, the major pheromone of *Dacus oleae*. *Journal of Chemical Ecology*, **12**, 1559–68.

Hardege, J. D., Müller, C. T., Bentley, M. G. & Beckmann, M. (1996) Chemical signaling in marine polychaetes: structure and function of homo- and heterospecific sex pheromones. Presented at 13th annual meeting of international society of chemical ecology, August 18–22, Prague, Czech Republic.

Hecker, E. & Butenandt, A. (1984) Bombykol revisited—reflections on pioneering period and on some of its consequences, in *Techniques in Pheromone Research* (Ed. by H. E. Hummel & T. A. Miller), Springer, New York, pp. 1–44.

Iwaki, S., Marumo, S., Saito, T., Yamada, M. & Katagiri, K. (1974) Synthesis and activity of optically active disparlure. *Journal of the American Chemical Society*, **96**, 7842–4.

Jactel, H., Menassieu, R, Lettere, M., Mori, K. & Einhorn, J. (1994) Field response of maritime pine scale, *Matsucoccus feytaudi* Duc. (Homoptera: Margarodidae), to synthetic sex pheromone stereoisomers. *Journal of Chemical Ecology*, **20**, 2159–70.

James, D. G. & Mori, K. (1995) Spined citrus bugs, *Biprorulus bibax* Breddin (Hemiptera: Pentatomidae), do not discriminate between enantiomers in their aggregation pheromone. *Journal of Chemical Ecology*, **21**, 403–6.

Kafka, W. A., Ohloff, G., Schneider, D. & Vareschi, E. (1973) Olfactory discrimination of two enantiomers of 4-methylhexanoic acid by the migratory locust and the honeybee. *Journal of Comparative Physiology*, **87**, 277–84.

Karlson, P. & Lüscher, M. (1959) Pheromones: a new term for a class of biologically active substances. *Nature*, **183**, 55–6.

Kodama, H., Mochizuki, K., Kohno, M., Ohnishi, A. & Kuwahara, Y. (1987) Inhibition of male response of drugstore beetles to stegobinone by its isomer. *Journal of Chemical Ecology*, **13**, 1859–69.

Kraemer, M., Coppel, H. C., Matsumura, R, Kikukawa, T. & Mori, K. (1979) Field responses of the white pine sawfly, *Neodiprion pinetum*, to optical isomers of sawfly sex pheromones. *Environmental Entomology*, **8**, 519–20.

Kuwahara, Y., Matsumoto, K., Wada, Y. & Suzuki, T. (1991) Aggregation pheromone and kairomone activity of synthetic lardolure $(1R,3R,5R,7R)$-1,3,5,7-

tetramethyldecyl formate and its optical isomers to *Lardoglyphus konoi* and *Carpoglyphus lactis* (Acari: Astigmata). *Applied Entomology and Zoology*, **26**, 85–9.

Leal, W. S. (1996). Chemical communication in scarab beetles: reciprocal behavioral agonist–antagonist activities of chiral pheromones. *Proceedings of the National Academy of Sciences, USA*, **93**, 12112–5.

Leitereg, T. J., Guadagni, D. G., Harris, L, Mon, T. R. & Teranishi, R. (1971) Evidence for the difference between the odours of the optical isomers (+)- and carvone. *Nature*, **230**, 455–6.

Lensky, Y. & Blum, M. S. (1974) Chirality in insect chemoreceptors. *Life Sciences*, **14**, 2045–9.

Lindgren, B. S., Gries, G., Pierce Jr, H. D. & Mori, K. (1992) *Dendroctonus pseudotsugae* Hopkins (Coleoptera: Scolytidae): Production and response to enantiomers of 1-methylcyclohex-2-en-1-ol. *Journal of Chemical Ecology*, **18**, 1201–8.

McDowell, P. G., Hassanali, A. & Dransfield, R. (1985). Activity of the diastereoisomers of 13,23-dimethylpentatriacontane, the sex pheromone of *Glossina pallidipes* and comparison with the natural pheromone. *Physiological Entomology*, **10**, 183–90.

Millar, J. G., Pierce Jr, H. D., Pierce, A. M., Oehlschlager, A. C. & Borden, J. H. (1985). Aggregation pheromones of the grain beetle, *Cryptolestes turcicus* (Coleoptera: Cucujidae). *Journal of Chemical Ecology*, **11**, 1071–81.

Miller, J. R., Mori, K. & Roelofs, W. L. (1977) Gypsy moth field trapping and electroantennogram studies with pheromone enantiomers. *Journal of Insect Physiology*, 23, 1447–54.

Mori, K. (1973) Absolute configurations of (–)-14-methyl-*cis*-8-hexadecen-1-ol and methyl (–)-14-methyl-*cis*-8-hexadecenoate, the sex attractant of female Dermestid beetle, *Trogoderma inclusum* LeConte. *Tetrahedron Letters*, 3869–72.

Mori, K. (1974a) Absolute configurations of (–)-14-methylhexadec-8-*cis*-en-1-ol and methyl (–)-14-methylhexadec-8-*cis*-enoate, the sex pheromone of female Dermestid beetle. *Tetrahedron*, **30**, 3817–20.

Mori, K. (1974b) Synthesis of *exo*-brevicomin, the pheromone of western pine beetle, to obtain optically active forms of known absolute configuration. *Tetrahedron*, **30**, 4223–7.

Mori, K. (1975a) Synthesis of optically active forms of frontalin. *Tetrahedron*, **31**, 1381–4.

Mori, K. (1975b) Synthesis of optically active forms of sulcatol, the aggregation pheromone in the scolytid beetle, *Gnathotrichus sulcatus*. *Tetrahedron*, **31**, 3011–2.

Mori, K. (1992) In: *The Total Synthesis of Natural Products*, Vol. 9 (Ed. by J. ApSimon), John Wiley, New York.

Mori, K. (1997) Pheromones: synthesis and bioactivity. *Chemical Communications*, 1153–1158.

Mori, K. & Kuwahara, S. (1986) Synthesis of both the enantiomers of lardolure, the aggregation pheromone of the acarid mite, *Lardoglyphus konoi*. *Tetrahedron*, **42**, 5539–44.

Mori, K. & Seu, Y.-B. (1986) Synthesis of both the enantiomers of 7-ethyl-5-methyl-6,8-dioxabicyclo[3.2.1]oct-3-ene, the *Mus musculus* (house mouse) pheromone. *Tetrahedron*, **42**, 5901–4.

Mori, K., Takigawa, T. & Matsui, M. (1976) Stereoselective synthesis of optically active disparlure, the pheromone of the gypsy moth (*Porthetria dispar* L.). *Tetrahedron Letters*, 3953–6.

Mori, K., Takigawa, T. & Matsui, M. (1979) Stereoselective synthesis of the both enantiomers of disparlure, the pheromone of the gypsy moth. *Tetrahedron*, **35**, 833–7.

Mori, K., Masuda, S. & Suguro, T. (1981) Stereocontrolled synthesis of all of the possible stereoisomers of 3,11-dimethylnonacosan-2-one and 29-hydroxy-3,11-dimethylnonacosan-2-one, the female sex pheromone of the German cockroach. *Tetrahedron*, **37**, 1329–40.

Mori, K., Uematsu, T., Yanagi, K. & Minobe, M. (1985a) Synthesis of the optically active forms of 4,10-dihydroxy-1,7-dioxaspiro[5.5]undecane and their conversion to the enantiomers, of 1,7-dioxaspiro[5.5]undecane, the olive fly pheromone. *Tetrahedron*, **41**, 2751–8.

Mori, K., Watanabe, H., Yanagi, K. & Minobe, M. (1985b) Synthesis of the enantiomers of 1,7-dioxaspiro[5.5]undecane, 4-hydroxy-1,7-dioxaspiro[5.5]undecane, and 3-hydroxy-1,7-dioxaspiro[5.5]undecane, the components of the olive fruit fly pheromone. *Tetrahedron*, **41**, 3663–72.

Mori, K., Amaike, M. & Watanabe, H. (1993) New synthesis and revision of the absolute configuration of the hemiacetal pheromone of the spined citrus bug *Biprorulus bibax*. *Liebigs Annalen der Chemie*, 1287–94.

Mori, M., Mochizuki, K., Kohno, M., Chuman, T., Ohnishi, A., Watanabe, H. & Mori, K. (1986) Inhibitory action of (4*S*,6*S*,7*R*)-isomer to pheromonal activity of serricomin, (4*S*,6*S*,7*S*)-7-hydroxy-4,6-dimethyl-3-nonanone. *Journal of Chemical Ecology*, **12**, 83–9.

Nishida, R. & Fukami, H. (1983) Female sex pheromone of the German cockroach, *Blattella germanica*. *Memoirs of College of Agriculture, Kyoto University*, **No. 122**, 1–24.

Norin, T. (1996) Chiral chemodiversity and its role for biological activity. Some observations from studies on insect/insect and insect/plant relationships. *Pure and Applied Chemistry*, **68**, 2043–9.

Novotny, M. V., Xie, T.-M., Harvey, S., Wiesler, D., Jemiolo, B. & Carmack, M. (1995) Stereoselectivity in mammalian chemical communication: male mouse pheromones. *Experientia*, **51**, 738–43.

Olaifa, J. I., Matsumura, F., Kikukawa, T. & Coppel, H. C. (1988) Pheromone-dependent species recognition mechanisms between *Neodiprion pinetum* and *Diprion similis* on white pine. *Journal of Chemical Ecology*, **14**, 1131–44.

Payne, T. L., Richerson, J. V., Dickens, J. C., West, J. R., Mori, K., Brisford, C. W., Hedden, R. L., Vité, J. P. & Blum, M. S. (1982) Southern pine beetle: olfactory receptor and behavior discrimination of enantiomers of the attractant pheromone frontalin. *Journal of Chemical Ecology*, **8**, 873–81.

Riley, R. G., Silverstein, R. M. & Moser, J. C. (1974) Biological responses of *Atta texana* to its alarm pheromone and the enantiomer of the pheromone. *Science*, **183**, 760–2.

Rodin, J. O., Silverstein, R. M., Burkholder, W. E. & Gorman, J. E. (1969) Sex attractant of female Dermestid beetle *Tragoderma inclusem* Le Conte. *Science*, **165**, 904–6.

Russel, G. F. & Hills, J. I. (1971) Odor differences between enantiomeric isomers. *Science*, **172**, 1043–4.

Seybold, S. L, Ohtsuka, T., Wood, D. L. & Kubo, I. (1995) Enantiomeric composition of ipsdienol: a chemotaxonomic character for North American populations of *Ips* spp. in the *pini* subgeneric group (Coleoptera: Scolytidae). *Journal of Chemical Ecology*, **21**, 995–1016.

Suzuki, T., Kozaki, J., Sugawara, R. & Mori, K. (1984) Biological activities of the analogs of the aggregation pheromone of *Tribolium castaneum* (Coleopera: Tenebrionidae). *Applied Entomology and Zoology*, **19**, 15–20.

Szöcs, G., Tóth, M., Francke, W., Schmidt, R, Philipp, P., König, W. A., Mori, K., Hansson, B. S. & Löfstedt, C. (1993) Species discrimination in five species of winter-flying geometrids (Lepidoptera) based on chirality of semiochemicals and flight season. *Journal of Chemical Ecology*, **19**, 2721–35.

Tamaki, Y., Noguchi, H., Sugie, H., Kariya, A., Arai, S., Ohba, M., Terada, T., Suguro, T. & Mori, K. (1980) Four-component synthetic sex pheromone of the smaller tea tortrix moth: field evaluation of its potency as an attractant for male moth. *Japanese Journal of Applied Entomology and Zoology*, **24**, 221–8.

Tumlinson, J. H., Klein, M. G., Doolittle, R. E., Ladd, T. L. & Proveaux, A. T. (1977) Identification of the female Japanese beetle sex pheromone: inhibition of male response by an enantiomer. *Science*, **197**, 789–92.

Vité, J. P., Klimetzek, D., Loskant, G., Hedden, R. & Mori, K. (1976) Chirality of insect pheromones: response interaction by inactive antipodes. *Naturwissenschaften*, **63**, 582–3.

Vité, J. P., Ohloff, G. & Billings, R. F. (1978) Pheromonal chirality and integrity of aggregation response in southern species of the bark beetle *Ips* sp. *Nature*, **272**, 817–8.

Williams, H. J., Silverstein, R. M., Burkholder, W. E. & Khorramshani, A. (1981) Dominicalure 1 and 2: Components of aggregation pheromone from male lesser grain borer *Rhyzopertha dominica* (R) (Coleoptera: Bastrichidae). *Journal of Chemical Ecology*, **7**, 759–80.

Wood, D. L., Browne, L. E., Ewing, B., Lindahl, K., Bedard, W. D., Tilden, P. E., Mori, K., Pitman, G. B. & Hughes, P. R. (1976) Western pine beetle: specificity among enantiomers of male and female components of an attractant pheromone. *Science*, **192**, 896–8.

Wright, R. H. (1963) Odour of optical isomers. *Nature*, **198**, 782.

Wright, R. H. (1964) *The Science of Smell*, Allen & Unwin, London.

Zeeck, E., Hardege, J. D., Bartels-Hardege, H. & Wesselmann, G. (1988) Sex pheromone in a marine polychaete: determination of the chemical structure. *Journal of Experimental Zoology*, **246**, 285–92.

Zeeck, E., Hardege, J. D., Willig, A., Krebber, R. & König, W. A. (1992) Preparative separation of enantiomeric polychaete sex pheromones. *Naturwissenschaften*, **79**, 182–3.

Chapter 10
Chirality in the Natural World—Odours and Tastes

Wilfried A. König
Institut für Organische Chemie, Universität Hamburg, 20146 Hamburg, Germany

INTRODUCTION

In our age of 'chemophobia' it seems archaic when we read about the analytical methods of early generations of chemists—it was common to *analyse* chemicals by smelling and tasting a small sample. Odour and taste still play an important role in characterizing a specimen of an unknown substance and, despite enormous advances in instrumental analysis, the selectivity and sensitivity of the human nose as a detector of volatile odorous compounds is superior to that of most technical equipment. Experienced flavourists and perfumers are rare and usually highly acknowledged (and paid) individuals. Their ability to recognize, differentiate and classify many different odours makes them valuable to any perfume company.

The sensation, transmission and transformation of signals caused by volatile chemicals is widespread among living systems and the basis of chemical communication between insects and between insects and plants and other members of the ecological system, which are usually more sensitive than man to odours, by orders of magnitude.

The physiology of the human *sense of smell* is quite complex. It is normally combined with at least one other sense. e.g. taste or touch. It is a fact that without smelling, a complete sensation of taste is impossible. Both senses are combined in the term 'chemical sense', which is correct, because both, smelling and tasting, are molecular processes initiated by the association of odorants or taste compounds with a specific receptor. The stimulation of a receptor cell by an odorant initiates a signal which is transmitted to the brain by nerve fibres. These are combined to form 'nerve fascicles' of 20–30 nerve fibres each ending into the brain in the so called 'bulbus olfactorius', which consists of approx. 27000 to 30000 olfactory nodules (*glomerula olfactoria*). From there the signals are distributed to many subordinate brain regions where specific sensations are induced.

Ten to twenty-five million olfactory cells (the olfactory epithelium) located in the upper region of the nose are responsible for olfactory sensations when they make contact with odorous molecules which are inhaled. These molecules must be sufficiently volatile and should not exceed a certain molecular mass (according to Bernreuther et al. (1997) the highest molecular mass of an odorant is 294). It is estimated that humans have approx. 1000 different receptors which can differentiate between 10000 different odours (Bernreuther et al. 1997). For odour perception depolarization of the olfactory cell membrane is the basis of signal transmission. A critical membrane threshold potential must be exceeded by a sufficient number of odorous molecules to cause an odour sensation ('sensation threshold') (VDI-Richtlinie 1986).

Similar 'sensors' are responsible for the perception of taste. They are located in different areas of the surface of the tongue (*papillae*) and can differentiate between four basic qualities of taste—sweet, salty, sour, and bitter. Both odour and taste sensations co-operate (the terms 'aroma' and 'flavour' combine both) and excite a cascade of secondary reactions and associations such as digestive secretions and reflexes.

THE INFLUENCE OF CHIRALITY ON BIOLOGICAL ACTIVITY

Biological systems have a pronounced sense of chirality, and biological activity is always somehow combined with stereochemical properties. As discussed in detail elsewhere in this volume, the enantiomers of chiral pharmaceutical compounds often have distinctly different activity, and numerous examples are known of one enantiomer having beneficial pharmacological activity whereas the other is inactive, inhibitory, synergistic or, occasionally, toxic. To underline this, some of the more well known examples are given in Fig. 10.1.

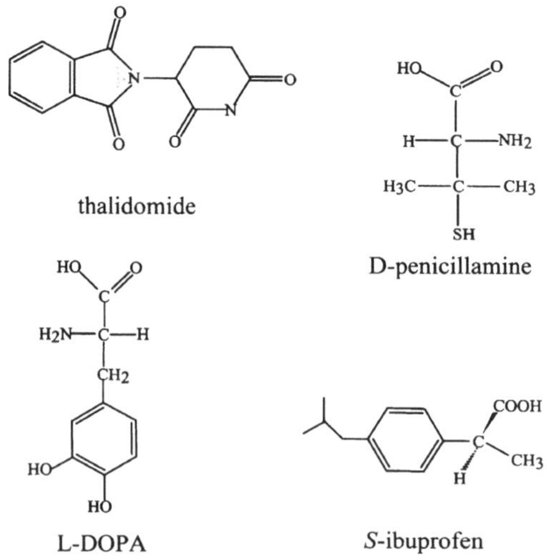

Fig. 10.1 Some chiral pharmaceutical compounds the enantiomers of which have different activity.

Enantioselective metabolism, pharmacokinetics and pharmacodynamics can also be observed for enantiomeric drugs when enzymes are involved (Wainer 1993). Only in recent years have pharmaceutical companies and legal authorities become aware of the importance of chirality. Now chiral drugs are either introduced as pure enantiomers or the racemic mixture is used only after thorough investigation of the activity and metabolic pathway of both enantiomers. Occasionally pure enantiomers of drugs are newly introduced to the market where the racemate has previously been used for a long time ('chiral switch')(Stinson 1993; Knabe 1995).

Stereochemistry also plays an important role in insect intraspecific chemical communication. In insects usually only *one* enantiomer is active, inducing a specific action, whereas the other is inactive or inhibitory. Occasionally both enantiomers are active in the same sense and sometimes even necessary for activity. The importance of stereochemistry in the pheromone field has been reviewed by K. Mori (1989) and is also revisited in this volume. The following examples demonstrate the complexity of structure–activity relationships (Fig. 10.2).

Chemical communication systems are also known for marine algae (Boland *et al.* 1984), fish (Brand *et al.* 1987), and even for mammals (Müller-Schwarze *et al.* 1978; Heth *et al.* 1992).

(a) Only one enantiomer is bioactive: *exo*-brevicomin (Western pine beetle)	
(b) All stereoisomers are bioactive: methyl ketone of the German cockroach	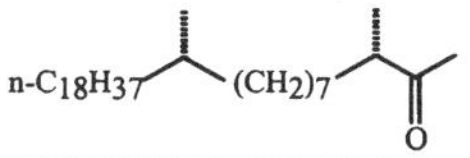 n-$C_{18}H_{37}$—$(CH_2)_7$
(c) Both enantiomers are required for bioactivity: sulcatol (ambrosia beetle)	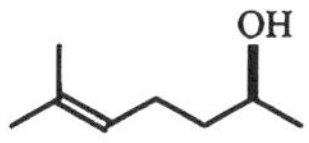
(d) One enantiomer is active on male insects, the other is active on females: 1,7-Dioxaspiro[5.5]undecane (Olive fruit fly)	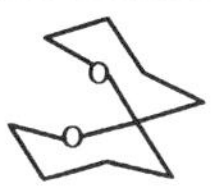
(e) One enantiomer is active, the other enantiomer (and diastereoisomers) are inhibitory: disparlure (gypsy moth)	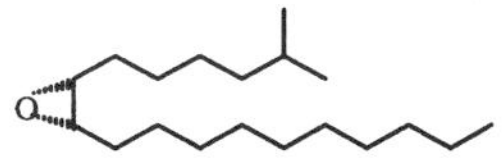
(f) In the same genus different species use different enantiomers: ipsdienol (bark beetles *Ips paraconfusus, Ips calligraphus*)	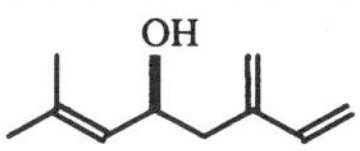
(g) Only one enantiomer is active, but its activity is enhanced by another stereoisomer: C_{12} aldehyde of the red flour beetle	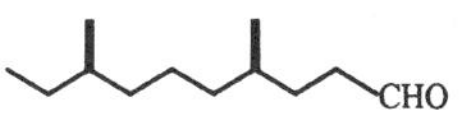
(h) Only the *meso*-isomer is bioactive: C_{37} hydrocarbon of the tsetse fly	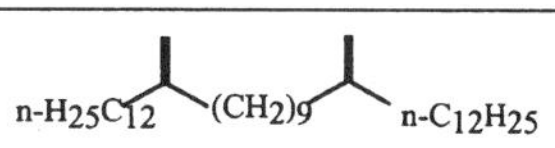 n-$H_{25}C_{12}$—$(CH_2)_9$—n-$C_{12}H_{25}$

Fig. 10.2 Relationship between the stereochemistry and biological activity of some pheromones.

THE RELATIONSHIP BETWEEN OLFACTION AND CHEMICAL STRUCTURE

Many theories about the correlation between sensory impression and the molecular shape of odorous compounds have been discussed in the literature (for a review see Bernreuther *et al.* 1997). Basically, it can be proved that molecules of similar size and three-dimensional shape cause similar olfactory sensations (Pauling 1946; Amoore 1970). This also explains why isomeric compounds (*cis* and *trans* isomers, chiral and non-chiral stereoisomers) result in extremely different sensory impressions (*stereochemical phenomenon of olfaction*) (Ohloff 1990).

As an example, the *E* isomer of 8-methyl-α-ionone **1** has a strong scent of violet whereas its *Z* isomer **2** has a pleasant woody, tobacco-like odour (Becker *et al.* 1974). As expected, the bicyclic analogue **3** has an odour which is extremely similar to that of **2** (Ohloff *et al.* 1986). A different orientation of only one substituent can result in drastic sensory changes. Thus, *Z*-4-*t*-butylcyclohexyl acetate (**4**) has a strong woody odour whereas the odour of the *E* isomer **5** is only very weak (Ohloff 1990) (Fig. 10.3).

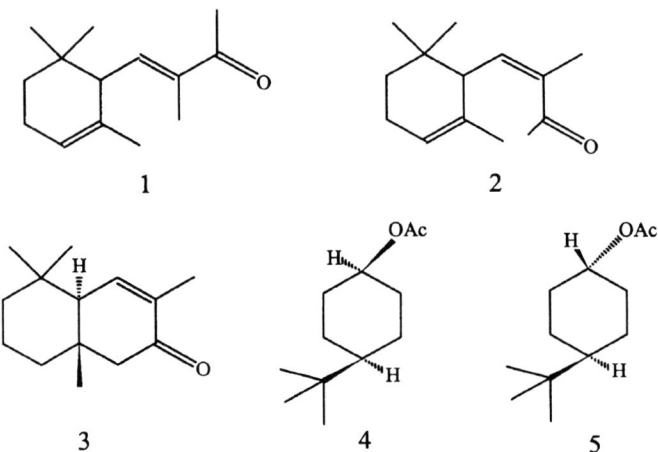

Fig. 10.3 Some examples of isomeric odorous compounds causing different odour impressions.

THE RELATIONSHIP BETWEEN ODOUR AND TASTE AND CHIRALITY

Pasteur was, in fact, the first to speculate that chiral compounds can be distinguished by the olfactory nerves (Pasteur *et al.* 1858). Although questioned for a long time (Wright *et al.* 1964), many odorous volatile chiral compounds may indeed cause different olfactory sensations depending on their configuration. This is because the receptors are proteins, themselves chiral compounds, which interact with enantiomers to form diastereomeric molecular complexes. These differ in their physical properties and energy content and, therefore, can result in different effects, i.e. olfactory sensations.

Emil Fischer's general 'key–lock' principle (Fischer et al. 1894) of the action of enzymes with their substrates implies that sensory perception is induced only by a molecule matching the active site of the chiral receptor, and this is only possible for one of two enantiomers. For the same reason enantiomeric compounds may taste different. The D enantiomers of many amino acids, including tryptophan (**6**), phenylalanine (**7**), tyrosine (**8**), leucine (**9**), and asparagine (**10**) have a sweet taste, whereas their L enantiomers taste bitter or neutral (Solms et al. 1965; Wiesner et al. 1977).

Asparagine was the first example of the proof of Pasteur's postulate (Pasteur et al. 1858), as early as 1886 (Piutti 1886). The unique property of the dipeptide derivative L-asp-L-phe-OMe (Aspartame, **11**) which tastes 200 times sweeter than sugar, makes it an important and profitable commercial product. Its enantiomer, however, tastes bitter (Mazur et al. 1969). The relationship between chemical structure and sweetness was reviewed by van der Hejden (van der Hejden 1993). Only the sodium salt of L-glutamic acid (**12**) is an important flavour enhancer and food additive—D-glutamate has no influence on taste (Solms 1967).

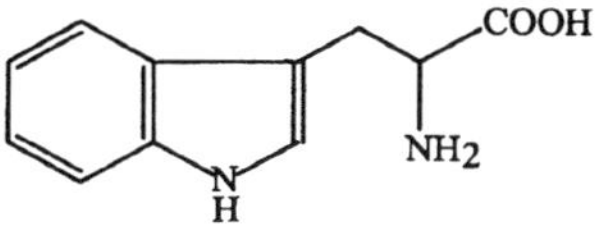

6

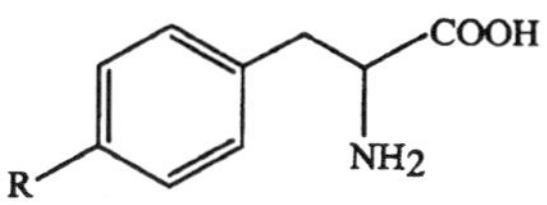

7, R = H; 8, R = OH

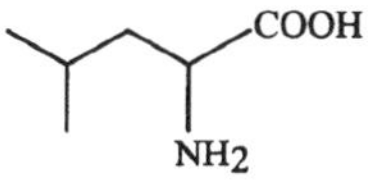

9

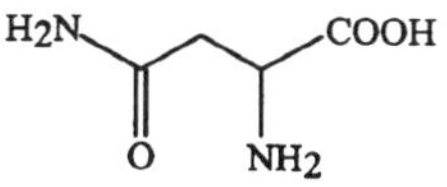

10

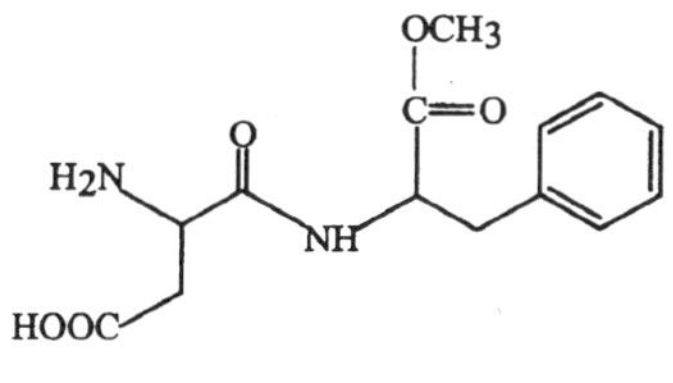

11

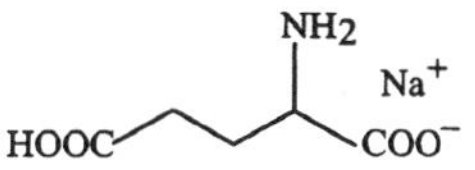

12

ENANTIOMERIC ODOUR DIFFERENCES AND THEIR QUANTITATIVE MEASUREMENT

Although examples of humans being able to differentiate between odours according to configuration are numerous, one should be aware that the detection limit (sensation threshold) for odours of the human *sense of smell* can be very different between individuals and can be subject to considerable fluctuations. It ranges from total deficiency of olfaction (*anosmia*) to hypersensitivity (*hyperosmia*).

Olfactometry, the controlled presentation of odorous substances and the recording of the resulting sensations in man, is an important method of measurement which is usually applied by a panel of trained individuals (with a 'normal' sense of smell) particularly suited to this task. When this technique is subject to many influences which are difficult to control, the quantitative determination of a certain odorous constituent of a complex mixture by a non-specific detector, as part of a physico-chemical analytical system (for example the flame-ionization detector of a gas chromatograph) should be more reliable in terms of accuracy and reproducibility.

Both methods have been combined in the gas chromatograph with 'sniffing port', which enables 'online' olfactory assessment of individual components of a complex mixture separated by a gas chromatographic column (Fuller et al. 1964). Humid and cooled air is added to the gas chromatographic effluent to improve the sniffing conditions for the human nose. Good ventilation is necessary to avoid sensory saturation. Sometimes this sniffing analysis is combined with a computer for real-time recording of odour impressions (CHARM analysis; Acree et al. 1984).

ANALYSIS OF THE CONFIGURATION AND ENANTIOMERIC COMPOSITION OF VOLATILE CHIRAL COMPOUNDS

During the past 30 years chiral flavour and fragrance compounds have been accessible to qualitative and quantitative enantiospecific analysis because of the introduction and optimization of enantioselective gas chromatography. E. Gil-Av and his associates were the first to resolve the enantiomers of amino acid derivatives by diastereoselective interaction with a chiral stationary phase derived from amino acids or peptides (Gil-Av et al. 1967). The interaction of enantiomers with the chiral stationary phase was predominantly based on hydrogen-bonding forces.

Compounds without hydrogen bonding sites were not suitable substrates and could not be separated. Although the thermal stability and the selectivity of these diamide chiral stationary phases could be considerably enhanced by covalent bonding of the chiral selector to a polysiloxane matrix (Frank et al. 1977; König et al. 1981) the range of applications was rather limited. Although improved by introducing appropriate derivatization procedures—reaction of chiral molecules with isocyanates (Benecke & König 1982), hydroxylamine (König et al. 1982), or phosgene (König et al. 1984), these chiral stationary phases enabled the separation of only a few classes of compounds relevant to flavour and fragrances (i.e. alcohols and ketones). A summary of the application of chiral diamide stationary phases is given elsewhere (König et al. 1987).

Extension of the range of application in the analysis of fragrance and flavour compounds was achieved when Schurig introduced *complexation gas chromatography* (Schurig 1977). Separation of the enantiomers of compounds lacking functional groups for derivative formation was a result of enantioselective interaction with chiral transition metal complexes added to a non-polar stationary phase in capillary columns. This technique enabled the separation of alcohols, ketones, spiroacetals, epoxides, and olefins (Schurig 1983).

A breakthrough in enantioselective gas chromatography was accomplished by introducing modified cyclodextrins as a new universal type of chiral selector in capillary gas chromatography (Koscielski et al. 1983; König & Lutz 1988; Schurig & Nowotny 1988). Enantiomers can be resolved by selective inclusion into the cavity of the chiral macrocyclic systems constructed from 6, 7, or 8 α-1→4-glycosidically connected glucose molecules (α-,β- and γ-cyclodextrins, respectively, Fig. 10.4).

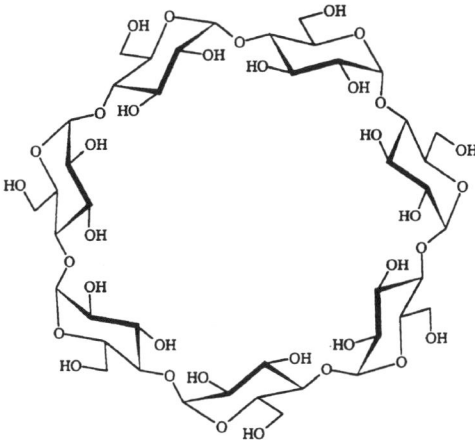

Fig. 10.4 The structure of β-cyclodextrin.

Cyclodextrins are obtained by enzymatic degradation of amylose and can be selectively alkylated, acylated, or silylated at the different hydroxyl groups at positions 2, 3, and 6 of the glucose unit to form a multitude of derivatives. The most important properties of these derivatives, in addition to their enantioselectivity, are their low melting points (sometimes below room temperature), their solubility in organic solvents, and their extraordinary thermal stability.

Enantioselectivity is not a property of specific functional groups, but is equally observed for saturated hydrocarbons and multi-functional molecules, as long as they are sufficiently volatile (König 1992). Each cyclodextrin derivative is selective for certain groups of compounds, but the selection of the most suitable cyclodextrin derivative for a certain separation problem is hard to predict, and still more or less empirical. This situation has been improved somewhat by publication of lists of separation factors for enantiomers in chromatographic journals (König 1993a, b, c; Maas *et al.* 1994) and by the establishment of a computer data base 'Chirbase/GC' where many published separations are cited (Koppenhöfer *et al.* 1993).

STEREOCHEMICAL ANALYSIS OF ODOROUS COMPOUNDS

MONOTERPENE HYDROCARBONS

For the analysis of odorous chiral compounds, for example the constituents of essential oils, other plant volatile compounds, or pheromones, selectively substituted cyclodextrins have been shown to have superior properties. Thus, the enantiomers of all monoterpene hydrocarbons usually present in essential oils can be resolved by a two-capillary-column system of heptakis(6-*O*-methyl-2,3-di-*O*-pentyl)-β-cyclodextrin and octakis(6-*O*-methyl-2,3-di-*O*-pentyl)-γ-cyclodextrin (König *et al.* 1992b). Samples are split by a Y-type device close to the injection port on to both columns, and two chromatograms are recorded simultaneously. The reliability in peak identification and quantitation is enhanced because the orders of elution of the enantiomers of most monoterpenes are reversed on changing from one chiral stationary phase to another (Fig. 10.5).

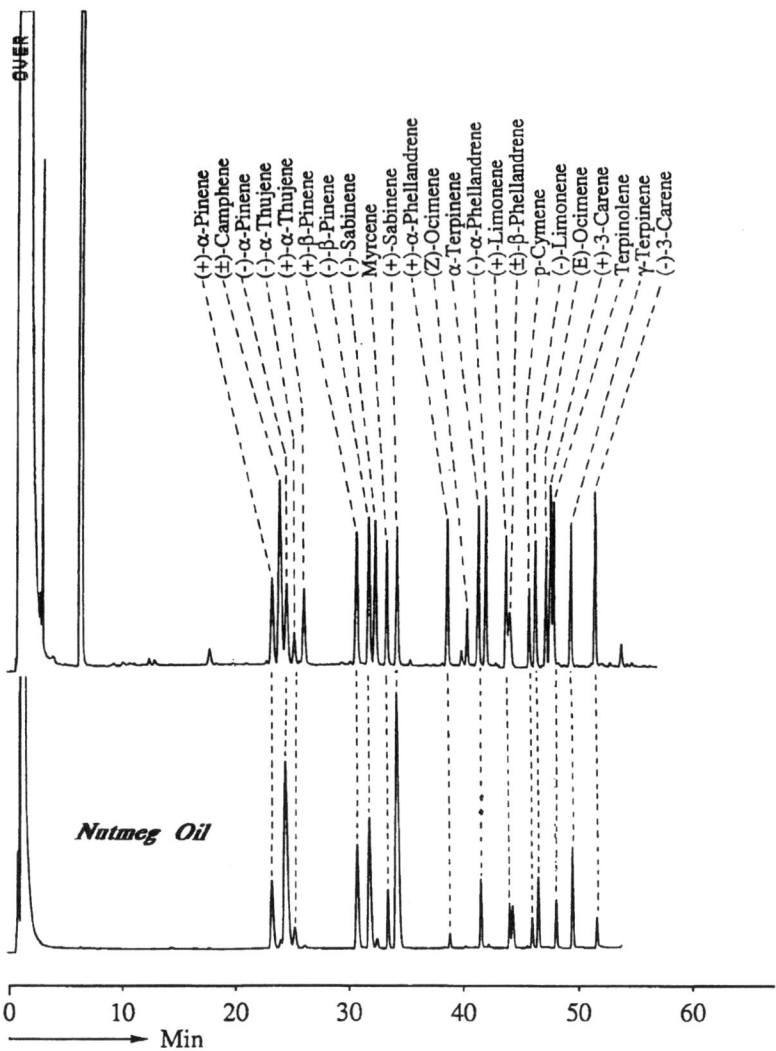

Fig. 10.5 Investigation of the enantiomeric composition of the monoterpene hydrocarbon fraction of Indonesian nutmeg oil. Upper: mixture of the enantiomers of standard monoterpene hydrocarbons; lower: fraction from nutmeg oil. The separations were performed on a 25-m fused-silica capillary column coated with octakis(6-*O*-methyl-2,3-di-*O*-pentyl)-γ-cyclodextrin; the column temperature was maintained at 30°C, isothermal, for 20 min then programmed at 1° min^{-1} to 160°C.

Because of their preponderance, monoterpene hydrocarbons make a significant contribution to the specific flavour of many essential oils. In contrast with the hypothesis, that nature prefers to synthesize pure enantiomers in enzymatically controlled processes, enantiomeric mixtures of monoterpene hydrocarbons are usually observed. The 'fingerprint' of the enantiomeric monoterpene hydrocarbon composition can be used to characterize many essential oils, and to prove their naturalness or adulteration (Dugo et al. 1992). As shown in Fig. 10.6, the enantiomeric composition of monoterpene hydrocarbons can differ significantly, even in related species.

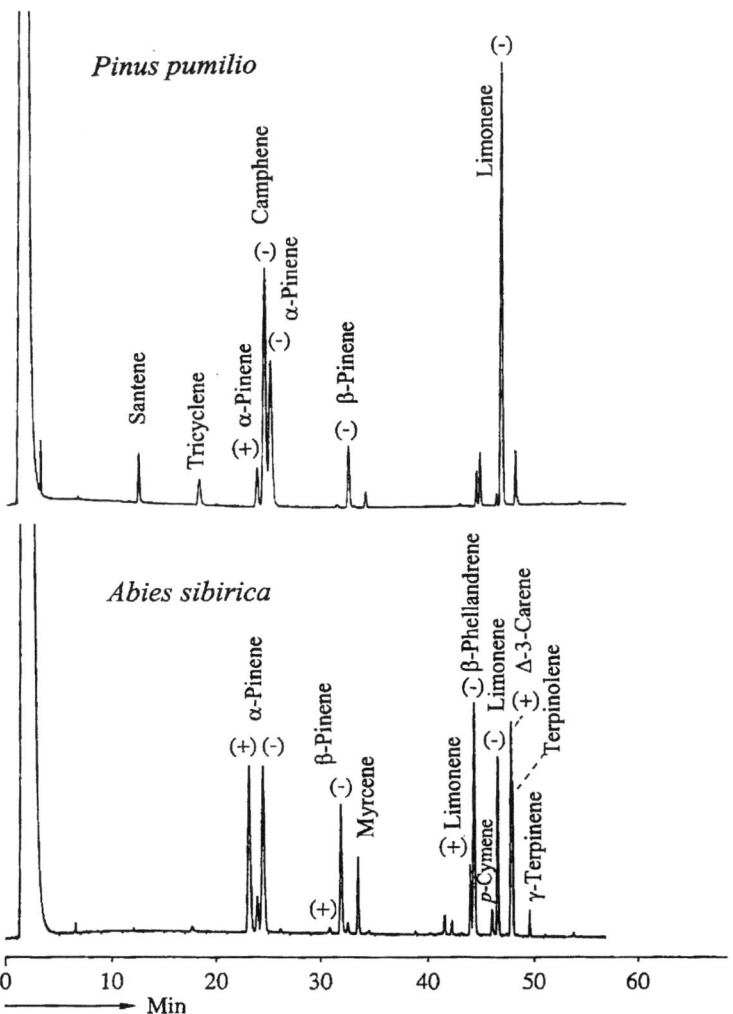

Fig. 10.6 Monoterpene hydrocarbon fraction from the needle oils of dwarf pine (Pinus pumilionis) and fir (Abies sibirica). The separation was performed on a 25-m fused-silica capillary column coated with octakis(6-O-methyl-2,3-di-O-pentyl)-γ-cyclodextrin (50%, w/w, in OV 1701). The column temperature was maintained at 30°C, isothermal, for 20 min then programmed at 1° min^{-1} to 160°C.

The odour of the enantiomers of limonene (**13**), although weak, is clearly different. The (+) enantiomer, the major enantiomer in most *citrus* oils (> 97%), has a more pleasant odour than the (–) enantiomer, which is present as the major enantiomer in most conifer needle oils (up to 90% in *Abies alba*) and has a turpentine-like odour. A slightly peppery flavour is reported for (+)-α-phellandrene (**14**) whereas that of the (–) enantiomer is weed- or dill-like (Blank & Grosch 1991). The enantiomers of α-pinene do not seem to smell differently but insects can probably distinguish between them. It has been shown that the defensive secretion of termites *Nasutitermes princeps* contain (+)-α-pinene (**15**) in high enantiomeric excess (Everaerts *et al.* 1990).

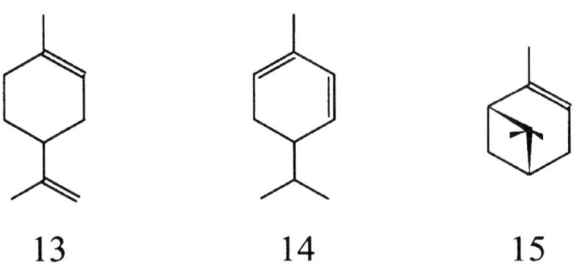

13   14   15

TERPENE ALCOHOLS

These oxygenated terpene derivatives usually have a high olfactory potential. Thus, linalool (**16**) is a widespread constituent of plant volatile compounds; it has a pleasant floral scent. The enantiomers differ in the intensity and tonality of their sensory impression (Ohloff & Klein 1962). The *R* enantiomer, which is a major constituent of the essential oils of lavender species, clary sage oil (*Salvia sclarea*) and bergamot oil (*Citrus bergamia*), has a more intense and lavender-type, woody scent whereas the *S* enantiomer, which is present in coriander (*Coriandrum sativum*, major constituent) and orange oil (minor constituent), is characterized by a sweeter, lavender-type smell. Both enantiomers of linalool in more or less equal proportions are present in Brazilian rosewood oil (major constituent), in rose oil, and in geranium oil (minor constituent). An example is given in Fig. 10.7.

The enantiomeric composition of linalool and linalyl acetate is an important criterion in the proof of adulteration by addition of synthetic (racemic) linalool and its acetate. Because natural, cold-pressed bergamot oil contains the *R* enantiomers only of both compounds (approx. 20% linalool) (König *et al.* 1997) adulteration is readily detected. Lavender and lavandin oils (approx. 30–50% linalool), prepared by hydrodistillation of the plants, contains approx. 5% *S*-linalool, formed by racemization during the hydrodistillation process.

Essential oils (e.g. rosewood oil, rosemary oil, neroli oil, etc.) very often contain another monoterpene alcohol, α-terpineol (**17**), which is readily formed by rearrangement of linalool under acidic conditions. The enantiomers are reported to make different sensory impressions, the *R* enantiomer having a strong, floral, sweet scent whereas the *S* enantiomer is associated with a tarry odour (Mosandl 1992a).

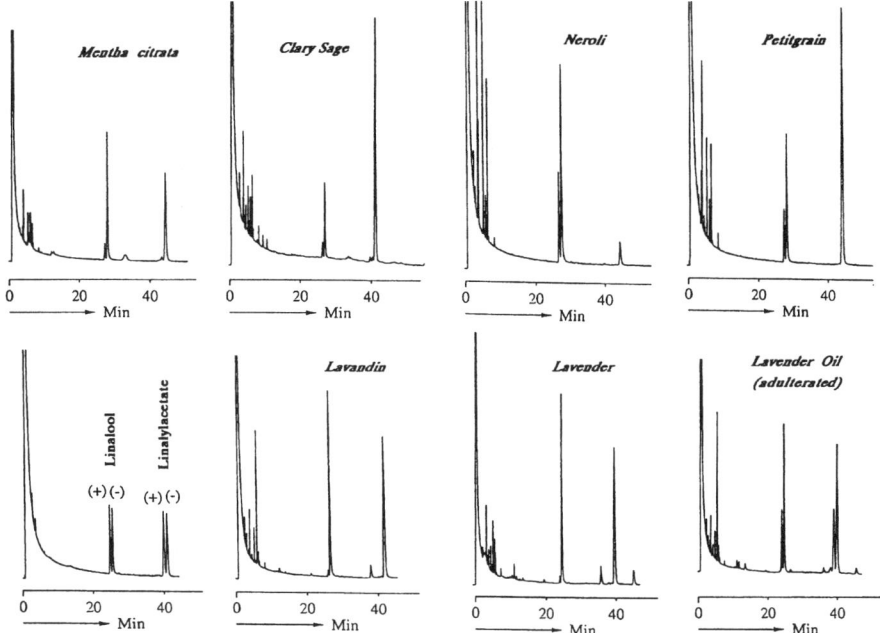

Fig. 10.7 Enantiomeric composition of linalool (**16**) and linalyl acetate in different essential oils together with a racemic standard of both compounds. The separation was performed at 70°C on a 25-m fused-silica capillary column coated with octakis(3-*O*-butyryl-2,6-di-*O*-pentyl)-γ-cyclodextrin (Lipodex E®).

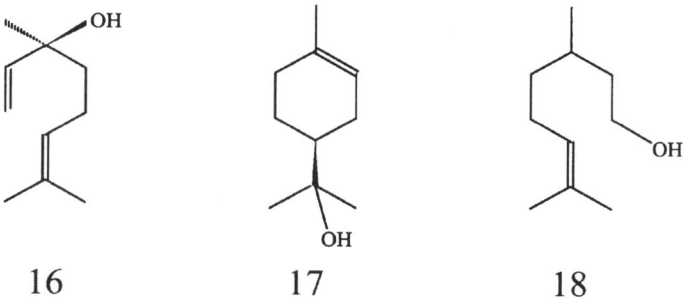

16   17   18

Another important alcohol with a sweet, rose-like scent, citronellol (**18**), is present in large amounts in rose oil (approx. 40%; >95% *S* enantiomer; Kreis & Mosandl 1992a), geranium oil (approx. 20–40%; both enantiomers; the *S* enantiomer is usually in excess; Ravid *et al.* 1992), and citronella oil (both enantiomers; *R* enantiomer always in excess; Kreis & Mosandl 1992b). The *S* isomer has a greater odour intensity and a finer rose odour than the *R* isomer (Rienäcker & Ohloff 1961).

Analysis of the enantiomeric composition of citronellol can most easily be achieved after trifluoroacetylation and separation of the enantiomers on a column coated with octakis(3-*O*-butyryl-2,6-di-*O*-pentyl)-γ-cyclodextrin (Lipodex E®; Macherey–Nagel, Düren, Germany; König *et al.* 1989b, 1990, 1997; Kreis & Mosandl

1992a). Resolution of underivatized citronellol is possible with heptakis(6-*O*-TBDMS-2,3-di-*O*-acetyl)-β-cyclodextrin (Kreis & Mosandl 1992a).

Borneol (**19**) is also a common constituent of essential oils. Although (−)-borneol is present in high enantiomeric excess (85–98%) in rosemary oil (Kreis *et al.* 1991) (approx. 3–6% borneol), different proportions were found in different commercial oils and also in oils prepared from fresh plants in the laboratory (König *et al.* 1997). In thyme oil (*Thymus vulgaris*, <5% borneol) the situation is similar (Kreis *et al.* 1991). A source of (−)-borneol is the essential oils of pine needles. As early as in 1874 (Plowman 1874) the literature reported differences between the odours of essential oils containing (+)-borneol (Borneo camphor oil) and (−)-borneol (Ngai camphor oil), the (−) isomer having a more pleasant smell (Bauer *et al.* 1990).

Enantioselective gas chromatographic analysis of borneol can be achieved on octakis(6-*O*-methyl-2,3-di-*O*-pentyl)-γ-cyclodextrin (König *et al.* 1997) or permethylated β-cyclodextrin (Kreis *et al.* 1991).

An interesting sensory evaluation was reported for the eight stereoisomers of menthol (**20**), the sensory impressions of which seem to be distinctly different (Emberger & Hopp 1985). (−)-Menthol is a major constituent of most *Mentha* essential oils, e.g. peppermint oil. The sensory intensity is greater for the (−) enantiomer (fresh, cool, minty) than for the (+) enantiomer. All eight stereoisomers of menthol can be resolved on a column coated with octakis(6-*O*-methyl-2,3-di-*O*-pentyl)-γ-cyclodextrin (König *et al.* 1997) (Fig. 10.8).

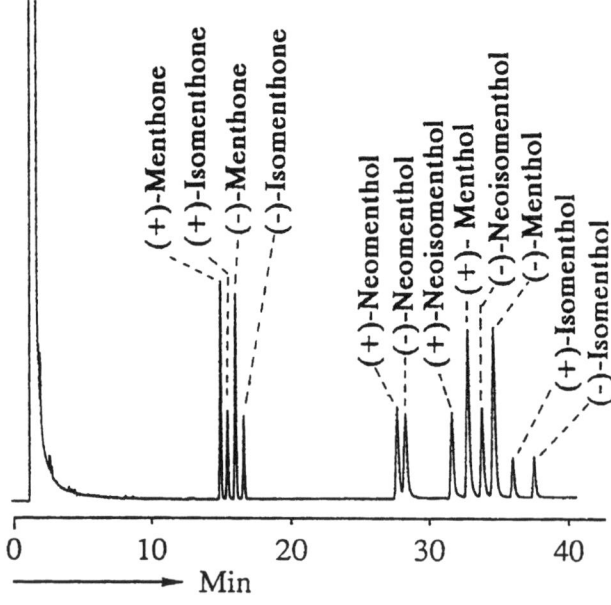

Fig. 10.8. Separation of the enantiomers of menthone and isomenthone, and of all the stereoisomers of menthol, on a 25 m fused-silica capillary column coated with octakis(6-*O*-methyl-2,3-di-*O*-pentyl)-γ-cyclodextrin (50%, w/w, in OV 1701). The column temperature was maintained at 80°C, isothermal, for 27 min then programmed at 1° min^{-1} to 140°C.

The enantiomers of 1-octen-3-ol (**21**), an important aliphatic flavour constituent, have very different olfactory properties. The *R* enantiomer has a much stronger mushroom flavour than the *S* isomer. Mushrooms usually contain a high excess of the *R* enantiomer (König *et al.* 1990b; Mosandl 1992a).

CARBONYL COMPOUNDS

The enantiomers of carvone (**22**) are present in many essential oils and have very different sensory properties. Whereas the (+) enantiomer is a major constituent of caraway with the specific flavour of this essential oil, the (–) isomer is the main component of spearmint oil with a specific spearmint odour. The difference between the flavours of this pair of enantiomers was observed as one of the first examples and was thoroughly investigated by comparison of enantiomerically pure synthetic compounds to prove that the organoleptic differences are not a result of impurities with low odour thresholds in the natural essential oils (Friedman & Miller 1971; Leitereg *et al.* 1971). In addition to this distinct difference in odour, the (–) enantiomer is a considerably stronger odorant than the (+) enantiomer. Enantioselective analysis of carvone enantiomers on hexakis(2,3,6-tri-*O*-pentyl)-α-cyclodextrin has been demonstrated (König 1990).

Citronellal (**23**), α-ionone (**24**), α-damascone (**25**), and *cis-* and *trans-*α- (**26**) and γ-irones (**27**) are very important chiral flavour compounds with carbonyl functions. Sensory differences between citronellal enantiomers have been reported but are questionable because the possibility of the presence of impurities could not be unambiguously eliminated (von Braun & Kaiser 1923). Citronellal is quite labile and is known to rearrange to several other fragrant products, including isopulegol, iso(iso)pulegol, menthone, isomenthone and menthol, when exposed to UV light in acidic media (Ziegler *et al.* 1991).

In essential oils citronellal can occur as the pure *R* enantiomer in balm oil (*Melissa offcinalis*), as the pure *S* enantiomer in the oil of *Nepeta citriodora* and as mixtures of the enantiomers in citronella oil, lemongrass oil and *Eucalyptus citriodora* oil. The enantiomeric purity of citronellal in balm oil is an important indicator in the authenticity control of this very expensive product (Kreis & Mosandl 1994; Schultze *et al.* 1995). Several cyclodextrin derivatives can be used to resolve the enantiomers, for example, heptakis(3-*O*-acetyl-2,6-di-*O*-methyl)-β-cyclodextrin (Lipodex B; Macherey–Nagel, Düren, Germany), heptakis(6-*O*-TBDMS-2,3-di-*O*-acetyl)-β-cyclodextrin (Kreis & Mosandl 1994), heptakis(2,6-di-*O*-methyl-3-*O*-pentyl)-β-cyclodextrin, and the corresponding γ-cyclodextrin derivatives.

α-Ionone (**24**) is an important fragrance compound with enantiomers of very different threshold values. It is a constituent of black tea (Bricout *et al.* 1967), raspberries (Winter & Sundt 1962), and many flower absolutes. Its natural configuration was first investigated by Eugster *et al.* (Eugster *et al.* 1969) and it was found to be the *R* enantiomer, which has a far lower olfactory threshold than the *S* isomer. The enantiomers can be separated by enantioselective gas chromatography with cyclodextrin derivatives (König *et al.* 1989a; Werkhoff *et al.* 1991).

α-Damascone (**25**), another constituent of black tea, is a compound with a high sensory impact. Although odour differences are clearly detectable, contradictory reports about the olfactory threshold have been reported in the literature (Werkhoff *et al.* 1991). α-Damascone is an important perfumery compound.

The irones (**26**, **27**) are natural constituents of *Iris* rhizomes and their essential oil (Orris oil) and concrete ('Orris butter') have a very pleasant violet-like scent. The irones are, therefore, highly esteemed as precious raw materials in the perfumery and fragrance industry. The relative composition of the ten different enantiomers and isomers of these C_{14} ketones is highly dependent on the *Iris* species and provenance, and has a great influence on sensory quality.

Again enantioselective gas chromatography has enabled precise analysis and quality control of *Iris* oils (Marner *et al.* 1990). Capillary gas chromatography using mainly octakis(6-*O*-methyl-2,3-di-*O*-pentyl)-γ-cyclodextrin (König *et al.* 1990a) combined with the GC-sniffing technique has enabled olfactory evaluation of all isomers (Galfre *et al.* 1993). (+)-*cis*-α-Irone and (+)-*cis*-γ-irone, produced mainly by the species *Iris pallida*, were found to have the most interesting sensory properties and the lowest threshold of perception, whereas their enantiomers are odourless (Galfre *et al.* 1993) (Fig. 10.9).

It is recognized that the human nose can differentiate clearly between the enantiomers of the irones. Generally, only the isomers with the 2*R* configuration have the typical violet- or iris-like scent. Although perfumers have been well aware of the different olfactory properties, and have preferred Iris oils prepared from *Iris pallida*, only after recent thorough investigation by enantioselective gas chromatography have the odour differences been explained. With this new analytical tool natural Iris oils can now be easily distinguished from non-natural (reconstituted) products.

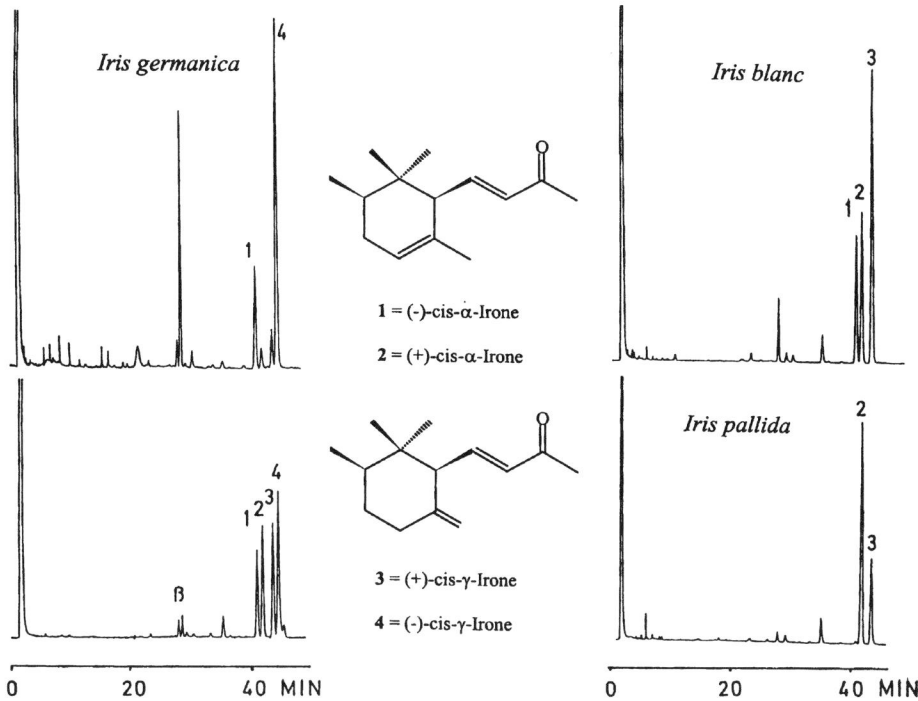

Fig. 10.9 Resolution of the enantiomers of *cis*-α-irone and *cis*-γ-irone, and investigation of irone isomers in different *Iris* species. The separation was performed at 85°C on a 25 m Pyrex glass capillary column coated with octakis(6-*O*-methyl-2,3-di-*O*-pentyl)-γ-cyclodextrin.

Another natural product highly appreciated by perfumers and flavourists is the essential oil and concrete of jasmine (*Jasminum grandiflorum*). The main constituents, in terms of the importance of their organoleptic properties, are (–)-methyl jasmonate (**28**), (+)-*epi*-methyl jasmonate (**29**) and δ-jasmolactone (**30**). Remarkable differences between the olfactory threshold values were observed for the stereoisomers of methyl jasmonate. According to Acree *et al.* (1985) the threshold value for *epi*-methyl jasmonate (1*R*,2*S* configuration) is over 400 times lower than that for methyl jasmonate (1*R*,2*R* configuration). The isomers are in 1:9 thermodynamic equilibrium, because of an acidic proton at the chiral center next to the carbonyl group.

The enantiomers of methyl jasmonate can be resolved on cyclodextrin phases, e.g. octakis(2,6-di-*O*-methyl-3-*O*-pentyl)-γ-cyclodextrin (König *et al.* 1992a). Two-dimensional gas chromatography is necessary for quantitation of the stereoisomers in jasmine concrete, because of the complexity of the mixture of volatile constituents (König *et al.* 1992a).

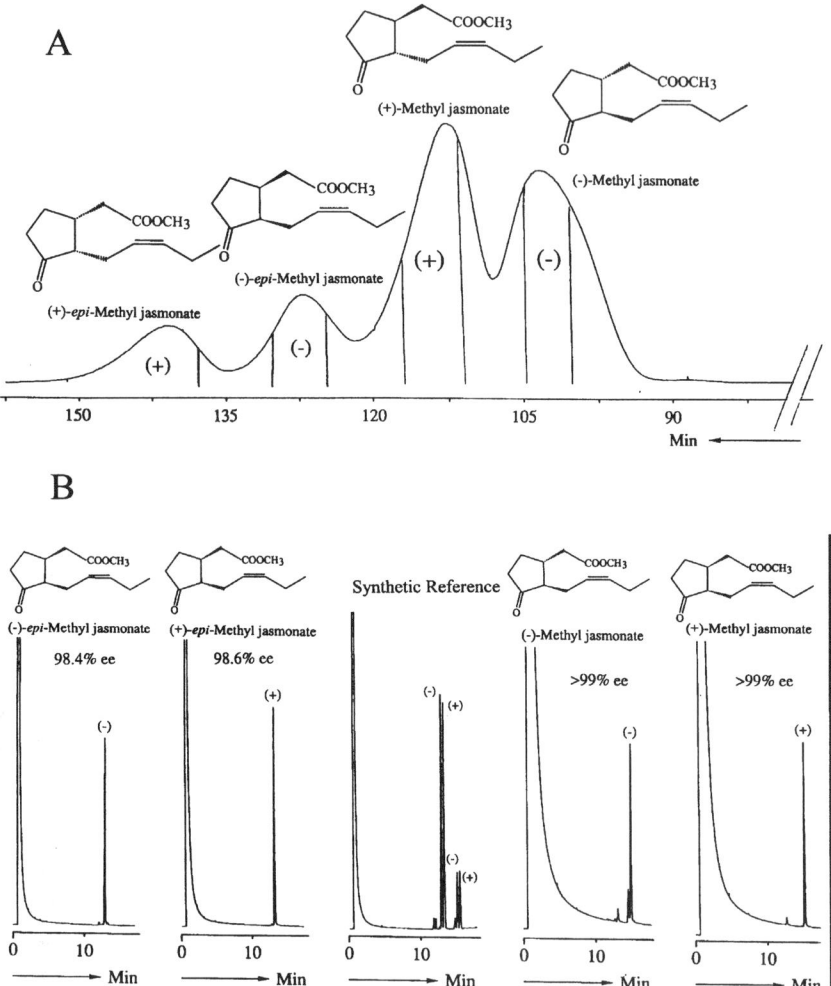

Fig. 10.10 A. Preparative separation, at 115°C, of the stereoisomers of methyl jasmonate (3 mg sample injected) on a 2 m × 5.3 mm gas chromatographic column packed with 2.5% heptakis(2,6-di-*O*-methyl-3-*O*-pentyl)-γ-cyclodextrin (50%, *w/w* in OV 1701). B. Analysis of the fractions by enantioselective gas chromatography at 145°C on a 25-m capillary coated with the same phase as for A.

Micro-preparative enantioselective separations of jasmonates can be performed by use of packed gas chromatographic columns to isolate single stereoisomers (Hardt &

König 1994), which can be investigated by NMR (to confirm their structure), by polarimetry (to determine the order of elution), and for evaluation of their organoleptic properties. This is demonstrated in Fig. 10.10 for the stereoisomers of methyl jasmonate. In addition to its interesting sensory properties jasmonic acid and its methyl esters are active as plant hormones and have a variety of functions in plant development (inhibition of plant growth, acceleration of ageing) (Ueda & Kato 1982; Sembdner & Parthier 1993). The induction of biosynthesis and emission of flavour compounds in plants by jasmonic acid and its methyl ester have recently been discovered as another interesting biological function (Boland *et al.* 1995).

Only the (–) enantiomer of δ-jasmolactone (**30**) is present in jasmine essential oil and in *Gardenia* blossoms, whereas the (+) enantiomer predominates in *Tuberose,* peach and mango fruit flavour (Werkhoff *et al.* 1993). The enantiomers differ in their sensory properties. The organoleptic intensity of the (–) enantiomer is stronger than that of the (+) isomer (Werkhoff *et al.* 1993).

SESQUITERPENES AND RELATED COMPOUNDS

Many examples are known of organoleptic and, more generally, biological differences between the enantiomers of sesquiterpene derivatives. Thus, of the four stereoisomers of α-bisabolol, (–)-α-bisabolol (**31**), the natural constituent of *chamomile* oil, is superior to the others in terms of its antiphlogistic properties. The gas chromatographic separation of all stereoisomers of α-bisabolol has been demonstrated (König *et al.* 1992a). Preparative resolution of the stereoisomers is possible by liquid chromatography with tri-*O*-benzoylcellulose as chiral selector (Günther *et al.* 1993).

Many sesquiterpene alcohols are important for their sensory properties; occasionally they are the flavour-determining 'impact' compounds in an essential oil. Although synthesis of the different stereoisomers of sesquiterpenes can be tedious and complicated, some examples of the enantiomers having different sensory properties have been reported. One such example is patchouli alcohol (**32**) (Näf *et al.* 1981). Only the natural, (–), enantiomer has the typical strong 'patchouli' scent. The (+) enantiomer has a weaker, less characteristic odour (Näf *et al.* 1981).

Nerolidol (**33**), an intermediate in an early stage of sesquiterpene biosynthesis, is very frequently observed in plants. Of the four stereoisomers, which can be easily resolved by enantioselective gas chromatography (König *et al.* 1992a), both enantiomers of the *E* isomer are common plant constituents. Whereas higher plants predominantly contain the (+) enantiomer, the (–) enantiomer was found as a major constituent of mushrooms (König, unpublished work). Mosandl has isolated all the stereoisomers, evaluated their organoleptic properties, and assigned different sensory characteristics to all four (Mosandl 1992b).

Different odour profiles and large differences between the threshold values were also reported for both enantiomers of nootkatone (**34**) and α-vetivone (**35**), on the basis of olfactometric experiments by Haring *et al.* (1972). The (+) enantiomer of **34**, which is a natural flavour constituent of grapefruit and has a bitter taste. Its odour (described as fresh, green, sour) is 2200 times more intense than that of the (–) enantiomer, which has a woody, spicy scent. (+)-α-Vetivone (**35**), a constituent of vetiver oil, has a woody, floral, balsamic scent and a threshold value (in aqueous solution) of 0.3 ppm, whereas the (–) enantiomer has a higher threshold value of approx. 10 ppm and a weaker and different scent. Equivalent differences were observed for the taste of the enantiomers of **34** and **35** in soft drinks.

Macrocyclic ketones such as (−)-muscone (**36**) have been found to be the sensorily active principles of musk, which occurs naturally as a constituent of the musk glands of *Moschus moschiferus*, a deer native to Asian countries. Separation of the enantiomers of muscone has been achieved on heptakis(2,6-di-*O*-methyl-3-*O*-pentyl)-β-cyclodextrin (König *et al.* 1992a; Fig. 10.11); the order of elution was assigned by polarimetric measurements after resolution of the enantiomers by preparative gas chromatography. Because natural musk is not easily accessible artificial musk flavour compounds have been synthesized. When the organoleptic properties of one of these, galaxolide (**37**), a compound occurring as four stereoisomers, were investigated it was found, that only the 4*S*,7*R* and 4*S*,7*S* isomers have a powerful musky scent (Frater *et al.* 1995).

A very informative study of the relationship between stereochemistry and odour was performed with di- and tricyclic compounds having the odour of ambergris, a metabolic product of the blue sperm whale (*Physeter macrocephalus*) with a fine amber-type fragrance highly appreciated by perfumers for its unique olfactory and fixative properties.

One of the important ingredients of natural ambergris is ambrox (ambroxan) (**38**). It was found that an amber-type fragrance of bicyclic or tricyclic decalin systems is always associated with the *trans* configuration. Systems with a *cis*-linked decalin structure are odourless. In addition a 1,2,4-triaxial arrangement of the substituents is necessary for a good substrate–receptor interaction ('three-point-interaction') and a strong amber-like scent ('triaxial rule of odour perception'; Ohloff *et al.* 1973). This is so for (−)- ambrox (**38**), which has an intense odour; (+)-isoambrox (**39**) is described as having a threshold value 100 times higher (Ohloff *et al.* 1985) or even as odourless (Brun 1989).

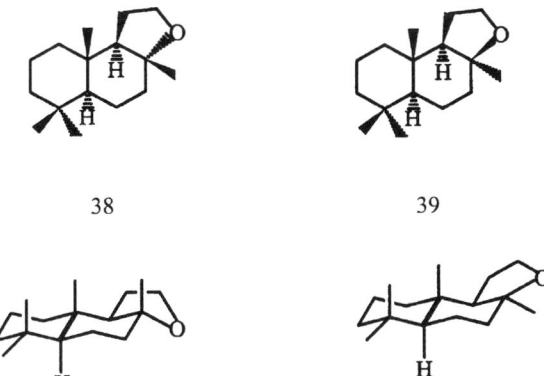

Similar correlations were found for several analogous synthetic compounds (Ohloff 1990; Salido 1995). Although only small odour differences were observed for (+)- and (−)-**38**, the (+) enantiomer with its higher threshold value and accentuated woody odour lacks the strong and warm animal smell and has therefore been called the 'poor man's ambrox' by perfumers (Salido 1995).

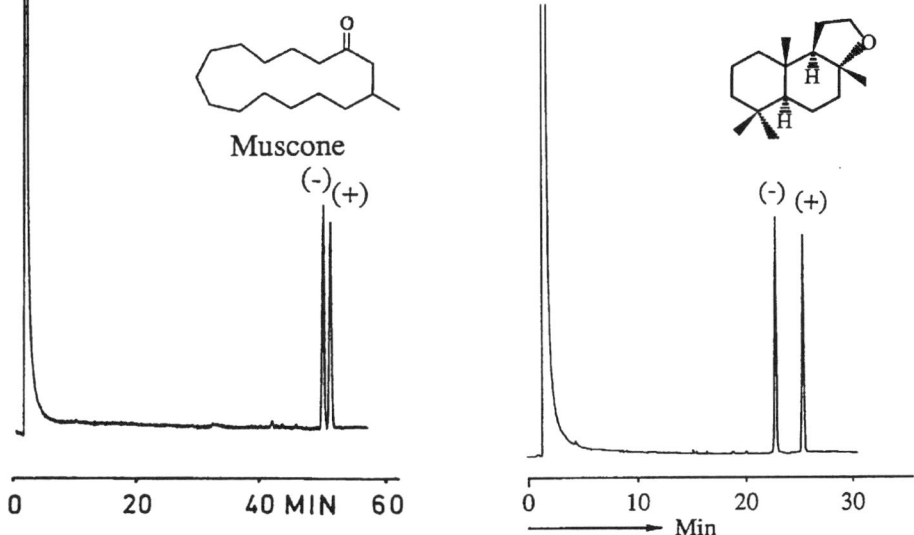

Fig. 10.11. Left. Separation at 135°C of the enantiomers of synthetic muscone (**36**) on a 25 m capillary column coated with heptakis(2,6-di-*O*-methyl-3-*O*-pentyl)-β-cyclodextrin (50%, w/w, in OV 1701). Right. Separation at 145°C of the enantiomers of ambrox (**38**) on a 25 m capillary column coated with heptakis(6-*O*-TBDMS-2,3-di-*O*-methyl)-β-cyclodextrin (50%, w/w, in OV 1701).

SULPHUR COMPOUNDS

Naturally occurring sulphur analogues of terpenoid compounds usually have a unique sensory impression at very low concentration. (−)-8-Mercapto-*p*-menthan-3-one (**40**) is a flavour-impact compound of *Buchu* leaf oil with a very low odour threshold value Sundt *et al.* 1971) which at very high dilution causes a fruity blackcurrant flavour (cassis). It is stereochemically related to (−)-isomenthone (**41**) and can be prepared by addition of hydrogen sulphide to (−)-pulegone (**42**) (Kaiser *et al.* 1975).

The sensory impressions of the four stereoisomers were investigated by Mosandl *et al.*, who also resolved the enantiomers by enantioselective gas chromatography (Köpke & Mosandl 1992). The natural $1S,4R$ and $1S,4S$ isomers are responsible for the typical cassis flavour. The $1R$ isomers, however, seem to have a lower odour threshold. The corresponding S-acetyl derivatives were also identified as trace components in *Buchu* leaf oil (Kaiser *et al.* 1975) and their sensory properties were investigated by Köpke *et al.* (1992). Again the four stereoisomers can be distinguished by their different odour impressions.

The related R-1-p-menthen-8-thiol (**43**) was identified as an important flavour constituent of grapefruit juice and has also a very low olfactory threshold (Demole *et al.* 1982). The S enantiomer is almost odourless (Lehmann *et al.* 1995).

'ELECTRONIC NOSE', SENSOR TECHNOLOGY

For several reasons the unambiguous evaluation of odorous compounds is still a great problem. Firstly, it is associated with the odour perception- capability of the individual panellist; secondly, despite of numerous attempts (Jaubert *et al.* 1995) there is no universal language for the description of approximately 40000 different odours; and, thirdly, absolute exclusion of low-odour-threshold impurities from test samples is difficult. It is, therefore, an ultimate challenge to develop an independent physical system for measuring odour intensity and character.

This seems possible as a result of the invention of mass-sensitive chemical sensors based on selective adsorption of a volatile compound on to the surface of oscillating quartz crystals (quartz micro-balance) (Sauerbrey 1959; King 1964), dielectric surfaces (Endres & Drost 1991), or calorimetric transducers (Lerchner *et al.* 1996). The selective interaction of a surface can be enhanced by coating it with a gas chromatographic stationary phase or a more specific layer of 'host' molecules which can form 'host–guest' complexes with volatile compounds in the gas phase of an environment. This complex formation depends on the steric properties and functionalities of the interacting molecules, which simulates the interaction of a substrate with a specific receptor in odour or taste perception. The extent of complex formation can be monitored by the increase in mass (quartz microbalance) or by enthalpy changes (calorimetric detection).

With an array of a multi-sensor system combined with a computer it is possible to classify mixtures of volatile compounds with a specific vapour profile characteristic of a particular product (essential oils, spices, coffee brands, etc.) (Horner & Vonach 1995; Hodgins 1995). Cyclodextrin derivatives were recently coated on to sensor surfaces as selective host compounds for enantiomers (Bodenhöfer et al. 1997; May et al. 1997) and statistically significant differences in response were measured; this is equivalent to the 'chiral discrimination' obtained in enantioselective gas chromatography.

CONCLUDING REMARKS

The interrelation of chirality and organoleptic perception is a complex and fascinating research field combining physiological, chemical and biophysical phenomena. A selection of examples of enantiomeric molecules inducing very different organoleptic impressions has been described. Although many theories of odour–molecular structure relationships have been established there is still much empiricism in the systematic search for new fragrances. The exciting history of this search has recently been documented by G. Ohloff (1992a, b).

Enantioselective gas chromatography has contributed tremendously to precise stereochemical analysis, to determination of enantiomeric composition, and to correlation of configuration with olfactory properties. Recent developments in sensor technology promise a new dimension in unambiguous measurement of the flavours which are an essential part of the world of our senses.

REFERENCES

Acree, T. E., Barnard, J. & Cunningham, D. C. (1984) In: *Analysis of Volatile Compounds*, (Ed. by P. Schreier), p. 251. de Gruyter, Berlin.
Acree, T. E., Nishida, R. & Fukami, H. (1985) *J. Agric. Food Chem.*, **33**, 425.
Amoore, J. E. (1970) *Molecular Basis of Odor*. C. C. Thomas, Springfield, Ill, USA.
Bauer, K., Garbe, D. & Surburg, H. (1990) *Common Fragrance and Flavor Materials*. VCH, Weinheim.
Becker, J., Ehrenfreund, J., Jeger, O., Ohloff, G. & Wolf, H. R. (1974) *Helv. Chim. Acta*, **57**, 2679.
Benecke, I. & König, W. A. (1982) *Angew. Chem. Int. Ed. Engl.*, **21**, 709.
Bernreuther, A., Epperlein, U. & Koppenhöfer, B. (1997). In: *Techniques for Analyzing Food Aroma*, (Ed. by R. Marsili). Marcel Dekker, New York.
Blank, I. & Grosch, W., (1991) *J. Food Sci.*, **56**, 63.
Bodenhöfer, K., Hierlemann, A., Seemann, J., Gauglitz, G., Christian, B., Koppenhöfer, B. & Göpel, W. (1997) *Nature*, **387**, 577.
Boland, W., Jaenicke, L., Müller, D. G. & Peters, A. (1984) *Eur. J. Biochem.*, **144**, 169.
Boland, W., Hopke, J., Donath, J., Nüske, J. & Bublitz, F. (1995) *Angew. Chem. Int. Ed. Engl.*, **34**, 1600.
Brand, J. G., Bryant, B. P., Cagan, R. H. & Kalinoski, D. L. (1987) *Ann. N. Y. Acad. Sci.*, **510**, 193.

Bricout, J., Viani, R., Müggler-Chavan, F., Marion, Reymond, D. & Egli, R. H. (1967) *Helv. Chim. Acta*, **50**, 1517.

Brun, K. (1989). In: *Henkel-Berichte, Int. Ed.* **25**, 118.

Demole, E., Enggist, P. & Ohloff, G. (1982) *Helv. Chim. Acta*, 65, 1785.

Dugo, G., Lamonica, G., Cotroneo, A., d'Alcontres, S. I., Verzera, A., Donato, M. G. & Dugo, P. (1992) *Perfumer & Flavorist*, 17, September/October, p. 57.

Emberger, R. & Hopp, R. (1985). In: *Topics in Flavour Research*, p. 201. Eichhorn, Marzling-Hargenham.

Endres, H. E. & Drost, S. (1991) *Sensors and Actuators*, B, 4, 95.

Eugster, C. H., Buchecker, R., Tscharner C., Uhde, G. & Ohloff, G. (1969) *Helv. Chim. Acta*, **52**, 1729.

Everaerts, C., Bonnard, O., Pasteels, J. M., Roisin, Y. & König, W. A. (1990) *Experientia*, **46**, 227.

Fischer, E. (1894) *Ber. Dtsch. Chem. Ges.*, 27, 2985.

Frank, H., Nicholson, G. J. & Bayer, E. (1977) *J. Chromatogr. Sci.*, 15, 174.

Frater, G., Bajgrowicz, J. A. & Petrzilka, M. (1995) *Proc. 13th Intern. Congress Flav. Fragr. Ess. Oils*, (Ed. by K. H. C. Baser). AREP, Istanbul.

Friedman, L. & Miller, J. G. (1971) *Science*, **172**, 1044.

Fuller, G. H., Steltenkamp, R. & Tisserant, G. A. (1964) *Annals. New York Acad. Sci.*, **116**, 711.

Galfre, A., Martin, P. & Petrzilka, M. (1993) *J. Essent. Oil Res.*, **5**, 265.

Gil-Av, E., Feibush, B. & Charles-Sigler, R. (1967). In: *Gas Chromatography 1966*, (Ed. by A. B. Littlewood), p. 227. Institute of Petroleum, London, 1967.

Günther, K., Carle, R., Fleischhauer, I. & Merget, S. (1993) *Fresenius J. Anal. Chem.*, **345**, 787.

Hardt, I. & König, W. A. (1994) *J. Chromatogr. A*, **666**, 611.

Haring, H. G., Rijkens, F., Boelens, H. & van der Gen, A. (1972) *J. Agr. Food Chem.*, **20**, 1018.

Heth, G., Nevo, E., Ikan, R., Weinstein, V., Ravid, U. & Duncan, H. (1992) *Experientia*, **48**, 897.

Hodgins, D. (1995) *Perfumer & Flavorist*, 20, Nov./Dec., 1.

Horner, G. & Vonach, B. (1995) *Labor-Praxis*, April 1995, 28.

Jaubert, J.-N., Tapiero, C. & Dore, J.-C. (1995) *Perfumer & Flavorist*, 20, May/June, 1.

Kaiser, R., Lamparsky, D. & Schudel, P. (1975) *J. Agric. Food Chem.*, 23, 943.

King, W. H. (1964) *Anal. Chem.*, 36, 1735.

Knabe, J. (1995) *Pharmazie in uns. Zeit*, 324.

König, W. A. (1987) *The Practice of Enantiomer Separation by Capillary Gas Chromatography*. Hüthig, Heidelberg.

König, W. A., (1990) *Kontakte* (Darmstadt), (2), 3.

König, W. A. (1992) *Gas Chromatographic Enantiomer Separation with Modified Cyclodextrins*, Hüthig Buch-Verlag, Heidelberg.

König, W. A. (1993a) *J. High Resol. Chromatogr.*, **16**, 312.

König, W. A. (1993b) *J. High Resol. Chromatogr.*, **16**, 338.

König, W. A. (1993c) *J. High Resol. Chromatogr.*, **16**, 569.

König, W. A. & Lutz, S. (1988). In: *Chirality and Biological Activity*, (Ed. by B. Holmstedt, H. Frank, & B. Testa), p. 55. A. R. Liss, New York, 1990.

König, W. A., Benecke, I. & Sievers, S. (1981) *J. Chromatogr.*, **217**, 71.

König, W. A., Benecke, I. & Ernst, K. (1982) *J. Chromatogr.*, **253**, 267.

König, W. A., Steinbach, E. & Ernst, K. (1984) *Angew. Chem. Int. Ed. Engl.*, **23**, 527.

König, W. A., Evers, P., Krebber, R., Schulz, S., Fehr, Ch. & Ohloff, G. (1989a) *Tetrahedron*, **45**, 7003; *Tetrahedron* **48**, 1741.

König, W. A., Krebber, R. & Mischnick, P. (1989b) *J. High Resol. Chromatogr.*, **12**, 732.

König, W. A., Icheln, D., Runge, T., Pforr, I. & Krebs, A. (1990a) *J. High Resol. Chromatogr.*, **13**, 702.

König, W. A., Krebber, R., Evers, P. & Bruhn, G. (1990b) *J. High Resol. Chromatogr.*, **13**, 328.

König, W. A., Gehrcke, B., Icheln, D., Evers, P., Dönnecke, J. & Wang, W. (1992a) *J. High Resol. Chromatogr.*, **15**, 367.

König, W. A., Krüger, A., Icheln, D. & Runge, T. (1992b) *J. High Resol. Chromatogr.*, **15**, 184.

König, W. A., Fricke, C., Saritas, Y., Momeni, B. & Hohenfeld, G. (1997) *J. High Resol. Chromatogr.* **20**, 55.

Köpke, T. & Mosandl, A. (1992) *Z. Lebensm. Unters. Forsch.*, **194**, 372.

Köpke, T., Schmarr, H.-G. & Mosandl, A. (1992) *Flavour & Fragr. J.*, **7**, 205.

Koppenhöfer, B., Nothdurft, A., Pierrot-Sanders, J., Piras, P., Popescu, C. & Roussel, C. (1993) *Chirality*, **5**, 213.

Koscielski, T., Sybilska, D. & Jurczak, J. (1983) *J. Chromatogr.*, **280**, 1.

Kreis, P. & Mosandl, A. (1992a) *Flavour & Fragr. J.*, **7**, 199.

Kreis, P. & Mosandl, A. (1992b) *Flavour & Fragr. J.*, **9**, 257.

Kreis, P. & Mosandl, A. (1994) *Flavour & Fragr. J.*, **9**, 249.

Kreis, P., Juchelka, D., Motz, C. & Mosandl, A. (1991) *Dtsch. Apothek. Ztg.*, **131**, 1984.

Lehmann, D., Dietrich, A., Hener, U. & Mosandl, A. (1995) *Phytochem. Anal.*

Leitereg, T. J., Guadagni, D. G., Harris, J., Mon, T. R. & Teranishi, R. (1971) *Nature*, **230**, 455.

Lerchner, J., Seidel, J., Wolf, G. & Weber, E. (1996) *Sensors and Actuators* B, **32**, 71.

Maas, B., Dietrich A. & Mosandl, A. (1994) *J. High Resol. Chromatogr.*, **17**, 109.

Marner, F. J., Runge, T. & König, W. A. (1990) *Helv. Chim. Acta*, **73**, 2165.

May, I. P., Byfield, M. P., Lindström, M. & Wünsche, L. F. (1997) *Chirality*, **9**, 225.

Mazur, R. H., Schlatter, J. M. & Goldkamp, A. H. (1969) *J. Am. Chem. Soc.*, **33**, 2684.

Mori, K. (1989) *Tetrahedron*, **45**, 3233.

Mosandl, A. (1992a) *Kontakte* (Darmstadt), (3), 38.

Mosandl, A. (1992b) *J. Chromatogr.*, **624**, 267

Müller-Schwarze, D., David, U., Claesson, A., Singer, A. G., Silverstein, R. M., Müller-Schwarze, C., Volkman, N. J., Zemanek, K. F. & Butler, R. G. (1978) *J. Chem. Ecol.*, **4**, 247.

Näf, F., Decorzant, R., Giersch, W. & Ohloff, G. (1981) *Helv. Chim. Acta*, **64**, 1387.

Ohloff, G. (1986) *Experientia*, **42**, 271.

Ohloff, G. (1990), *Riechstoffe und Geruchssinn,* Springer, Berlin.

Ohloff, G. (1992a) *Helv. Chim. Acta*, **75**, 1341.

Ohloff, G. (1992b) *Helv. Chim. Acta*, **75**, 2041.

Ohloff, G., Klein, E. (1962) *Tetrahedron*, **18**, 37.

Ohloff, G., Näf, F., Decorzant, R., Thommen, W. & Sundt, E. (1973) *Helv. Chim. Acta*, 56, 1414.

Ohloff, G., Giersch, W., Pickenhagen, W., Furrer, A. & Frei, B. (1985) *Helv. Chim. Acta*, 68, 2022.

Pasteur, L. (1858) *C. R. Acad. Sci. Paris*, **46**, 615.

Pauling, L. (1946) *Chem. Eng. News*, **24**, 1375.

Piutti, A. (1886) *Ber. Dtsch. Chem. Ges.*, **19**, 1691.

Plowman, S. (1874) *Arch. Pharm.*, **205**, 237.

Ravid, U., Putievsky, E., Katzir, I., Ikan, R. & Weinstein, V. (1992) *Flavour & Fragr. J.*, 7, 235.

Rienäcker, R. & Ohloff, G. (1961) *Angew. Chem.*, **73**, 240.

Salido, S. (1995). In: *Proc. 13th Intern. Conf. Flav. Fragr. Ess. Oils*, (Ed. by K. H. C. Baser). AREP, Istanbul.

Sauerbrey, G. (1959) *Z. Physik.*, **155**, 206.

Schultze, W., König, W. A., Hilkert, A. & Richter, R. (1995) *Dtsch. Apothek. Ztg.*, **135**(7), 17.

Schurig, V. (1977) *Angew. Chem. Int. Ed. Engl.*, **16**, 110.

Schurig, V. (1983). In: *Asymmetric Synthesis*, Vol. 1, (Ed. by J. D. Morrison). Academic Press, New York.

Schurig, V. & Nowotny, H.-P. (1988) *J. Chromatogr.*, **441**, 155.

Sembdner, G. & Parthier, B. (1993) *Annu. Rev. Plant Physiol. Plant Mol. Biol.*, **44**, 569.

Solms, J. (1967) *Chimia*, **21**, 169.

Solms, J., Vuataz, L. & Egli, R. H. (1965) *Experientia*, **21**, 692.

Stinson, S. C. (1993) *Chem. Eng. News*, Sept. 27, p. 38.

Sundt, E., Willhalm, B., Chappaz, R. & Ohloff, G. (1971) *Helv. Chim. Acta*, **54**, 1801.

Ueda, J. & Kato, J. (1982) *Agric. Biol Chem.*, **46**, 1975.

van der Hejden, A. (1993). In: *Flavor Science*, (Ed. by T. E. Acree & R. Teranishi), ACS Prof. Ref. Book. Am. Chem. Soc., Washington, DC.

VDI-Richtlinie 3881, *VDI-Handbuch Reinhaltung der Luft*, Vol. 1, 1986. VDI, Düsseldorf.

von Braun, J. & Kaiser, W. (1923) *Chem. Ber.*, **56**, 2268.

Wainer, I. W. (Ed.) (1993) *Drug Stereochemistry*. Marcel Dekker, New York.

Werkhoff, P., Bretschneider, W., Güntert, M., Hopp, R. & Surburg, H. (1991) *Z. Lebensm. Unters. Forsch.*, **192**, 111.

Werkhoff, P., Brennecke, S., Bretschneider, W., Güntert, M., Hopp, R. & Surburg, H. (1993) *Z. Lebensm. Unters. Forsch.*, **196**, 307.

Wiesner, H., Jugel, H. & Belitz, H.-D. (1977) *Z. Lebensm. Unters.-Forsch.*, **164**, 277.

Winter, M. & Sundt, E. (1962) *Helv. Chim. Acta*, **45**, 2195.

Wright, R. H., (1964) *The Science of Smell*. Allan & Unwin, London.

Ziegler, M., Brandauer, H., Ziegler, E. & Ziegler, G. (1991) *J. Ess. Oil Res.*, **3**, 209.

Chapter 11
Chirality in the Natural World: Life Through the Looking Glass

Christopher J. Welch
Merck & Co., Inc., Rahway, NJ 007065, USA

INTRODUCTION

Although most organisms have a symmetric exterior form, there are some well known exceptions. For example, snails, flounders, fiddler crabs, narwhals and twining vines are all chiral, because they cannot be superimposed on their mirror images. The existence of this 'handedness' in the exterior form of some organisms was recognized by Aristotle, and examination of even earlier artworks and artefacts from a variety of cultures suggests that humankind has in some way been aware of chirality and enantiomerism for millennia (Washburn & Crowe 1988; Hargittai & Hargittai 1994). Although Kant (1783) provided a concise description of chirality nearly a century before Pasteur's discoveries:

> What can more resemble my hand or my ear, and be in all points more like, than its image in the looking glass? And yet I cannot put such a hand as I see in the glass in the place of its original ...

It was, however, Pasteur who was first to understand the central importance of chirality in the biological world (Dubos 1950):

> ... all living species are primordially, in their structure, in their external forms, functions of cosmic asymmetry.

Pasteur came to the conclusion that enantioenrichment:

> ... forms perhaps the only sharply defined boundary which can be drawn at the present time between the chemistry of dead and that of living matter.

Pasteur's 'hunch' about the importance of asymmetry led him to many truly remarkable discoveries concerning molecular chirality and stereochemistry. These discoveries laid the foundation for many of the chemical developments of the 20th century (Jacques 1993), and Pasteur's enthusiasm about the importance of chirality is today shared by thousands of chemists.

In this chapter asymmetry in the exterior forms of various living organisms is reviewed. Although this field has received only a small fraction of the attention devoted to the study of molecular chirality, recent advances in the field of developmental biology are bringing questions concerning animal asymmetry into the arena of chemistry, as some of the genes which govern the development of asymmetric body forms are now being discovered. It is hoped that this review will make the existence of this interesting topic better known to the stereochemistry community.

BACKGROUND

Many examples of asymmetry in the animal world were compiled by A. C. Neville (1976) more than twenty years ago in the book *Animal Asymmetry*. The subject has also been addressed by Martin Gardner (1990) in his wonderful book on the subject of asymmetry, *The New Ambidextrous Universe*. There is some discussion of the subject in the classic work by D'Arcy Thompson (1992), *On Growth and Form*, and, finally, several interesting examples are included in an internet web site (http://www.biology.ualberta.ca/palmer.hp/asym/asymmetry.htm) maintained by Professor Richard Palmer at the University of Alberta.

Information from a number of sources, including personal observations and communications with other scientists has been compiled. Nearly every stereochemist has some favourite teaching anecdotes concerning enantiomerism in the natural world. The book by Hargittai & Hargittai (1994) on symmetry contains many good photographs, and Henri Brunner from the University of Regensberg has collected many interesting examples of asymmetric organisms and objects. These observations and anecdotes are, however, rarely published. In this review a survey of the literature will be augmented with some personal observations.

SPECIAL PROBLEMS IN THE STUDY OF ASYMMETRIC ORGANISMS

At this point it might be important to emphasize some special problems in the study of asymmetric organisms. For example, if a diner in a restaurant orders escargot, he is likely to receive a plate containing several individuals of the snail *Helix pomatia* prepared with butter and garlic. Closer examination of the snail shells will probably show all to be dextral, i.e. having right-handed helicity (Fig. 11.1).

From this limited data set it might be incorrectly concluded that all shells of *Helix pomatia* are dextral, yet the sinistral form of the species does indeed exist, albeit rarely. Professor Brunner (personal communication) informs me that workers at an escargot processing facility in Dijon use a specially designed chiral knife to extract the snails from their shells, and occasionally encounter a sinistral shell into which this knife does not 'fit'. The occurrence of such individuals is approximately 1 in 20000 individuals, which represents an enantiomeric excess of 99.99% for the dextral form.

Fig. 11.1 The two enantiomeric forms of the escargot snail, *Helix pomatia*. The sinistral form, shown left, is quite rare, whereas the dextral form, shown right, is commonplace. The occurrence of the sinistral form has been calculated by Brunner (personal communication) to be approximately 1 in 20000. (Photograph by the author.)

It is clear that such data will not be available for many chiral organisms where enantiopurity might be high, but still less than 100%. Thus it can be seen that measuring the enantiopurity of organisms can be very difficult, and labour intensive, especially when the enantioenrichment is quite high. Needless to say, the requirement to observe large numbers of organisms is sometimes impossible to meet for rare species or for those which are difficult to observe.

Another problem confronting those trying to collect data on asymmetric organisms relates to populations. Brunner's data imply that the *Helix pomatia* at this one particular processing plant in Dijon have an enantiomeric excess of 99.99% for the dextral form, but care must be taken in extrapolating this finding to the same species in other locations within its geographic range or in places where it is introduced or cultivated.

Considerable variation in enantioenrichment has been shown to be possible for some species. For example, on the Polynesian island of Moorea, the direction of coiling of populations of the land snail *Partula suturalis* have been shown to differ substantially depending upon collection location (Johnson *et al.* 1987). The sinistral form of the species predominates in the northern regions of the island, the dextral form in the south, and intermediate regions are populated by mixtures of the two forms.

Similarly, starry flounders (*Platichthys stellatus*; Fig. 11.2) from the coast of California are found as a nearly equal mixture of left-eyed and right-eyed individuals, whereas the same species in Japanese waters consists almost exclusively of the left-eyed variety (Policansky 1982).

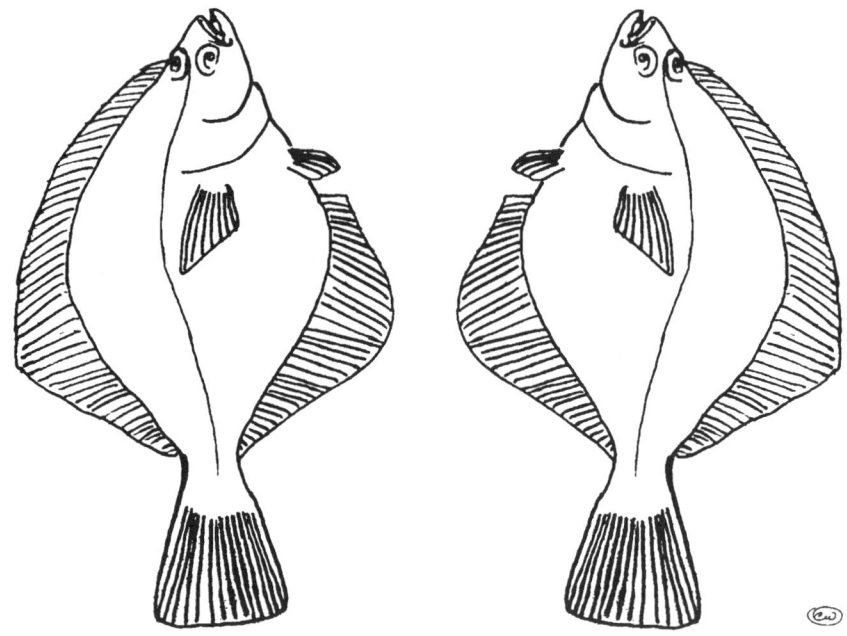

Fig. 11.2 The starry flounder, *Platichthys stellatus*, is found off the California coast as a nearly equal mixture of left-eyed and right-eyed individuals, whereas the same species in Japanese waters consists almost exclusively of left-eyed individuals.

Although, when organisms are not readily available for direct observation, photographs, prints and other such published media can be useful, such sources must be used with great caution, because the photographs and prints found in books and magazines are frequently reversed. Although such reversals are almost always detected before publication if the error results in the reversal of written letters or some other easily recognized chirality inversion, these errors often go undetected when more subtle types of asymmetry are involved.

For example, reversal of the published images of snails is a frequent problem, and in a recent article, Steven Jay Gould (1995) traces the history of this problem for more than three centuries. A letter to the editor from the February 1990 issue of *Scientific American* points out such a reversal error for a photograph of an ancient Greek kouros sculpture published in the June 1989 issue. In this instance a careful observer noted that the published photograph of the nearly symmetric statue showed the right testicle hanging lower than the left, whereas the opposite situation normally prevails. Doubting that the ancient Greeks would have tolerated such a flagrant error in human anatomy, the writer proposed that the statue must be either counterfeit or (as proved to be the case) that the negative had been reversed.

Further problems are encountered when using depictions of chiral objects in non-photographic prints made by a negative image process, for example lithographs, etchings, or engravings. In some of these techniques the image is drawn directly on a stone, block of wood or metal plate, and a reversed image is then printed. Obviously, this introduces a mechanism whereby chirality information can be reversed. Again, when these errors of inversion are obvious, for example with printed text, the printmaker takes care to prepare the printing medium with a reversed images, perhaps by sketching the object as its mirror reflection. When the asymmetry is more subtle, the printmaker might or might not have inverted the chirality of the object being depicted. This ambiguity makes the use of lithographs, etchings and similar prints nearly unusable for determining information about chirality. Paintings or drawings, which entail no such reversals, are, however, still useful.

Finally, as if this were not already enough, caution must be used when dealing with micrographs. The image observed in a compound microscope is typically reversed left to right. Thus, for example, a helical microorganism which appears under the microscope to have (P) helicity would actually have the opposite (M) helicity. When this is compounded with the fact that such microscopic images are typically photographed, and the resulting photographic images may or may not be reversed, a potential for great confusion can be seen to exist.

THE MACHINERY OF LIFE: SYMMETRIC ORGANISMS FROM ASYMMETRIC BUILDING BLOCKS

Considering that living creatures are constructed almost entirely from enantiopure molecular building blocks it is somewhat surprising that most organisms have a symmetric exterior form. The apparent bilateral symmetry of an organism such as a fish is, of course, an illusion, because both the left and right sides of the fish contain D sugars, L amino acids, α-helices with (P) helicity, starch molecules with (M) helicity, *etc.* Thus the apparent symmetry of many animals or plants is actually 'pseudosymmetry'.

Because proteins are produced from L amino acids, it follows that all proteins are inherently chiral. One of the most interesting recent developments in the study of the chirality of folded protein structures is the discovery by Mislow and others that some proteins can fold in such a way as to form a topological knot, thus adding yet another form of chirality (Zurer 1996). Other protein folding motifs, such as the β-sheet and the α-helix are also chiral.

The asymmetric α-helix coils with (P) helicity and occurs frequently in protein structure. For example, many transmembrane proteins are based on a structure containing seven membrane-spanning α-helices, which forms a tube-like supramolecular structure. Such proteins can have an appearance which looks quite symmetric. It is interesting to note that in protein structure, and in the form of animal bodies, a situation is frequently encountered in which enantiopure building blocks have been repeated to form a supramolecular structure with high symmetry. Thus, many enzymes have structures which are dimers, tetramers, *etc.*, giving an overall appearance which looks quite symmetric, although in the final analysis, they are all chiral because they are composed of L amino acids (Fersht 1985).

A further example of this trend can be seen in the immunoglobulin family of proteins (Eisen 1980), in which an assembly containing a single antigen recognition site has been dimerized to form the class of IgG immunoglobulins. Similarly, IgA immunoglobulins exist as monomers, dimers, or higher oligomers of an IgG-like structural unit, and IgM immunoglobulins are C_5 pentameric structures containing five joined IgG-like structural units.

A variety of cellular components with interesting asymmetric shapes play structural roles in the cell. Many structural proteins have a helical shape. Whenever a tube or a rod or a strut or a spar is required at the subcellular level, nature frequently turns to a helical solution. Examples include the muscle proteins actin and myosin, the 'stalk' of tobacco mosaic virus, collagen fibrils, microtubules, and prokaryotic flagella.

The structures of several rotary proteins have recently been determined. For example, the catalytic domain of adenosine triphosphate (ATP) synthetase was recently shown to rotate in the anticlockwise direction in the presence of ATP (Brennan 1997), and the structures of functional kinesin motor proteins on microtubules have been imaged by electron cryo-microscopy (Arnal *et al.* 1996). One of the most remarkable and longest studied protein 'motors' is the bacterial flagellar motor (Macnab & Aizawa 1984). The motor runs on a proton gradient, and is used for propulsion by some prokaryotic bacteria, such as *E. coli*. The flow of protons through the motor results in rotation, which turns the flagellum. In essence this motor is a wheel, a chiral device central to many of mankind's inventions, but which is rarely found in nature.

When thinking about the relative scarcity of asymmetric body forms, it is, perhaps, pertinent to begin with the question of why most organisms have a symmetric body plan. Martin Gardner (1990) speculates that whereas early life was probably spherically symmetric, "once a living form anchors itself to the bottom of a sea or to the land, a permanent up-down axis is created".

He goes on to state that most plants and animals which live such a rooted existence, for example trees, sea anemones, tube worms, *etc.* generally have a conic symmetry in which the top is readily distinguished from the bottom, but front and back and left and right are often indistinguishable. Although a few motile animals, for example starfish and jellyfish, maintain this body plan with conical symmetry, most motile animals have developed a body plan with bilateral symmetry in which, in addition to the top and bottom, the front and back can be distinguished. Mouth and eyes or other sensory organs make more sense on the front end of a moving organism. There is virtually nothing in the environment of a swimming fish, a flying bird or a running mammal which can distinguish between right and left, thus there is little incentive for such animals to deviate from a body plan with bilateral symmetry.

Nevertheless, a variety of chiral body plans have developed in many plants and animals, and the causes of these departures from symmetry, be they a chance event or some environmental asymmetry, are a fascinating subject for study. The study of such anomalies has largely been relegated to the individual disciplines; ichthyologists study the asymmetry of flounders, malacologists study gastropod shell asymmetry, and botanists study the asymmetry of twining vines and tendrils. Recent advances in the field of developmental biology are bringing such questions into the arena of chemistry, and some of the genes which govern the development of asymmetric body forms are now being discovered.

THE ASYMMETRY OF MICROORGANISMS

Microscopic organisms with asymmetric body forms are actually fairly common (Sagan & Margulis 1988). Gardner's aforementioned reasoning concerning why most swimming organisms have symmetric bodies seems not to apply to microorganisms, a substantial number of which have helical body shapes which enable them to swim by 'screwing' their way through their highly viscous surroundings. Examples are found among the bacteria, for example the syphilis spirochete, *Trepanoma pallidum*, is shaped like a (P) helical corkscrew. In a review entitled "The Handedness of the Universe" Hegstrom and Kondepudi (1990) point out that the bacterium *Bacillus subtilis* normally forms spiral colonies with (P) helicity, but when heated, these change to (M) helicity (Potera 1997).

The familiar ciliated protozoan, *Paramecium*, has a body plan which resembles a right footed shoe or slipper; this also leads to a helical swimming motion. Neville (1976) mentions a study of ciliated protozoans which found that 102 species traced an (M) helix as they swam, whereas 62 spiralled the other way. This result is undoubtedly a result of some asymmetry of the body shape of the organisms. Dinoflagellates, a family of algal protists famous for causing 'red tides' have a flagellum protruding from the posterior which is responsible for forward propulsion, and another flagellum which coils around the midsection, which is responsible for spinning the animal as it moves forward (Sagan & Margulis 1988). Quite a few single-celled algae, such as the familiar *Spirogyra*, contain helical elements. Helical structures are also found among the cyanobacteria.

ASYMMETRY IN THE PLANT WORLD

Many kinds of asymmetry can be observed in the plant world. A familiar example is twining vines, such as bindweed or pole beans which generally twine to form a (P) helix, although several species such as hops and honeysuckle twine so as to form an (M) helix. Gardner (1990) has pointed out that the intertwining of vines of opposite helicity is a theme which has been explored by poets including Shakespeare and Johnson.

Another kind of asymmetry among the plants involves the placement of flowers or buds along a stem to trace a helical path. This subject, referred to as phyllotaxis, was studied by da Vinci, Kepler and Goethe, and became very popular during the nineteenth century, where it became permeated with mysticism and pseudoscience. Much of this early work is critically reviewed by Thompson (1992) and this history, and more recent developments, are treated in a recent book by Jean (1994). Many plants have chirality as a result of phyllotaxis. Notable examples include the flower spikes of the orchid genus *Spiranthes*, and the placement of scales on pinecones.

A spiral arrangement of fronds or branches is found on many palms and other trees. One example is provided by *Pandanus forsteri*, a species of *Pandanus* or screw-pine native to New Zealand (Fig. 11.3), in which both enantiomeric forms are found.

Another type of plant asymmetry involves chirality of flowers. Many flowers such as the hibiscus shown below have a sweeping arrangement of petals and show a preferred handedness. Interestingly, the female flowers of the papaya show a clockwise swirl of petals, whereas the male flowers show the opposite (Fig. 11.4).

Fig. 11.3 The New Zealand 'screwpine' species, *Pandanus forsteri*, comes in two enantiomeric forms. (Photographs by the author; Royal Botanical Gardens, Kew).

Fig. 11.4 Many flowers such as this hibiscus are chiral owing to a sweeping of the petals in one direction or the other. (Photograph by the author; Chicago Botanical Gardens, Glencoe, IL.)

The asymmetry of orchid flowers is a subject which has been of interest to the author for some time (Welch 1998). Whereas the overwhelming majority of orchid species have flowers with bilateral symmetry, it has been known for some time that a few orchid flowers are chiral (Dressler 1993). For example, species such as *Ludisia discolour* and *Mormodes colossus* depart dramatically from bilateral symmetry owing to bending of various flower parts. Some species with twisted petals, which seem at first glance to have bilateral symmetry, are in fact chiral, however, because the petals twist with the same helicity on either side of the plane of symmetry. The author has conducted a survey of such twisted petal species with interesting results (Welch 1998).

One of the most dramatic examples of this type of chirality is the species *Tricopilia tortilis* from the new world tropics, in which both petals and all three sepals have a strong (M) helical twist. A few related members of the neotropical genus *Encyclia*, also have (M) helical petals, as do most members of section *Spatulata* of the huge Asia–Pacific genus, *Dendrobium*. This group is known as the 'antelope *Dendrobiums*' because the twisted petals stand upright like two horns. Unlike their namesakes, however, the 'horns' of antelope *Dendrobiums* both twist with the same (M) helical sense.

The lady slipper orchids (Subfamily *Cypripediodeae*; Fig. 11.5) often have helical petals. *Phragmipedium*, a genus restricted to the tropical Americas contains several species with twisted petals, invariably with (P) helicity. Similarly, the genus *Cypripedium*, which has a circumboreal distribution, occurring in North America, Asia and Europe also has a number of species with twisted petals, all with (P) helicity.

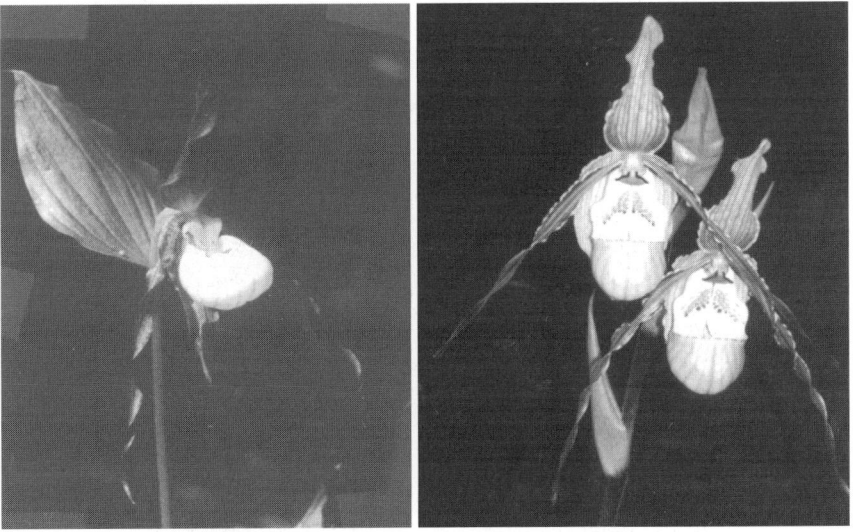

Fig. 11.5 Representative species from two 'slipper' orchid genera in which helical petals invariably twist with (P) helicity. *Cypripedium montanum*, from the Rocky Mountains is shown on the left and *Phragmipedium pearcei* from Ecuador on the right. The related Asian genus, *Paphiopedilum*, contains some species with (P) helical petals and some with (M) helical petals. (Photograph of *Cyp. montanum* by Ron Coleman, used with permission. Photograph of *Phrag. pearcei* by the author.)

The genus *Paphiopedilum*, however, native to Asia and the Pacific, contains some species with (M) and some with (P) helical petals. For example, *Paphiopedilum glanduliferum* (New Guinea) and *Paphiopedilum philippinensis* (The Philippines) are similar in appearance, although they can easily be distinguished by comparing petal twist direction.

Although botanical descriptions of these plants contain hundreds of terms describing leaf shape, growing habit, and other distinguishing features, these descriptions make reference solely to the spiral shape of the petals, without any mention of twist direction. It is clear that information about the absolute sense of chirality should be included in species descriptions, and it seems probable that this information could prove useful for taxonomic segregation, or could offer some insight into evolutionary relationships.

ASYMMETRY OF INVERTEBRATES

One of the most studied areas of animal asymmetry is the asymmetry of gastropod or snail shells (Robertson 1993). It has long been noted that whereas most snail species have shells with dextral twist some species have shells which coil in the opposite, sinistral direction (Fig. 11.6; see Fig. 11.1 for the convention of dextral and sinistral). Furthermore, many species which are normally dextral occasionally have sinistral mutants, and *vice versa*. This topic has been studied by malacologists for many years, and has a rich history. For example, the sacred Indian Chank shell, *Turbinella pyrum*, normally a dextral species, is occasionally found in the very rare sinistral form. Such rarities have been sought after for millennia, and in ancient sculptures, the god Vishnu is often depicted with a sinistral form of this shell (Welch 1998).

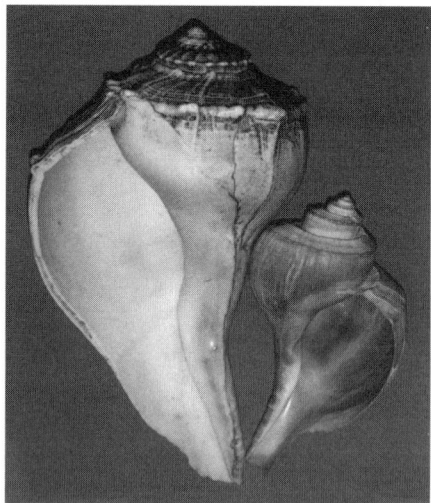

Fig. 11.6 Two species of American whelk. The lightning whelk (*Busycon contrarium*) shown on the left is a normally sinistral species, whereas the channelled whelk (*Busycon canaliculatum*) shown on the right is a normally dextral species. (Photograph by the author; invertebrate collection, Field Museum of Natural History, Chicago.)

The overwhelming majority of gastropod species are normally dextral, with only a few normally sinistral species and a few species where significant numbers of both dextral and sinistral forms are found. This preponderance of dextral species has led many to theorize about a possible evolutionary advantage in dextrality. Although no such advantage need exist to account for the present state of affairs, Gould and others have noted slight differences in the proportions of rare sinistral forms compared with the common dextral Gould *et al.*

Given that the cellular machinery which is responsible for the biomineralization process is inherently chiral, it might be possible that sinistral and dextral forms are not truly enantiomeric, but differ in some way. One can liken the situation to the building of a ziggurat from enantiomeric bricks. The bricks naturally fit together to make one enantiomeric form of the helical tower, and fit poorly when the opposite form is attempted.

Although it is currently impossible to say what, if any, reason underlies the preponderance of dextral forms, it has been suggested that individuals which differ in chirality from the majority of the population might be at a disadvantage when it comes to mating (Asami 1993). Consequently, in a normally dextral population, a sinistral mutant is at a competitive disadvantage, providing an evolutionary mechanism which may prevent the change from dextrality to sinistrality within a species. Asami has pointed out successful mating between enantiomeric forms is most difficult for snails with globose or discoidal shells whereas snails with high-spired shells can engage in inter-enantiomeric mating, albeit with some difficulty (Asami 1993). Interestingly, almost all examples of species with significant numbers of both enantiomeric forms are high-spired pulmonate land snails.

The inheritance of coil direction has been studied in some gastropods. In *Helix*, sinistrality is recessive, whereas in *Partula*, dextrality is recessive (Robertson 1993). Interestingly, the factor which controls coil direction is already present in the unfertilized egg, thus coil direction is determined not by the individual's genotype, but by the genotype of the mother. This is an example of 'maternal inheritance' or delayed segregation, and complicates the study of coil direction genetics considerably, requiring that several generations be studied to ascertain the genotype of an individual.

Although gastropods are the best known asymmetric molluscs, the bivalve molluscs, or pelecypods, also have a subtle form of chirality. Close examination of the hinge region of a clam will reveal that the two shells are not identical. Most cephalopods such as the squid or the chambered nautilus have a bilaterally symmetric body plan.

There are many famous examples of asymmetric crustaceans, for example, the unequal claw sizes of lobsters and fiddler crabs (Fig. 11.7). This unequal development of the chelae (claws) is termed heterochely, and is most pronounced in the male fiddler crab, *Uca pugnax*. According to Neville (1976), in this species the large claw can be as much as twenty times heavier than the smaller, comprising as much as 70% of the total body weight. Not surprisingly, such heterochely is rarely observed in free-swimming decapod crustaceans.

Crustaceans displaying heterochely typically show a roughly 1:1 mixture of the two enantiomeric forms. Young male fiddler crabs begin life with two equal claws. If by chance one of these claws is lost, it is regenerated as a small, female type claw. Loss of both claws leads to an adult with two female type claws, and an individual which loses neither claw develops two enlarged male type claws.

Fig. 11.7 Enantiomeric fiddler crabs, onetime residents of the author's laboratory aquarium. Photo by author.

The hermit crabs, which make their homes in gastropod shells, often have asymmetric bodies which correspond to their shell homes. Neville (1976) reports that asymmetric hermit crabs are designed to fit into dextral gastropod shells, and when presented with a sinistral shell, reject it. The author has, however, observed hermit crabs inhabiting the sinistral shell of the lightning whelk, *Busycon contrarium* with no apparent problem.

A well known example of asymmetry among insects is the way the wings overlap (Neville 1976). In the cockroach the preferred direction is left over right, whereas in the cricket the opposite pattern is preferred

Other examples of asymmetric invertebrates include the Portuguese man-o'-war, *Physalia physalis*, which Neville (1976) reports has a sail which is set at an angle to the body axis.

> The progeny of one parent are polymorphic, some individuals having the sail from right to left, other from left to right. They are thus blown in opposite directions, reducing the risk of all offspring being washed ashore.

ASYMMETRY OF FISHES

Among the fishes the asymmetry of flatfish such as flounders, halibuts and soles is very pronounced and has been much studied (Neville 1976; Policansky 1982; Gardner 1990). Interestingly, these fish begin life with bilateral symmetry and swim freely. After a few weeks they settle to the bottom, where they lie on one side. The eye facing the bottom then begins to migrate upwards and forward, and after a few weeks the transformation is complete.

Other changes accompany this transformation, including the loss of pigmentation from the underside and some desymmetrization of the jaw, scales, fins and lateral line. Some flatfish species are normally right-eyed, some species are normally left-eyed, and some species such as the starry flounder, mentioned previously, have both right- and left-eyed individuals.

Policansky (1982) has conducted some interesting breeding experiments in an effort to elucidate the inheritance of asymmetry in the starry flounder. These studies are hampered by the need to raise the young to the stage where they begin to show asymmetry, which can be as long as several months. The results of these experiments are somewhat confusing, and suggest that environment or chance, as well as genetics, might play some role in the development of flounder asymmetry.

Another famous case of fish asymmetry related by Neville involves the species *Anableps anableps*, a South American live-bearing tooth-carp (Neville 1976). The males of this species have a modified anal fin which is used as a gonopodium, a tubular extension of the sperm channel. This gonopodium can be moved to the right side by about half the males and to the left by the others. Similarly, the female anableps has the sex opening on either the right or left side, owing to partial occlusion by a large scale. The two enantiomeric female forms occur in roughly equal numbers. Mating between anableps occurs when the male and female are swimming side by side, thus a male with a gonopodium which can move to the right can only mate with a female with the sex opening on the left and *vice versa*.

Finally, an interesting example of asymmetry has been reported for some African cichlids of the genus *Perissodus*. The African rift lake cichlids are a much studied group of fish well known to the aquarium hobby. Competition among these fish for limited food resources has led to a remarkable diversity of species, with many exploiting unusual food sources. One of the most remarkable examples involves a group which has become adapted to eating the scales of other fishes. Of these, perhaps the most specialized example involves the genus *Perissodus* which is native to Lake Tanganyika. According to Axelrod & Burgess (1988) some members of this genus have the head and jaws skewed to one side to facilitate the scale biting but are limited to attacking from one particular side only.

It is interesting to note that whereas many microorganisms have developed helical swimming strategies, often with accompanying asymmetric bodies, no fish seem to utilize this method of swimming.

ASYMMETRY OF BIRDS

Asymmetry is something of a rarity among birds. The crossbill, *Loxia curvirsotra*, is a well known exception. The crossed beak of this bird is a specialization which helps it to remove the seeds from pine cones. According to Neville (1976), the birds are born with a symmetric bill, which begins to cross at approximately one month of age. Gardner (1990) states that the majority of crossbills in the United States have the upper bill crossing to the bird's left whereas the opposite arrangement is more common in Europe (Neville (1976) reports that a study in Norway showed this form to be in 16% enantiomeric excess).

Another interesting asymmetric bird mentioned by Neville (1976) is the New Zealand wry billed plover, *Anarhyncus frontalis*, which has a beak which is always bent to the right side. This presumably helps the bird in its task of turning over stones in the search for food.

The barn owls (family Tytonidae) have an interesting asymmetry of the skull in which one ear opening is displaced relative to the other. This improves the owl's hunting ability by enabling it to locate the source of a sound more precisely (Hume 1991). Neville (1976) notes that:

> Goslings hatch from the egg by making an anti-clockwise perforation as seen from the outside. This correlates with the position of the head since the neck always coils to one side.

Finally, Gardner (1990) reports that:

> ... the female birds of all genera, with few exceptions, exhibit a curious left–right asymmetry with respect to their ovaries and oviducts. In young birds, both the left and the right ovaries and their ducts are equal in size; as the bird matures, the organs on the right degenerate and become useless. Only the left oviduct, which greatly enlarges during the egg-laying season, remains functional.

ASYMMETRY OF REPTILES AND AMPHIBIANS

Few examples of asymmetry among reptiles and amphibians have been reported, although Neville (1976) has noted that many salamanders and frogs show asymmetric patterning.

ASYMMETRY OF MAMMALS

Asymmetry among mammals is also rather rare. Gardner (1990) mentions the tail of the Japanese akita dog, which has a tail "that curls one way on males, the other way on females". The porcine penis is also corkscrew shaped. Although the tusks of the wild boar and the horns of an antelope are individually chiral, an overall *meso* structure is maintained by the presence of a tusk or horn of the opposite chirality on the other side of the organism. An exception to this rule is the very interesting case of the narwhal, *Monodon monoceros*.

The narwhal is a whale which is found today in the Greenland area, but which once ranged throughout the Arctic. The male narwhal has a single elongated pulp-filled hollow ivory tooth with a highly visible spiral pattern with (M) helicity. This tooth projects from the left side of the narwhal's mouth and the corresponding tooth from the other side ordinarily does not become elongated. Interestingly, there are reports of occasional narwhals (about 0.2% of males) in which both left and right teeth are elongated (Bruemmer 1993).

In contrast with what one might expect by analogy with the teeth and tusks of other mammals, both of these tusks have the same (M) helicity. The skull of the narwhal also has marked asymmetry, as does the blowhole, which usually is to the left of the midline. The mechanism by which the narwhal tusk acquires its helicity is a topic which Thompson (1992) has treated at length, proposing an unusual theory having to do with the torque forces produced by an asymmetry in the swimming pattern in the animal. Neville (1976) discounts this hypothesis, pointing out that skull asymmetry is already apparent in the narwhal embryo.

The trade in narwhal ivory has a long and interesting history (Bruemmer 1993). The Vikings established a trading business bringing narwhal ivory to Europe, where it was much sought after as the horn of the unicorn. Many magical properties were attributed to unicorn horn, especially the ability to neutralize or counteract the effect of poisons. Consequently, much unicorn horn was used in the manufacture of goblets and other receptacles for food and drink. The powdered horn was widely used as a pharmaceutical for variety of ailments until well into the 18th century. During most of this period unicorn horn was worth considerably more than its weight in gold.

A cursory survey suggests that depictions of unicorn from this early period usually show the animal with an (M) helical horn, which might suggest they were modelled from 'authentic unicorn horn'. In contrast, modern depictions of unicorns seem to be nearly equal mixtures of the two enantiomeric forms. Interestingly, a depiction of a unicorn from Leonardo da Vinci's notebooks shows an animal with a (P) helical horn. Because these notebooks are written in an unusual mirror image enantiomeric script, however, in which the writing is read from right to left, perhaps the twist direction in this illustration should also be reversed.

ASYMMETRY IN HUMANS

The human body is, to a first approximation, bilaterally symmetrical. There are several well known exceptions, for the hair on the back of the head can form a 'cowlick' which spirals in either direction. The best known example of human asymmetry involves the strong preference for right handedness. Approximately 95% of humans are right handed, a bias which is observed in all countries and cultures. Lefthanders are often at a disadvantage, because many everyday items are designed for use by right-handers. Examples include scissors, baseball gloves, can openers, and wristwatches (Gardner 1990).

An enormous amount of material has been written about asymmetry of the human brain, some of it verging on pseudoscience. A critical review of the left brain–right brain topic can be found in an article by Saravi (1993). Despite the abundance of nonsense on this topic, the facts are quite interesting. The left and right hemispheres of the brain function more or less independently, and are specialized for certain tasks. For example, the right hemisphere is generally more adept at spatial perception and recognition of faces whereas the left hemisphere is generally better suited to verbal and mathematical functions.

Much of our knowledge on this subject comes from studies of subjects in which the *corpus callosum*, a large bundle of nerves connecting the two hemispheres, has been severed. This procedure is sometimes used for treatment of patients who experience severe epileptic seizures. Other studies involve patients in which a portion or even an entire hemisphere has been removed.

Recently, the use of the techniques of positron emission tomography (PET scan) and magnetic resonance imaging (MRI) have furnished interesting information on brain asymmetry in normal individuals. One such recent study of the brains of musicians shows that the *planum temporarale* region is larger in the left hemisphere, and that this asymmetry is most pronounced in musicians with 'perfect pitch' (Schlaug *et al.* 1995). This finding contrasts with many previous reports which had suggested that music and other 'artistic' functions are primarily processed by the right hemisphere.

The internal organs of humans are also asymmetrically disposed, the heart, stomach and pancreas are normally on the left side, whereas the liver, appendix and gall bladder are normally on the right. In an unusual medical condition known as *situs inversus*, which affects approximately 0.01% of the population, this asymmetry in the placement of the internal organs is reversed. Interestingly, *situs inversus* is always observed in one of a pair of Siamese twins. Recently, a developmental gene which governs a familial form of this condition has been mapped to the X chromosome (Casey *et al.* 1993).

SOME RECENT ADVANCES IN THE DEVELOPMENTAL BIOLOGY OF ANIMAL ASYMMETRY

A very active area within the discipline of developmental biology involves the discovery and study of genes which control *situs inversus* and other developmental asymmetries. One such gene, referred to as sonic hedgehog, has been identified as playing a role in the asymmetric development of the internal organs of the chicken, and closely related genes might play similar roles in such diverse organisms as fruit flies, zebra fish, mice, and even humans (Dickman 1996). The current pace of discovery in developmental biology is staggering, and it seems clear that some sort of 'chemical' understanding of the basis of animal asymmetry is not very far off (Levin *et al.* 1995; Isaac *et al.* 1997).

BEHAVIOURAL ASYMMETRY

Several types of behaviour in humans and animals are asymmetric. Gardner (1990) mentions the example of a colony of bats which invariably fly out of their cavern and up into the sky in a (P) helical path. One wonders whether there might be some similar preference among soaring birds or coil direction among snakes. Neville (1976) discusses some interesting studies in which blindfolded subjects attempt to walk in a straight line and invariably end up moving around in circles.

Similar results were obtained when blindfolded subjects swam, rowed a boat, and even when they drove a car! Gardner (1990) reports that dolphins confined to a circular tank tend to swim in a counter-clockwise direction, and a study has found that cats treated with *S*-amphetamine tend to circle in an anticlockwise direction (Glick *et al.* 1981) (no information was provided about cats treated with *R*-amphetamine).

SUMMARY AND CONCLUSION

It is clear that asymmetry in the exterior form of organisms is an important topic which warrants more careful attention. Asymmetry of various organisms might be of some importance in taxonomy, and descriptions of species should include information relating to the sense of chirality when asymmetry is present. Care must be take to use unambiguous and correct terms in descriptions of chirality. Caution should be exercised when dealing with photographic negatives and compound microscopic images or in other instances in which the sense of chirality can be inadvertently inverted. This is a interesting topic for further study and hopefully a source of some amusement for stereochemists.

ACKNOWLEDGEMENTS

Many people have helped me over the past few years with valuable discussions and communications on this topic. I would especially like to thank Michael Cornwell, John Slapcinski, Henri Brunner, Takahiro Asami, Pedro Lehmann, Koji Nakanishi, Mark Zaffron, Linda Wish, Eric Muehlbauer, and my wife, Renee.

REFERENCES

Arnal, I., Metoz, F., DeBonis, S., & Wade, R. (1996) Three-dimensional structure of functional motor proteins on microtubules. *Current Biology*, **6**, 1265–1270.

Asami, T. (1993) Genetic variation and evolution of coiling chirality in snails. *Forma*, **8**, 263–276.

Axelrod, H. R. & Burgess, W. E. (1988; originally published 1973) African Cichlids of Lakes Malawi and Tanganyika, 12th Edition, TFH Publications, Neptune City, NJ, 12th Edition,.

Brennan, M. (1997) Rotary catalysis of ATP imaged. *Chemical & Engineering News*, **27**, March 31.

Bruemmer, F. (1993) *The Narwhal: Unicorn of the Sea*. Key Porter, Toronto,.

Casey, B., Devoto, M., Jones, K. L., & Ballabio, A. (1993) Mapping a gene for familial situs abnormalities to human chromosome Xq24-q27.1. *Nature Genetics*, **5**, 403–407.

Dickman, S. (1996) The left-hearted gene. *Discover*, 71–75, August.

Dressler, R. L. (1993) *Phylogeny and Classification of the Orchid Family*. Dioscorides Press, Portland, OR.

Dubos, R. (1950) *Louis Pasteur: Free Lance of Science*. Charles Scribner's Sons, New York.

Eisen, H. N. (1980) *Immunology*. Harper & Row, Philadelphia.

Fersht, A. (1985) *Enzyme Structure and Mechanism*. W. H. Freeman & Company, New York.

Gardner, M. (1990; originally published 1964), *The New Ambidextrous Universe: Symmetry and Asymmetry from Mirror Reflections to Superstrings*, revised edition. W. H. Freeman, New York.

Glick, S. D., Weaver, L. M. & Meibach, R. C. (1981) *Brain Research*, **208**, 227–229.

Gould, S. J. (1995) Left snails and right minds. *Natural History*, 10–18, April.

Gould, S. J., Young, N. D., & Kasson, B. (19??) The consequences of being different: sinistral coiling in *Cerion*. *Evolution*, **39**, 1364–1379.

Hargittai, I. & Hargittai, M. (1994) *Symmetry: A Unifying Concept*. Shelter Publications, Bolinas, CA.

Hegstrom, R. A. & Kondepudi, D. K. (1990) The handedness of the universe. *Scientific American*, **262**, 108–115, January.

Hume, R. (1991) Owls of the World. Running Press, Philadelphia.

Isaac, A., Sargent, M. G. & Cooke, J. (1997) Control of vertebrate left–right asymmetry by a snail-related zinc finger gene. *Science*, **275**, 1301–1304.

Jacques, J. (1993) *The Molecule and its Double*. McGraw–Hill, New York.

Jean, R. V. (1994) *Phyllotaxis: A Systematic Study in Plant Morphogenesis*, Cambridge University Press, New York.

Johnson, M. S., Murray, J. & Clarke, B. (1987) Independence of genetic subdivision and variation for coil in *Partula suturalis*. *Heredity*, **58**, 307–313.

Kant, I. (1783; 12th printing 1995) *Prolegomena to any Future Metaphysics That Can Qualify as a Science*. Paul Carus Translation. Open Court, LaSalle, IL.

Levin, M., Johnson, R. L., Stern, C. D., Kuehn, M., & Tabin, C. (1995) A molecular pathway determining left–right asymmetry in chick embryogenesis. *Cell*, **82**, 803–814.

Macnab, R. M. & Aizawa, S. (1984) Bacterial motility ant ht bacterial flagellar motor. *Ann. Rev. Biophys. Bioeng.*, **13**, 51–83.

Neville, A. C. (1976) *Animal Asymmetry*, Studies in Biology No. 67. Edward Arnold Publishers, London.

Policansky, D. (1982) The asymmetry of flounders. *Scientific American*, 116–122.

Potera, C. (1997) Physics, biology meet in self assembling bacterial fibers. *Science*, **276**, June 6.

Robertson, R. (1993) Snail Handedness. *National Geographic Research & Exploration*, **9**, 104–119.

Sagan, D. & Margulis, L. (1988) *Garden of Microbial Delights*. Harcourt Brace Jovanovich, Boston.

Saravi, F. D. (1993) The right hemisphere: an esoteric closet? *Skeptical Inquirer*, **17**, 380–387.

Schlaug, G., Jäncke, L., Huang, Y., & Steinmetz, H. (1995) *In vivo* evidence of structural brain asymmetry in musicians, *Science*, **267**, 699–701.

Thompson, D. W. (1992; originally published 1917) *On Growth and Form, The Complete Revised Edition*. Dover Publications, New York.

Washburn, D. K. & Crowe, D. W. (1988) *Symmetries of Culture: Theory and Practice of Plane Pattern Analysis*. University of Washington Press, Seattle.

Welch, C. (1998) Some observations concerning the asymmetry of orchid flowers. *Malayan Orchid Review*, **32**, 86–92.

Welch, C. J. (1998) Vishnu's Trumpet. *Enantiomer*, **3**, 491–493.

Welch, C. J. (1998) Outfitting the sinistral minority. *Enantiomer*, **4**, 485–486.

Zurer, P. (1996) Much ado about knotting. *Chemical & Engineering News*, December 9, 43–44.

Concluding Remarks
Chirality, Chemistry, the Future

The Editors

In 1815 Jean-Baptiste Biot discovered the ability of crystals to rotate the plane of polarized light, thereby initiating the study of molecular asymmetry, or chirality. This book is a celebration of the diversity of this phenomenon and its varied impact on the natural and physical sciences.

The book began with an Introduction that dealt with the work of Louis Pasteur, the scientist credited with changing the study of chirality from an abstract question to a concrete scientific endeavour. Pasteur accomplished this by the manual resolution of the enantiomeric crystals of ammonium tartrate. When this experiment was repeated for Biot, Biot took Pasteur by the arm and said, "My dear child, I have loved science so much throughout my life that this makes my heart throb" (Wainer & Drayer 1988). The beauty and wonder of chirality is indeed one of the marvels of nature. Accordingly, at the end of the book it is appropriate to dwell on how far we have come in our appreciation of this phenomenon and how far there is still to go. Clearly, in some areas we have come a long way whereas in others we have only begun our journey.

Pasteur's first success in the field of chirality was as a separation scientist when he resolved the enantiomeric crystals of ammonium tartrate with a pair of tweezers. This first success was followed by a second method of enantiomer separation that involved the formation of diastereomeric salts. In this manner Pasteur was acting as a separation scientist. It is fitting, then, that the separation of enantiomers is one area of chirality that has made great progress towards maturity. This is particularly true for the decade 1980–1990 during which liquid chromatographic chiral stationary phases became commercially available and widely used.

With the separation of racemic tartaric acid into its enantiomorphs Pasteur showed that whereas molecules made by ordinary laboratory processes did not rotate the plane of polarized light, those produced naturally did so. From this critical observation two very different consequences followed. The first was that molecules had a shape in three-dimensional space, the second that Nature must set about making molecules differently from the chemists of that time (Bernal 1971). The observations foreshadowed the development of spatial chemistry by van't Hoff and Le Bel and led to great progress in organic and inorganic chemistry.

In mainstream organic chemistry there has been much progress in stereochemical synthesis and structural identification. Sufficient work had been done by the early sixties for Eliel to be able to produce his seminal text on stereochemistry (Eliel 1962, later updated Eliel & Wilen 1994).

Almost all the work that had been done at that time was on natural compounds. The stereochemistry of natural products, the determination of absolute configuration, and many aspects of the chemistry of carbohydrates, peptides, and proteins had all been well explored. In the pharmaceutical arena, single-isomer natural products and chiral synthetic compounds (usually as racemates) had been extensively used as drugs. Regulatory pressures (e.g. FDA policy statement 1992), however, resulted in a situation where it is almost always necessary to develop a single-enantiomer drug candidate both for synthetic and natural compounds. Thus, synthetic methods to produce single-enantiomer drug candidates without a resolution step became a necessity rather than just a useful academic pursuit.

Many elegant synthetic routes have been devised on the basis of single-enantiomer, usually natural, starting materials and there have been some stunning developments in asymmetric synthesis. As Sanders (1998) so eloquently pointed out, however, even with recent advances in supramolecular chemistry using assemblies of molecules held together by non-covalent interactions, man-made catalysts are still well short of the "astonishing selectivities and catalytic efficiencies" of enzymes. With regard to chirality in organic chemistry, then, there is still more to be achieved.

In addition to the manual and chemical separation of ammonium tartrate, Pasteur also achieved a biological separation. He added the racemic salt to a solution of yeast and the fermentation process consumed the dextro-rotatory isomer leaving only the levo-rotatory tartrate. Pasteur had demonstrated the inherent enantioselectivity of biological processes and opened the door to our understanding of ligand–receptor and substrate–enzyme interactions. As Pasteur noted (Wainer and Drayer 1988):

> ... thus we find introduced into physiological principles and investigations the idea of the influence of the molecular asymmetry of natural organic products, of this great character which establishes perhaps the only well marked line of demarcation that can at present be drawn between the chemistry of dead matter and the chemistry of living matter.

Chirality remains both a powerful tool and an elusive goal in modern biological and pharmacological sciences.

In identifying the areas in the study of chirality where there is the most still to do we can look again to Pasteur. Pasteur searched in vain for the cosmic force of dissymmetry. Today we are still a long way from fully understanding the reasons for dissymmetry in flounders (to name but one species), the origins of chirality in the Universe, and, related to that, the origins of the Universe.

What the book has hopefully demonstrated is that chirality is within us and is all around us. The day we stop studying what is within us and what is all around us will be the day we stop studying. Indeed it will be the day we stop. Let us hope, then, that there is still a century or two left in the field of chirality!

REFERENCES

Bernal, J.D. (1971) *Science in History, Volume 2: The Scientific and Industrial Revolutions*, The MIT Press, Cambridge, MA, p. 630.
Eliel, E.L. (1962) *Stereochemistry of Carbon Compounds*. McGraw–Hill. New York.

Eliel, E.L. & Wilen, S.H. (1994) *Stereochemistry of Organic Compounds*, Wiley–Interscience, New York.
FDA Policy Statement (1992) *Chirality* **4**, 338.
Sanders, J.K.M. (1998) *Chem. Eur. J.*, **4**, 1378.
Wainer, I.W. & Drayer D.E. (1988) *Drug Stereochemistry: Analytical Methods and Pharmacology*, Marcel Dekker, New York (first citation, p. 12; second, p. 17).

Index

α_1-acid glycoprotein, 117, 186, 187, 192
α-bisabolol, 277–278
α-bungarotoxin, 153
α-Burke 2 CSP, 190
α-damascone, 273–274
α-ionone, 273–274
α-methyldopa, 141–142
α-methyldopamine, 141–142
α-methylnoradrenaline, 141–142, 157, 162–163
α-phellandrene, 270
α-pinene, 216, 251, 269–270
α-terpineol, 270–271
α-(2,4,5,7-tetranitro-9-fluorenylidineaminooxy)-propionic acid, 181
α-vetivone, 277
β-blocker drugs, 189, 193
β-decay, 17, 31–33, 70–71
β-emitters, 17, 32–33, 41
β-Gem 1 CSP, 190
β-haloalkylamines, 167
β-rays, 17, 32–33, 37, 41
δ-jasmolactone, 275–277

1,2-diarylethane-1,2-diols, 228
1,4-dihydropyridines, 127
1-(4-nitrophenyl)-2-amino-1,3-propanediol, 184
1,7-dioxaspiro[5,5]undecane, 246, 263
1-(9-fluorenyl)ethyl chloroformate, 184
14-methylhexadecen-1-ol, 242–243
14-methylhexadecenal, 243
1-chloro-4,4-dimethylpenta-1,2-diene, 227, 232
1-methoxy-2-propanol, 180
1-octen-3-ol, 273
1-phenylethylamine, 184
1-*p*-menthen-8-thiol, 280
1*R*,2*R*-cyclohexanediol, 229
2,2,2-trifluoro-1-(9-anthryl)ethanol, 181–182
2-(3,4-dihydroxy)cyclobutylamines, 158
[2-[4-(3-ethoxy-2-hydroxypropoxy)phenyl-carbamoyl]ethyl]dimethylsulphonium ions, 180
2-methyl-1-butanol, 242–243
2-methyl-4-pentenoic acid, 244
2-naphthalene acetic acid, 187
2-naphthylethyl isocyanate, 184
2-octanol, 244
3,4,4-trimethyl-1-pentyn-3-ol, 232
3,4-dichloroisoproterenol, 168
3,4-dihydroxyphenylserine, 141–142
3,4-dihydroxy-α-methylpropiophenone, 168
3-methyl-1-pentene, 44
3-methylcyclohexanone, 216
4-carboxy-4-pentanolide, 245
4-methyl-3-heptanone, 244
4-methylhexanoic acid, 242, 244
5-methyl-3-heptanone, 254
5-methylcyclopentadiene, 225–226
8-mercapto-*p*-menthan-3-one, 279–280
8-methyl-α-ionone, 264

(+)-*cis*-2-hydroxy-2-phenylcyclohexane-carboxylic acid, 229

absorption, drug, 116–117
acarid mite, 249
acebutolol, 115
aceclidine, 147–149
acetylcholine, 139–149
acetylcholine esterase, 146, 150–151
achiral derivatisation, 197
adrenaline, 140–141, 145, 157, 161–162, 171
adrenoreceptors, 145, 164–170
adsorptive bubble separation, 199
agonists, 144–149
AIDS, 97, 103
alanine, 30
albumin, 117
albumin CSP, 183, 188, 192
albuterol, 115
alcohols
    acetylenic, 232
    benzoates of, 232
algae, 291
alkaloids, 2–3, 139, 179
    aporphine, helicity rules for, 224
    allylic, axial rule, 226
alprenolol, 121
ambergris, 278
ambroxan, 278–279
amides, 180
amino acids, 17, 23–24, 33, 37, 46, 88, 110
    cyanohydrin, from, 43
    derivatives of, 266
    GC stationary phases from, 266
    α-methyl, 18–19, 26, 43
amosulalol, 166–167
amphetamine, 154–156, 171, 300
amphibians, asymmetry of, 298
amplification mechanisms, 35–37
amyl alcohol, 8
amylose, 189
anaesthetics, 117, 128
anatoxin-a, 149
aniline, dye, 181
anosmia, 265
antelope, 298
antibiotics, 117
anti-fermion, 31–32

antihistamines, 116
anti-hypertensive drugs, 189
anti-inflammatory drugs, 116, 117, 120
anti-matter, 18, 46–48
antimolecules, 76
antineutrinos, 71, 74, 81–82
antiparallel, 70
antiparticles, 73–74
antipions, 74
antiworld, 71, 73
anyon theory, 83
aporphine, 224
asparagine, 265
Aspartame, 265
aspartic acid, 10–11
asymmetric synthesis, absolute, 65–70
asymmetry
    behavioural, 300
    development biology of animal, 300
atropine, 110, 151–152
aziridine, 217

bacteria
    asymmetry of, 291
    prokaryotic, 290
Barron–Thiemann effect, 29
beetles, pheromones from, 241, 244, 245, 249, 251–255, 263
benzene, 11, 12
benzodiazepines, substituted, 221
benzylnaphazoline, 164
bethanechol, 147
biaryl derivatives, 217
bifurcations, hypersensitive, 36, 41
biosynthesis, 141–144
Biot, 53, 303
biphenyl systems, 222
    helicity rules for, 224
birds, asymmetry of, 297–298
boar, wild, 298
bombykol, 241–242
borneol, 272
boson, 31–35, 83
bovine serum albumin, 160, 183
brain, human, 299
Brewster Rule, 227
'bridging studies', 130
bupranolol, 157–158, 169

C_2, symmetric chromophore chirality rule, 226
camphorsulphonic acid, 184, 189
capillary GC, 181, 193–194, 266–280
capillary electrochromatography, 196
capillary electrophoresis, 194–195
caraway, 243
carbohydrates
    structural analysis of, 231
carbon tetrabromide, 12
carbon tetraiodide, 12
cardiovascular drugs, 115

carvedilol, 170
carvone, 242, 244, 273
cassis, 280
catalysis, 40, 42, 75
catecholamines, 110, 120, 122
catecholimidazolines, 163, 171
catechol-O-transferase, 140–142
cellulose CSP, 186, 188, 194
cellulose triacetate, 183
cephalopods, 295
chamomile oil, 277
charge conjugation, 60, 75–76
charge transfer, 181
chiral
    centres, 101–105
    influence, 2, 24–27
    influences, falsely, 66–70
    influences, truly, 66–70
    inversion, 118
    memory, 113
    polymers, 25–26
    recognition, conformationally-driven, 94
    stationary phases, 181–183, 184–192
    'switch', 115–116, 130, 263
    synthesis, stereoselective, 11–15
chirality
    biomolecular, iii, 15–17, 23–48
    content, 81
    dimensions, in two, 82–83
    cf. dissymmetry, 55
    'false', 27–33, 55, 65, 83–84
    heavy atoms, of, iii
    homo-, iii, 15–17, 23–24, 42
    new definition of, 62
    relativity, and, 81–82
    rules, 226
    symmetry violation, and, 70
    'true', 27, 33–35, 55, 61–65, 72–74
Chiralpak AD, 189
Chiralpak AS, 189
'Chirbase', 267
chlorophyll, 45
chloroquine, 120
chlorostrychnine, 6
choleretic agent, 229
chromophore
    benzene, 227, 230, 233
    benzoate, 213
    carbonyl, 212–213, 218, 219, 224–226
    diene, cis-, 213
    dissymmetric, 216
    extended conjugated system, 213
    ketone, 218, 219
    naphthalene, 213
    olefin, 212
    tropolone, 222
    twisted enol ether, 217
cichlids, 297
cicloprofen, 116
cimetidine, 118–119

cinchonine, 2–3, 87
cinchotoxine, iv
circular birefringence (see also optical rotation), 6, 16
circular dichroism
    discovery of, 16
    electronic, 203–239
    fluorescence, 203
    luminescence, 203
    near-infrared, 203
    Raman, 203
circular dichroism spectra, 207–208
    β-adrenoreceptor-containing membranes, of, 160
    applications of, 234
    band maximum, 29–30
    bovine serum albumin, of, 160
    couplet in, 228
    empirical correlation of, 221–227
    exciton couplet in, 228, 229
    non-empirical correlation of, 227–233
    pheromones, of, 242
    structural information from, 220–233
circular polarisation, ii, 6, 16, 30
citraconic acid, 12–13
citronellal, 273–274
citronellol, 271–272
clay minerals, 37, 40, 42
CNDO molecular orbital calculations, 216–217
CNS drugs, 115
cocaine, 155–156, 171
cockroach, 251, 263
colchicine, 222
comets, 30
cones, 63–64
contingency, principle of, 45
Cotton effects, 16, 203, 208–209, 223, 224
    intensity of, 231
coumarins, 217
    dihydroiso-, 221
crabs, 295–296
crossbills, 297
crown ethers, 183, 186, 188
Crownpak CR, 188
crystallisation, 2–11, 35, 179–181
curare, 139
Cyclobond® CSP, 188
cyclobutenes, 28
cyclodextrins, 183, 186, 188, 190, 193, 194, 196–198, 266–280, 281
    p-dibromobenzoate of, 229
cyclohexanone, 3-substituted, 219
cyclopentyl methylphosphoryl thiocholine, 150–151
cyclopeptides, 112
cycloxilic acid, 229
cytisine, 149
cytochrome P450, 95

D lines, sodium, 206
of chiral allenes, 226
Darwin telescope, 45
DeVoe calculations, 233
debrisoquine, 119
decarboxylation, enzymatic, 91–92
dehydro-*exo*-brevocomin, 249
dexetimidine, 152–153
diamagnetic systems, 209
diastereomers, 180, 183–184
dibozane, 158, 164
dichrograph, 208
diene, 1,3-, helicity rules for, 224
diene, *cis*, distorted, 216
difluoromuscarine, 147–148
dihydroergoorptine, 164
dinitrobenzoylphenylglycine, 182, 185–187
diols, 98–105
di-3-pinanylborane, 216–217
dipolar strength, 214
dipyridines, 217
disopyramide, 115
disparlure, 247, 250, 255, 263
displacement chromatography, 198
dissymmetric force, cosmic, 3, 11, 83–84
dissymmetry, i, 54–55
disulphide, helicity rules for, 224
DNA, 25, 38–41, 46
dobutamine, 115, 120–121, 145
dominicalure, 250–251
DOPA, 141–142, 262
dopamine, 117, 140, 142–143
double well model, 76–78
dwarf pine and fir, 269

Easson–Stedman hypothesis, 91, 105, 161–164
eigenfunctions, 58
eigenstates, 63
electric field, 64–65, 205, 209
electroantennographic studies, 244, 247, 251
electromagnetic field, 209
'electronic nose', 280–281
electronic transitions, the nature of, 211–215
electrons
    β-, 32
    longitudinally spin-polarised, 66
    spinning, translating, 63–64
'electroweak', 18, 33, 35, 55, 72
ellipticity, 207
Enantiopac, 187
enantiomeric
    antagonism, 36
    cross-inhibition, 24
    excesses, 26, 40
    microscopic reversibility, 28, 68
    particles, 67–68
    reactions, 68
enantioselectivity, enzymatic, 89–90
enzymes, 24, 37, 88–90, 140
ephedrine, 155, 157, 170
epibatidine, 149

epi-methyl jasmonate, 275–277
epinephrine, 121, 145, 162
essential oils, 267–277
esters, 180
ethambutol, 130
ethylene, 12
European corn borer, 248
excretion, drug, 116, 120
exo-brevocomin, 242, 244, 249, 263

Faraday effect, 54
felodipine, 89–90
fenofibrate, 119
fermions, 31–32, 46–48, 82
Fischer's convention, 14
fish, asymmetry of, 296–297
flagellum, 291
'flash chromatography', 197
flounders, 287–288, 296–297
fluoromuscarine, 147–148
fluoxetine, 90, 115
Food and Drug Administration, 128–130, 304
Fresnel, 53–54
Fresnel equation, 205–207
frontalin, 245, 252
frogs, asymmetry of, 298

galaxies, 19
galaxolide, 278
gallopamil, 126
gastropods, 294–295
glaucine, 222
glucal, 217
gluons, 31
glutamic acid, 244, 246, 265
glyceraldehyde, 13–14
glyceramide, 30
glyceric acid, 30
glycerol, 30, 38
glycine, 91–92
glycopeptides, 191–192
grain borer, lesser, 250–251
Grand-Unified theories, 47
grapefruit, 277, 280
gravitational field, 64–65
gravitol, 165
gravitons, 31

Hamiltonian, 58, 79
halibut, 296
halothane, 128
hibiscus, 291–292
helicenes, 29, 181–182
helicity, 289–295
helicity rules, 224
helix model, 209–211
hemihedral correlation, law of, 7–8
hexahelicene, 216
histamine, 165
Hofmann degradation, 230

Höhn–Weigang coupling, 227
homatropines, 110
homochirality, 23–24
honeybee drone, 244
hormone, luteinising hormone-releasing, 97
host–guest chromatography, 183
house mouse, 249
Hückel molecular orbital calculations, 216
Hückel, extended, molecular orbital calculations 216
human immune deficiency virus (HIV), 97
    protease, 97–105
humans, asymmetry of, 299–300
huperzine, 150
hydrogen cyanide, 43–44
hydrogen-bonding, 100
hydroxylamine, 266
hydroxytryptamine, 165
Hypercarb®, 197
hyoscyamine, 110, 151–152, 170
hyperosmia, 265

ibuprofen, 116, 118, 120, 262
ifosfamide, 94–96
immunoglobulins, 290
indancrinone, 130
independent systems approach, 217–219
INDO molecular orbital calculations, 216–217
inhibitors, 99–105
ion-channels, 123–127
ion-exchange, 192
ion-pairing agents, 189
ipsdienol, 252, 263
Iris oils, 274–275
irones, 273–275
isoambrox, 278–279
isocyanates, 266
isoflurane, 128
isomenthone, 272
isomorphism, 4–6
isopropylmethoxamine, 169
isoproterenol, 142–144, 160, 168

Japanese akita dog, 298
japonilure, 247, 250, 255
jasmine, 275–277
jellyfish, 290, 296

ketamine, 130
ketone, α,β-unsaturated
    helicity rules for, 224
ketoprofen, 116
kinetics, of enzymatic reactions, 88
Kondepudi effect, 39–40
Kondepudi mechanism, 35–38, 41, 44
Kronig–Kramers transform, 232
Kuhn–Condon sum rule, 29, 41

labetolol, 114, 121–123, 170
lactonic acid, 245

lardolure, 249–250
lemon, essence of, 53
leptons, 47, 81
leucine, 265
lidocaine, 123–124
ligand exchange, 181, 185, 194
limonene, 269–270
linalool, 270–271
linalyl acetate, 270–271
Lipodex®, 271
lobeline, 149
lobsters, 295–296
lock-and-key concept, 23, 92, 243, 265
locust, 244
loganine, 225–226
lorazepam, 115
Lowe Rule, 226, 232

macrocyclic antibiotic CSP, 191–192
'magic mobile phase', 190
magnetic field, 28–29, 64–65
magnetic resonance imaging, 299
malic acid, 246, 247
mammals, asymmetry of, 298–299
maritime pine scale, 251
Maxwell equations, 56, 209–210
meclizine, 115
medetomide, 159
mediocrity, principle of, 45
menthol, 272
menthone, 272
mesons, 31
metabolism, drug, 116, 118–120
metamphetamine, 154–156
meteorites, 18–19, 27, 42
methacholine, 146
methyl jasmonate, 275–277
methylcyclopentanones, 225–226
methylmalic acid, 13
metocurine iodide, 154
metoprolol, 120–121
Michaelis–Menten theory, 88–90
'microfossils', 27
microscopic reversibility, 67–69
microorganisms, 291
microsomes, human liver, 90, 95
Miller–Urey experiment, 37
mobile phase additives, 188–189, 197
modulator, photoelastic, 208, 209
molecular orbital theory, 216–217
mono-ols, 98–105
monoamine oxidase, 140–141
monochromator, 204
Mösher's ester, 246
moths, 244, 247, 253, 255, 263
motors, protein, 290
muscarine, 144, 146–149
muscarinic receptors, 144–149, 151–153
muscone, 278–279
musk, 278

myrcene, 245

naphthalene ring systems, 223
naphthanyl, analogues of medetomide, 159
naproxen, 196
narwhal, 298–299
nautilus, chambered, 295
N-benzoxy-glycyl-proline, 185, 186, 189, 193
N-dechloroethylation, 94–96
neostigmine, 150
nereid marine polychaetes, 254
nerolidol, 277
nervous system, 140
neuroeffector junction, 154–155
neurones, 140
neutral current, 33, 72–73, 83
neutrinos, 71, 74. 81–82
neutron stars, 30, 42
Newman projection, 230
Newton's equations, 56
nicardipine, 115
nicotine, 139–140, 149
nicotinic agonists, 149
nicotinic blockers, 153–154
nifedipine, 127
nitrendipine, 127
N-methylcathinone, 155
nodal planes, 225
non-observables, 56–57
nootkatone, 277
noradrenaline, 140, 141–144, 154–156, 158–159
norepinephrine, 121, 155–156
norfluoxetine, 90
nuclear magnetic resonance spectroscopy (NMR), 33, 97, 181–182, 183, 192–193, 246, 276
nucleic acids, 37, 233
nutmeg oil, 268

octant, representation of cyclohexanones, 219
Octant Rule, 224–226
Ogston's model, 92–93
olean, 254
olfaction, 243, 261–262, 264, 265–266
O-methylation, 143
optical resolution, 1–11
optical rotation, 53, 57
optical rotatory dispersion, 53, 203, 211
   spectra, 207–208, 242
optically active media
   macroscopic characteristics of, 209–210
optical activity, natural and magnetic, 61–62
orchids, 291–294
organophosphates, 145, 150–151
organs, human, 299–300
oscillators, coupled, 219–220
ovomucoid CSP, 192
owls, barn, 297–298
oxirane, 217

palms, 291
pancuronium bromide, 154
papaya, 291–292
paracyclophanes, 228
parallel, 70
paratartaric acid, 1–11, 54, 87, 109, 179, 245, 246, 303
parity, 17–19, 31, 33–34, 46–48, 58–59, 63–65, 70–84
partial-filling technique, 195
Pasteur, Louis, i–iv, 1–19, 54, 61, 83–84, 87–88, 109–110, 139, 179, 285–286, 303–304
Pasteur's Law, ii
patchouli alcohol, 277
pelecypods, 295
pepsin CSP, 192
peptides, 96–101, 110–112
penicillamine, 262
penis, porcine, 298
peroxide, 217
phenoxybenzamine, 167–168
phentolamine, 166
phenylalanine, 265
phenylephrine, 165
phenytoin, 119
pheromones, 194, 241–255
phosgene, 266
positrons, 17, 32–33
photomultiplier, 208
photons
    circularly polarised, 66
    spinning, translating, 63–64
phyllotaxis, 291
physostigmine, 150, 170–171
picenadol, 114
piezoelectric device, 208
pilocarpine, 147–148
pindolol, 120, 169
pine sawfly, 253
pions, 74, 81–82
Pirkle 1-J CSP
plover, New Zealand wry billed, 297
polar-organic mode, 190
polarimetric, detection in HPLC, 196–197
polarimetry, 277
polarised light
    Biot, 303
    circularly, 203–207, 208
    elliptically, 206–207
    plane, 203–207, 208
polariser, 204
polycyclic aromatic hydrocarbons
    helicity rules for, 224
polyols, acyclic
    absolute configuration of, 231
polypeptides, 233
polysaccharide CSP, 186, 189, 194
polysiloxane, 266

porphyrins, 113, 231
positrons, 71
positron emission tomography, 299
postjunctional receptors, 157–160
PPP molecular orbital calculations, 216–217
prebiotic, origins of biological molecules, 26, 65
preparative HPLC, 197–199
preparative GC, 276
proline, 181
promethazine, 114
propoxyphene, 114
propranolol, 114, 117, 118, 121, 168–170, 193
proteins, 37, 96–101
protozoa, 291
pseudococaine, 155–156
pseudoephedrine, 155
pseudopeptides, 97
pseudoscalar, 57, 61
pseudoyohimbine, 164
pulegone, 280
pyroelectricity, 7

quantum field theory, relativistic, 75
quark, 31, 47
quartz, i, 5–6, 8–9, 35, 39, 40, 53, 181, 280
quartz cell, 208
quinotoxine, iv
quinuclidinyl benzylate, 160

racemisation, 18, 26
receptors, 110–113, 116, 120–127, 140
redbanded leaf roller, 248
refractive index, detection in HPLC, 197
regulatory requirements, for chiral drugs, 128–130
reptiles, asymmetry of, 298
reserpine, 170–171
respiratory drugs, 115
replicator, achiral, 24
retro-enantiomers, 111
retropeptides, 111–112
rhodopsin, bovine, 231
ritonavir, 103–105
RNA, 25, 37–39
'rocking tetrahedron' model, 93, 95–96
rotational strength, 214
    ab initio calculations of, 216–217
    symmetry properties of, 215
    non-vanishing, 216

saccharides, 13–14
salamanders, asymmetry of, 298
salicylideneamino, rule for amines, 226
salsoline, 144
salsolinol, 143
scalars, 57
scalene triangle, 82–83
sector rules, 224–226
serine, 30, 91–92, 150

serricornin, 250, 255
sesquiterpenes, 277–279
SETH, 42, 44
SEXSOH, 44–46
simulated moving bed chromatography, 198
snails, 286–287, 288, 294
'sniffing port', for GC, 266
sodium chlorate, 33
sole, 296
sotalol, 123, 169
spasmolytic agent, 229
spearmint, 243
specific rotation, 242
spined citrus bug, 251
squid, 295
starfish, 290
stars, 46
stegobinone, 250
stilbene oxide, *trans*-, 227
Strecker synthesis, 43
structure–activity studies, 161
strychnine, 6
succinic acid, 10–11
sulcatol, 246, 253
sulphur compounds, 279–280
superconductivity, high-temperature, 83
supercritical fluid chromatography (SFC), 194
supramolecular chemistry, 304
symmetry
    breaking, spontaneous, 80–81
    C2, 98–99, 101, 103
    operations, 56–57
    physical quantities, of, 57
    principles, 56–60
    quantum mechanics, in, 58–60
    violation, 80–81
sympathomimetic, class of drugs, 156
synthetic multiple-interaction CSP, 190–191

tacrine, 150
tartaric acid, i, iii, iv, 1–11, 54, 87, 109, 179, 245, 246, 303
tartramide CSP, 192
taste, 261–262
teicoplanin, 192
telenzepine, 153
tensors, 57
terbutaline, 115, 120
terfenadine, 116
terodiline, 130
terpenoids, 180, 267–273
tetrahydropyridine, 217
thalidomide, 24, 185, 262
thermodynamic equilibrium, 69–70
thermodynamics, of enzymatic reactions, 89
thin-layer chromatography (TLC), 194
three-point interaction rule, 91–93, 182, 183, 278
threo-methylphenidate, 155–156, 171
timolol, 169

'time reversal', 17–19, 56–57, 59–60, 63–65, 74–76
tolazoline, 166
tooth-carp, 297
tournaline, 5
trans-2-acetoxycyclopropyltrimethylammonium iodide, 148
trans-2-methylcyclopent-3-en-1-ol, 225–226
trans-2-phenylcyclopropylamine, 155–156
trans-3-methylcyclopent-4-ol, 225–226
tranylcypromine, 142
triage, 180
tribolure, 253–254
Tröger's base, 181
true enantiomers, 72–74
tryptophan, 265
tsetse fly, 254, 263
tubocurarine, 153–154
tunnelling splitting, 79
turpentine, 53
two-state systems, 78–80
tyramine, 142, 154
tyrosine, 141–142, 265

uncertainty principle, 78
unicorn horn, 299
unitarity, 69–70
universe, 46–48

vancomycin, 191–192
vancuronium bromide, 154
vectors, 57
verapamil, 115, 118–119, 124–126
vetiver oil, 277
vines, 291

warfarin, 118
weak force, 18, 48
whelk, 294
Whelk-O 1 CSP, 190–191
white dwarf, 30–31, 41
Wu experiment, 70–71, 73

X-ray
    analysis, 229
    crystallography, 97, 99
xenon arc lamp, 208

yohimbine, 164, 166

Z-11-tetradecenyl acetate, 248
Z-t-butylcyclohexyl acetate, 264